About Island Press

Since 1984, the nonprofit Island Press has been stimulating, shaping, and communicating the ideas that are essential for solving environmental problems worldwide. With more than 800 titles in print and some 40 new releases each year, we are the nation's leading publisher on environmental issues. We identify innovative thinkers and emerging trends in the environmental field. We work with world-renowned experts and authors to develop cross-disciplinary solutions to environmental challenges.

Island Press designs and implements coordinated book publication campaigns in order to communicate our critical messages in print, in person, and online using the latest technologies, programs, and the media. Our goal: to reach targeted audiences—scientists, policymakers, environmental advocates, the media, and concerned citizens—who can and will take action to protect the plants and animals that enrich our world, the ecosystems we need to survive, the water we drink, and the air we breathe.

Island Press gratefully acknowledges the support of its work by the Agua Fund, Inc., Annenberg Foundation, The Christensen Fund, The Nathan Cummings Foundation, The Geraldine R. Dodge Foundation, Doris Duke Charitable Foundation, The Educational Foundation of America, Betsy and Jesse Fink Foundation, The William and Flora Hewlett Foundation, The Kendeda Fund, The Andrew W. Mellon Foundation, The Curtis and Edith Munson Foundation, Oak Foundation, The Overbrook Foundation, the David and Lucile Packard Foundation, The Summit Fund of Washington, Trust for Architectural Easements, Wallace Global Fund, The Winslow Foundation, and other generous donors.

CLIMATE AND CONSERVATION

Climate and Conservation

LANDSCAPE AND SEASCAPE SCIENCE, PLANNING, AND ACTION

Edited by

Jodi A. Hilty, Charles C. Chester, and Molly S. Cross

Washington | Covelo | London

Library of Congress Cataloging-in-Publication Data

Climate and conservation : landscape and seascape science, planning, and action / edited by Jodi A. Hilty, Charles C. Chester, and Molly S. Cross.

p. cm.

Includes bibliographical references and index.

ISBN-13: 978-1-61091-170-2 (cloth : alk. paper)

ISBN-10: 1-61091-170-9 (cloth : alk. paper)

ISBN-13: 978-1-61091-171-9 (pbk. : alk. paper)

ISBN-10: 1-61091-171-7 (pbk. : alk. paper) 1. Biodiversity conservation. 2. Climate change mitigation. 3. Endangered ecosystems. 4. Ecosystem management. I. Hilty, Jodi A. II. Chester, Charles C. III. Cross, Molly S.

QH75.C56 2012

333.95'16—dc23

2012001573

Printed on recycled, acid-free paper 

Manufactured in the United States of America

10 9 8 7 6 5 4 3 2 1

KEYWORDS: Island Press, Landscapes, seascapes, conservation, climate change, climate stress, climate strategy, adaptation, conservation planning, biodiversity, resilience, resistance, transformation, conservation area networks, livelihoods, science, connectivity, corridor, ecosystem-based adaptation, wildlife, ecosystems, large-scale conservation.

To hope for our children

Samuel, Caleb, and Sophia Chester;

Jesse House and Remi Hilty;

Phoebe and Willa Cross;

and to the future of biodiversity

CONTENTS

FOREWORD

There has been a recent and worrying sea change in how people view climate change. For much of the past twenty years that I've been working in this realm, I've felt compelled to devote my public lectures to convincing audiences that climate change is really happening, and that humans are the cause. My talks to scientific audiences (mostly biologists) on the other hand, sought to convince them that climate change was a threat to biodiversity equal to or greater than the usual suspects such as habitat loss and invasive species. Then, seemingly in the middle of the night, the mood among both audiences swung to the polar opposite, to a fatalistic pessimism. Doomsday is coming and there is nothing to be done, so why even try?

This shift is perhaps the most disturbing experience I've had in my academic life. Even normally "can-do" Americans feel helpless in the face of such an overwhelming challenge. Classic psychological depression has set in: when disaster is perceived as inevitable, a resignation follows that prevents action being taken to avert calamity. Recently, I was chatting at a cafe with a member of Interfaith Power and Light, an interdenominational religious coalition that focuses on climate change, and learned that they had also witnessed this phenomenon among their congregants. Mass depression was catholic in its victims, with conservation practitioners just as susceptible as lay people.

This is a powerful trend that needs to be countered by well-thought-out, concrete, positive actions that offer hope of preserving biodiversity in the face of climate change. The case studies in this book give us the necessary prescriptions for such actions, from simple steps like getting relevant players on board and talking to one another, to more complex endeavors such as developing scientifically based scenarios of possible futures, to implementing transnational conservation strategies that incorporate climate change. It was heartening to read "best" cases where actions taken are already giving value for money, such as the Great Eastern Ranges Connectivity Conservation Corridor in Australia. But I was particularly impressed that the editors chose to include the hard cases as well, such as the Sundarbans of Asia, where mangrove forests are rapidly being inundated by sea-level rise. Using two native river/estuary dolphin species as sensitive indicators of the entire hydrological system from inland to coast is just the kind of creative thinking we need to tackle this challenge. We won't know for many years, perhaps decades, whether these efforts are truly successful in reducing biodiversity losses from climate change, but as the authors repeatedly point out, all are "win-win" in that they are good conservation practices even without climate change, and so their implementation can only be beneficial.

This volume is timely because it pushes against a growing depression, giving conservation practitioners not only hope but also suggestions for concrete actions that *will* help, even if we cannot precisely predict the future. The message of this work is more powerful than the nuts and bolts of specific recommendations; it embraces a much-needed philosophical premise that we *can do* something to help wildlife cope with the future. This is not to say the authors are Panglossian—they do not shy from detailing the obstacles to implementing climate change adaptation plans (be they political, economic, or social), nor do they present outcomes as known or inevitable. The case of the Amazon is particularly sobering; the authors of that chapter warn that "there is no indication of any ongoing action that can effectively tackle illegal land appropriation, limit deforestation to authorized sites, and stop anarchical logging and fires." On the other hand, these obstacles existed before climate change became a threat, and conservation has always suffered when enforcement of agreements and laws has been lacking. Climate change may be new, but such problems are very old.

Historically, the difficulty of obtaining agreements that operate across national boundaries has been a major bane of conservation biology, impeding action to address transboundary species and annual migrations. In this respect, the global nature of the climate-change conversation and the growing power of international policy forums may be helpful. Every case study in this volume has at the heart of its climate change adaptation strategy the connectivity of systems across large landscapes and seascapes. Conservation biologists have been extolling the virtues of this approach for decades, but international climate change initiatives have primed diverse sectors to understand, appreciate, and support the inherent need for protected areas across management and national boundaries. While the consequences of climate change are too grave to call this a silver lining, it is at least one positive side effect for conservation.

In sum, this volume provides much-needed hope and optimism to get people involved in proactive planning and action. As our Australian colleagues proudly assert in their case study chapter, the Great Eastern Ranges Connectivity Conservation Corridor "has helped introduce a new paradigm . . . one that has empowered individuals to make a difference when responding to climate change." Through a diverse set of case studies spanning the globe, this volume affirms that while climate change may be the great known unknown, there are already pathways to follow and robust role models for taking action against future uncertainties. With action, there is hope.

Camille Parmesan
Marine Institute, University of Plymouth, England,
and Section of Integrative Biology, University of Texas, Austin

PREFACE

Only a decade ago, most general treatments of biodiversity loss classified climate change as one of several threats to biodiversity, far behind the more tangible menaces of habitat loss and invasive species. This is not to say that climate change as a hazard to biodiversity went unrecognized; indeed, since at least the late 1980s there have been conferences dedicated to the subject, as well as multiple efforts to compile bibliographic databases of pertinent journal articles, books, and conference materials. Notably, the relevant body of literature is sufficiently vast that none of these attempts to catalog it have been comprehensive.

But even as climate change has been recognized as a significant threat to biodiversity, only over the past few years have a sizeable number of scientists and conservationists begun pointing to climate change as one of the great conservation challenges of the twenty-first century. Partly because the media has increasingly focused on climate change as a political problem, conservationists have had no small success in conveying the message that climate change could cause a species to go extinct. That this is no longer an obscure concern is evinced by the polar bear's current status as the "climate change poster child."

With the increased recognition of actual and potential climate change impacts on biodiversity, what solutions have conservationists put forth? At both the domestic and international levels, efforts to address climate change have focused on *mitigation*, or reducing our emissions of greenhouse gases and finding ways to store carbon in nonatmospheric sinks. No doubt, reducing carbon emissions today is critical to shaping what life on planet Earth will look like in the not-too-distant future. Yet even if we were to limit atmospheric greenhouse gas concentrations immediately—a highly idealistic hope given contemporary energy policies—the global climate will continue to change over the next century.

Along with mitigation, an increasingly common refrain has concerned the capacity of populations, species, and ecosystems to *adapt* to a changing climate. Again, this is not a new message. In 1992, the international community signed the United Nations Framework Convention on Climate Change, which specifically and prominently (article 2) called for avoiding disruption to the climate system "within a time frame sufficient to allow ecosystems to adapt naturally to climate change." Yet in the years following the 1992 Rio Earth summit, nearly all of the effort at an international scale was on mitigation, and negotiations leading up to the 1997 Kyoto Protocol were mostly about which countries would do what (and by how little) to lower their emissions. In many, if not most, countries, domestic attention to climate change followed suit.

It was not until a decade later, under the "Bali Roadmap," that adaptation finally came to the foreground in climate change negotiations. In the few years since the 2007 Bali Conference of the Parties, adaptation has become *de riguer* throughout the international system, all the way up to the United Nations General Assembly. Within the relatively compact arena of international biodiversity conservation, this was clearly in evidence by October 2008, when we found ourselves in Barcelona wandering the long hallways of the quadrennial Congress of the International Union for the Conservation of Nature and Natural Resources (IUCN). Due to extraneous reasons, most participants at this global conservation conference were looking somewhat shell shocked. For it was exactly then, just as we were settling into the collegiality of our conservation colleagues along the bright beaches of Barcelona, that the global economy took its initial nosedive from which we are still recovering. Over the ensuing days, you could read the same thought on nearly every face: *What does this mean for my grant funding?* Those with a more optimistic outlook observed that the crash would ease the rapid growth rate of anthropogenic carbon emissions—at least for a bit. On consideration, most of us remained pessimistic about the prospects of a global economic crash constituting a long-term solution to anything.

But it was not just the contortions of the global economy that had us wandering slack-jawed through the halls. Although we were fully expecting climate change to take center stage at the Congress—indeed, one of the formal themes was "A new climate for change"—the pervasiveness of the climate change theme throughout the presentations, exhibits, and hallway conversations was nonetheless highly notable. And what was remarkable, at least for those of us who had long been tracking climate change as a mitigation problem, was that mitigation was barely a secondary subject. The question everyone in Barcelona was asking was: *What do we actually do to protect biodiversity from inexorable climate change?* It wasn't the first time people were asking the question (hardly), but it did seem to be the first time when everybody was asking it.

The volume you now hold in your hands was born from that initial question. Having now grappled with it in both practitioner and academic settings, we are convinced that the question is one of the toughest facing the conservation community. It is tough for more reasons than we can count, but two related stand-out reasons are (1) that climate change will entail inexorable ecosystem change, which in turn demands us to rethink what it is we are actually trying to protect; and (2) that climate change will force a proactive break with our conservation heritage of "let it alone," requiring us to consider actively managing species if we are not to lose them.

As noted in a landmark study on climate change as a policy and management problem for biodiversity (Heller and Zavaleta 2009), one of the major recommendations has been to ensure that species can move across "the landscape." For anyone with a rudimentary understanding of our Pleistocene history, this finding did not come as a surprise. Over the millennia, the principal manner in which biodiversity has responded to changing climates has been to move across the landscape to more agreeable climates. Today, the capacity of biodiversity to make such peregrinations is threatened by human artifacts, e.g., cities, suburbs, and extensively modified agricultural areas. Simply put, by blocking the ability of biodiversity to move, anthropogenic changes to the landscape effectively vitiate the possibility of movement in response to climate change.

Consequently, scientists and conservationists explicitly argue that the current system of protected areas in most countries and transboundary regions is entirely inadequate to the long-term goal of biodiversity protection under the threat of climate change. Rather than a system of "postage stamp parks" left for native flora and fauna, they argue, we *have* to protect large landscapes in order to provide a stage for a successful response to climate change. One might interpret such a prescription as little more than a scaling up of the traditional conservation tactic of establishing formal protected areas. In practice, however, only a few conservationists argue that a strict protected status for broad regional landscapes is practical, feasible, or even imaginable. Nonetheless, conservation of some kind at the scale of large landscapes and seascapes—or "scapes" as we generically call them—will be necessary in protecting biodiversity from climate change.

Despite increasing evidence that the practice of biodiversity conservation will require a reframing of missions, goals, and objectives in order to match the scale of large scapes, current management practices in terrestrial, freshwater, and marine environments are, from a global perspective, only beginning to incorporate climate change considerations in order to facilitate long-term persistence of biodiversity. This volume can be considered an early survey of these beginnings. Admittedly, our initial hopes with this volume were to give the reader a definitive model for studying, planning, and implementing a scape-scale approach to conservation under conditions of climate change. Yet as the nineteen geographic case studies in this volume make clear, scape-scale responses to climate change are highly context dependent; we can assure the reader that you will not find *the* answer in this volume. Rather, what you will find is a range of approaches from which different lessons can be drawn.

In addition to the practical lessons that can be found throughout this volume, you will also find an engaging portrait of how much of inestimable value we have in the diversity of life on this planet. This kind of portrayal has been published an uncountable number of times, and such was not our original purpose in putting together this volume. But in reaching out to potential chapter authors (and then pleading with them for what we knew would be insightful contributions), we inadvertently chose a sizeable group of individuals who first and foremost care deeply about the places where they live and work. Although we can only take credit for bringing the case studies together, we are very proud to say that this love of place shines throughout this volume. It only makes the challenging work ahead of us all the more profoundly important.

We committed to writing this book so that all the pieces fit well together as one cohesive piece. Through multiple reviews and the process of consolidating sections, we each had substantial input on all chapters. Charles and Jodi both had the initial idea for the book and each took the lead at different stages, and in reality are co-lead editors. We are grateful to the organizations and people who have made this book possible. This includes the Wildlife Conservation Society; the staff at Island Press, especially our ever-cheerful and patient task masters Barbara Dean, Erin Johnson, and Sharis Simonian; the talented imagery whiz Andra Toivola; the helpful assistance of Eva Fearn; the book contributors who thankfully responded to our requests/nagging demands; and perhaps most of all, our supportive husbands, wives, and kids who gave us the love and time we needed to make it to the finish line.

PART I

Setting the Context

The vast scale of climate change impacts on species and ecosystems suggests the need for a bigger approach to conservation, both spatially (across large geographies) and temporally (across long periods of time). These impacts reveal both challenges and opportunities for investigating, conceptualizing, and implementing climate change adaptation at a landscape and seascape scale.

Chapter 1

Climate Change Science, Impacts, and Opportunities

CHARLES C. CHESTER, JODI A. HILTY, AND STEPHEN C. TROMBULAK

Across a vast array of human endeavors, climate change is demanding fundamental reconsideration of past approaches—be it in applied research on agricultural technologies, policy setting on water distribution, long-range planning for natural disasters, and so on. The list is long, and the arena of biodiversity conservation is no exception.

The practice of conservation has a long tradition of managing for historical reference points that range from particular dates (e.g., pre-Columbian in the Americas) or idealized conditions (e.g., "primitive wilderness") to quantitative assessments of particular species (e.g., fish stock and ungulate levels under "maximum sustained yield"). Even having long shed the naïve notion of a "balance of nature," biologists and conservationists have nonetheless incorporated cyclical and nonlinear dynamics within what are, ultimately, static models of whatever it is they are trying to protect. Yet under rapid climate change, many of these reference points may soon become unachievable, perhaps nonsensical. Climate change is forcing us to modify our approach to conservation, for climate change means that we can no longer manage for a historical reference point, but rather must manage for change. Barring all but the most unlikely of scenarios, this is not a choice: change is coming.

Conservation at the scale of landscapes and seascapes—hereafter referred to as "scapes"—is a necessary perspective and tool for achieving conservation goals in light of these inevitable changes. For this book we define a scape as a mosaic of local ecosystems interconnected by biophysical processes and allowing for populations of focal species to thrive over multiple generations (we come back to this in more detail later in this chapter). Increasingly over the past few decades, scientists, managers, and conservationists have come to acknowledge that working across scapes to conserve biodiversity is a prerequisite for effective biodiversity conservation. The threat of climate change has added further weight to the necessity of this approach. While scape-scale conservation is one of the most commonly recommended strategies for conserving biodiversity as the climate changes (Heller and Zavaleta 2009), those who have begun putting these concepts into practice have quickly realized that this is largely uncharted territory. As but one tangible example, in our experience of administering a climate change grant program under the auspices of the Wildlife Conservation Society, we found little agreement over what constitutes "applied, landscape-scale climate change adaptation." When we invited a panel of climate change professionals to help us review a number of short preproposals, there was unanimous consensus on whether or not to invite a longer second proposal only 22 percent of the time. Further, dialogue both within the grant program's advisory committee and across the science and conservation community has revealed a lack of consensus on terminology, priorities, and what constitutes an applied climate change-adaptation project.

Even as climate change can be detected and needs to be conceptualized and considered at a scape scale, on-the-ground management action generally occurs at more localized scales. Further, what is needed in one region of the scape might be very different from another. Additionally, land management agencies have their own mission and goals and lack mandate, experience, and mechanisms for effective cross-agency planning and implementation. Compound all this by working across international borders and the matter becomes even more challenging. The combination of these multijurisdictional challenges with the uncertainty inherent in projected impacts of climate change results in a variant of "analysis paralysis"—in this case, "climate paralysis."

All the same, there are emerging examples of large-scape conservation efforts incorporating climate change into their work. A number of initiatives around the world are working to implement conservation across large expanses of land and oceans, often working across multiple international boundaries. Many of these efforts have incorporated climate change adaptation—coping strategies in natural or human systems in response to actual or expected altered climate conditions—into their planning and action. In this volume, we capture nineteen examples of such efforts from around the globe. We hope that these case studies can help direct governments, nonprofits, industry, and others as they grapple with how to prioritize their planning and activities in light of climate change.

This chapter first offers a short overview of climate change's historical influence and projected effects on biodiversity. We use this section to summarize the context for this book, referring to key insights from the rapidly expanding literature on the relationship between climate change and biodiversity (and while there are now any number of books and articles providing extensive overviews of the issue, two highly recommended background publications are Hansen and Hoffman [2011] and Hannah [2011]). We then summarize why those impacts necessitate a large-scape perspective on biodiversity conservation, and outline a number of general principles for making conservation decisions in light of climate change. After highlighting several opportunities for conservation that climate change creates, we provide an overview of the remainder of the volume.

Climate Change and Biodiversity

The intertwined relationship between climate and biodiversity constitutes a fundamental tenet of biogeography (Lomolino et al. 2010). Specific concern over the effect of anthropogenic climate change on biodiversity dates to at least 1980, when Norse and McManus recognized the threat that natural or anthropogenic climatic changes may pose to biodiversity in protected reserves. Since then, a perpetual intertwining of high uncertainty and generalized concern has constituted a stable facet of the literature on climate change and biodiversity, causing Peters and Darling (1985) to recommend that conservation plans should be amended to consider the potential negative effects of climate change. Although we have learned a good deal about these effects in subsequent decades, the natural resource management and conservation community writ large has only just commenced to act on this grand recommendation.

Past and Future Changes

Researchers have examined the effects of climate change on biodiversity both through empirical and modeling studies. Regarding the former, an average global warming of roughly 0.74 degree Celsius over the last century has entailed a host of physical changes that includes decreased frequency of extreme cold events, increased frequency of extreme hot events, increased area affected by drought, decreased snow cover, melting of glaciers and ice in the arctic and Antarctica, warmer ocean temperatures, increased ocean acidity, sea-level rise, and changes to disturbance regimes such as wildfire, flooding, coastal storm surges, and pest outbreaks (IPCC-WG2 2007). In turn, these physical changes in climate are already influencing plants, animals, ecological processes, ecosystem structure and function, and ecosystem services around the globe. The Intergovernmental Panel on Climate Change (IPCC) has also cataloged evidence of long-term biological trends as evinced in studies lasting over twenty years and ending after 1990, concluding that out of over 28,000 data series, 90 percent indicated significant changes consistent with warming (IPCC-WG2 2007).

Speaking most broadly, biodiversity will respond to changes in climate in two ways. First, a species can adapt to the changes within its current range via the mechanisms of phenotypic plasticity and/or adaptive evolution. Phenotypic plasticity is simply the capacity of an individual to respond to changing conditions (e.g., phenological changes in the timing of activities such as mating and movement or behavioral changes such as altering food sources), whereas adaptive evolution constitutes genetic change brought about by the process of natural selection (Running and Mills 2009).

Second, a species can adapt by movement, be it shifting to a different aspect on a mountain (i.e., north, east, south, or west), or moving a hundred kilometers away. Research has shown that this latter type of movement is already happening with many species, and it is happening without the conscious help or interference of humankind (more on this later). However, the extant human footprint influencing more than 80 percent of the Earth's terrestrial surface (excluding Antarctica) will likely keep expanding and increasingly impede the ability of species to move (Sanderson et al. 2002; Hilty et al. 2006). The difference between past movements during paleoclimatic changes and those underway today is that species now need to negotiate not only the biophysical or natural components, but also an increasingly humanized scape.

The idea that conservation has to occur at a large scale did not arise recently. Well before the threat of climate change became widely recognized, the rise of landscape ecology and its associated tools (such as geographic information systems and global positioning system collars) had literally transformed our ability to understand and predict how species move across regions, thereby leading us to understand that very few protected areas will be large enough to protect all species within their boundaries (Hilty et al. 2006). The need for species to move beyond formal protected areas in order to survive constituted a central reason why natural resource managers and conservationists had begun to refocus their attention to the scape scale. Today, climate change analyses and projections have reinforced this need to look across large scapes.

Specific ecological responses to climate change have been reviewed extensively (see for example, Lovejoy and Hannah 2005; Parmesan 2006; IPCC-WG2 2007; Walther 2010; Walther et al. 2002). These included significant changes in the following:

- primary productivity and growing season length
- phenology (the timing of life-history events such as bud burst, flowering, denning, migrations, etc.)
- species assemblages and community composition
- trophic and other species interactions
- habitat availability and quality
- the distribution and abundance of terrestrial, aquatic, and marine plants, vertebrates, and invertebrates (including both native and nonnative species)
- the geographical ranges of pathogens and diseases

The complexity of these myriad potential changes lies in direct contrast to how the threat of climate change has been commonly perceived. Polar bears (*Ursus maritimus*) and American pikas (*Ochotona princeps*) are two species widely associated with the threat of climate change (Kovacs et al. 2011; Beever et al. 2011). For both species, the mechanism of the climate threat—viz., decreasing habitat due to warming conditions—seems intuitively straightforward. The reality, however, can be more complex. Pikas, for example, can be affected by *cooler* temperatures due to loss of snow pack. In other words, even where the overlying causal change is a warming climate, the actual mechanisms leading to a decline in the species can manifest themselves in unexpected ways. To this point, Parmesan et al. (2011) warn that it is difficult to attribute changes in any single particular species to climate change, not only because climate change operates at large spatial scales, but because a species' responses are particular to that species' unique biology. They further argue that "responses to climate are inextricably intertwined with reactions to other human modifications of the environment. Even where climate is a clear driver of change, little insight is gained by asking what proportion of the overall trend is due to greenhouse gases versus solar activity. From the perspective of a wild plant or animal, a changing climate is a changing climate, irrespective of its cause" (2). Although Parmesan and colleagues are making a relatively fine point here regarding how biologists should respond to guidance from the IPCC on distinguishing attribution, the broader cautionary point holds. Overall, as Willis and Bhagwat (2009) conclude in a review of several fine-scaled modeling studies, understanding the effects of climate change on biodiversity combined with human-induced habitat fragmentation constitutes "a serious undertaking fraught with caveats and complexities."

Despite these uncertainties, with anthropogenic greenhouse gas emissions continuing to increase toward the upper end of scenarios developed by the IPCC (IPCC-SRES 2000), further warming of the Earth's climate seems inevitable and ecological consequences will undoubtedly continue to manifest across the globe. Coupled atmosphere-ocean general circulation models generally concur that global average temperatures will likely increase a further 0.64–0.69 degree Celsius over the next twenty years, with the greatest temperature increases at high northern latitudes and over land, and less warming over the southern oceans and North Atlantic. Best estimates of global average temperature change by the end of the twenty-first century range from 1.8 to 4.0 degrees Celsius, depending on assumptions about future greenhouse gas emissions (IPCC-WG1 2007). Projected emission rates and the resultant warming will likely further exacerbate many of the observed ecological responses to recent climate changes, as well as result in environmental and ecological conditions that are difficult to anticipate. In addition to these changes in global averages, we can expect to see the magnitude and even direction of changes in temperature and precipitation vary across space and time (e.g., across seasons) with more extremes likely (Girvetz et al. 2009).

Consequences for Conservation

All of the changes described above dictate the need to incorporate climate change in our natural resource and conservation efforts. To be more specific: change is happening and inevitable such that it is not possible to move forward with effective conservation without explicitly considering climate change. Failure to do so in many cases may mean ultimate failure to achieve the conservation goals set forth. For example, in working to protect a species with a northward-shifting range due to climate change, designing a reserve around its current range could lead to a reserve that only hosts the target species in the short term. Instead, a reserve farther north or a network of reserves that facilitate the species' northward movement may be more appropriate. This inchoate practice of incorporating climate change considerations into natural resource planning has led to the proliferation of a number of labels ranging from "climate ready" and "climate resilient" to "climate smart" and "climate savvy." While we do not at all reject any of these, the term that we use throughout this book is *climate change adaptation*.

As noted earlier, climate change entails significant implications regarding the common use of historical reference points as management targets. This includes the manner in which conservation practitioners have been coming to grips with the problem of "shifting baselines." One need tread carefully with this useful concept, for there are at least two central conceptualizations of shifting baselines that differ in a nuanced but substantial way from each other (Papworth et al. 2009). First, there is the physical reality of shifting baselines—higher average temperatures, lower fish population counts, higher concentrations of CO_2 in the atmosphere, lower precipitation levels, greater extent of urban sprawl—all of which are readily measurable according to a baseline rooted to a particular time. Whether due to natural or anthropogenic causes, these will all vary over time, thus creating the *physical reality* of a shifting baseline. Then there is the *human perception* of shifting baselines—the common tendency of humans to perceive a baseline as inexorable, forgetting or ignoring that the baseline has shifted to something very different than what it was at an even earlier time. This social perception of the physical reality is generally referred to as "shifting baseline syndrome" (Weber and Stern 2011).

The conservation community has to grapple both with how baselines have shifted over time and with how the perceptions of those baselines have shifted. Natural resource managers ultimately must come to terms with some variant of the following question: What is it going to mean to "protect an ecosystem" in a particular area—be it Yellowstone National Park, the Alps, or the Great Barrier Reef—if climate change is inexorably shuffling and redistributing their biological communities? The answer will require empirical data, philosophical interpretation, and moral judgment, and consequently it is one of the most challenging questions facing the conservation community today. A central premise of this volume is that the difficulty of all of these questions can be ameliorated by focusing on a scape approach that allows for movement of species, be it across ranges of latitude, altitude, or microclimate.

Need for Scape-scale Management and Conservation

Climate science points to two critical and intertwined issues for biodiversity conservation. The first is that the scope and scale of climate change and associated ecological responses require examining large-scale or scape-scale regions. Second, even as increasing scientific evidence *without* regard to climate change has long pointed to the inadequacy of protected areas in protecting biodiversity, climate change makes it imperative that natural resource managers look beyond protected areas in order to achieve long-term conservation.

As mentioned earlier, employing a large-scape perspective on biodiversity conservation is one of the most frequently cited recommendations in the scientific literature for biodiversity conservation as climate changes (Heller and Zavaleta 2009). Indeed, some within the conservation community had identified the need for this approach before the advent of either the IPCC's First Assessment Report in 1990 or the 1992 UN Framework Convention on Climate Change. In 1988, the World Wildlife Fund held a conference entitled "Consequences of the Greenhouse Effect for Biological Diversity" in Washington, DC, the purposes of which were to stimulate research, to pull together existing information, to draw general conclusions about conservation consequences, and to bring these conclusions to scientists, policy makers, funders, and management agencies. Apparently the first of its kind, the conference highlighted the need for scape-scale conservation by focusing on the conundrums facing protected area management, ultimately recommending that individual parks and reserves be managed in the context of integrated regional or global conservation plans (Peters 1988).

There are several reasons why a focus on scape-scale conservation is essential for addressing the threat of climate change. First, climate change is a widespread stressor. Although the exact direction and magnitude of changes in climate and ecological conditions will vary from place to place, climate change is not an isolated concern for localized areas. Second, the ability of species and ecosystems to respond to climate change also necessitates a larger spatial perspective. Many species responded to paleoclimatic changes with dramatic range shifts (DeChaine and Martin 2004; Botkin et al. 2007), warranting the need to think bigger about the amount, location, and connectivity of core habitat areas. Third, effective conservation of ecosystem services and functions requires actions across spatial areas large enough to support sustainable ecological processes (e.g., hydrology, wildlife, native pest infestations, etc.) (Game et al. 2010). All of these factors justify an approach to conservation from a bigger-picture perspective.

Overall, the primary concern is the incompatibility between large-scale dynamic species range limits (which are likely to shift in response to climate change) and the comparatively small static boundaries of protected areas, the juxtaposition of which suggests that the current protected area network will be inadequate given climate-induced species range shifts (Burns et al. 2003; Hannah 2008; Krosby et al. 2010). Since it is not likely possible to add protected areas in many places around the world, shifting to a broader landscape perspective will require looking at protected areas in the context of the surrounding matrix and using other conservation tools on key non-protected lands.

It is important to acknowledge the spatial ambiguity inherent in the terms *landscape* and *seascape*. One prominent definition of landscape held it as "a mosaic where the mix of local ecosystems or land uses is repeated in similar form over a kilometers-wide area" (Forman 1995, 13). A more expansive encyclopedic characterization of landscapes described them as "complex phenomena comprising interconnected physical, biological and cultural systems" (LaGro 2001). And in an extensive treatment of "landscape-scale conservation planning," Trombulak and Baldwin (2010, 7) defined landscape as "a collection of habitat patches sufficient enough in size to allow population processes to take place at a multigenerational time scale. What constitutes a landscape for one organism may simply be a portion of a landscape for another." They add, however, that despite the resultant unavoidable ambiguity, "even such an amorphous concept as landscape can be used as an effective tool for conservation planning if the scale selected for planning is appropriate for the conservation goals to be achieved" (Trombulak and Baldwin 2010, 8). While the term *landscape* has received much more attention in the peer-review literature than the term *seascape*, conservation organizations also utilize the latter term with comparable definitions that reference large areas, multiple jurisdictions and uses, and connectivity (e.g., CI 2007; WCS 2011).

Responding to the Threat of Climate Change

The past decade has seen many entities in the natural resource arena put a great deal of thought into how to effectively respond to the challenges posed by current and future climate changes. Researchers have begun to converge on a set of widely accepted basic tenets (Hansen et al. 2010), and associated principles of climate change adaptation with a number of general principles for each (e.g., Game et al. 2010; Hansen and Hoffman 2011; Heller and Zavaleta 2009; Lawler 2009; Mawdsley et al. 2009; West et al. 2009). The following list summarizes these tenets:

- Protect appropriate and adequate space.
    1. Make conservation planning more dynamic by taking shifting species ranges into account when designing reserve networks.
    2. Increase the size and number of protected areas.

    3. Maintain/enhance connectivity between those areas.
    4. Protect climate refugia, those areas least likely to undergo significant climate-induced changes or that will likely house suitable climate conditions in the future but are not currently occupied.
- Reduce nonclimate stressors.
    1. Repulse invasive species, pests, and diseases.
    2. Sustain ecosystem processes and functions.
    3. Implement ecological restoration.
    4. Plan for human responses to climate change that may entail additional threats to biodiversity (e.g., renewable energy facilities).
- Adopt adaptive management and focus on ecosystem services.
    1. Study management interventions and species responses.
    2. Monitor changes and interventions.
    3. Implement ecosystem-based adaptation (EBA) practices.
    4. Ensure that decision making systems are able to incorporate and adjust to new information.
- Consider species-specific interventions.
    1. Protect, restore, and create critical habitat and connectivity areas.
    2. Consider translocation (also called assisted migration, assisted colonization, and managed relocation) and/or captive breeding.
- Reduce the rate and extent of climate change.

Given that many of these conservation strategies are at least a century old, the overall structure of this list will be familiar to natural resource managers and conservation practitioners. Many of the advances that have been made in conservation planning and action over the last several decades will continue to play a critical role in the conservation of species and ecosystems under a changing climate. Instead of replacing these tools, developing climate change-adaptation actions will require reprioritizing and rearranging the conservation toolbox, while also looking out for when we need new tools or to use existing tools in substantially different ways. For example, a small to moderate shift in the timing or sequencing of a conventional management action may dramatically improve its probability of success as the climate changes (e.g., shifting the timing window for the use of prescribed fire based on a trend toward earlier and longer fire seasons).

To reiterate, climate change adaptation actions are not necessarily brand-new additions to the conservation toolbox, but rather a collection of prioritized actions for achieving goals and objectives after a systematic consideration of the impacts of climate change and associated uncertainties. As Hansen and Hoffman (2011, 107) put it: "Addressing climate change in your work does not mean reinventing the wheel. It does mean making sure you have the right wheels for the terrain."

Although general adaptation principles are broadly accepted, exactly how to put those principles into practice remains murky in many (if not most) cases. Conservation practitioners need ways to translate these general principles into practical site- and target-specific actions. Accordingly, chapter 2 describes in greater detail the process of integrating climate change into scape-scale conservation planning and action, and the case study chapters then demonstrate how general adaptation principles are being interpreted and applied in particular scapes around the globe. Debate continues over how and when to adopt adaptation approaches. For example, some proponents suggest that we should conserve "the stage" rather than "the actors" (Anderson and Ferree 2010), suggesting that we should be conserving intact places and letting species arrive and depart as they will. Others feel that various levels of species-specific strategies are needed at least in some cases. Likewise, discussions increasingly revolve around when an invasive species should no longer be considered invasive, and when and how should translocations occur given the checkered past of many intentional species introductions (and sometimes reintroductions).

Examining priorities across a larger relatively intact scape may address some of the concerns around these debates. Large scapes in theory can conserve large enough stages for natural resource managers to allow for or even facilitate redistribution of species through time and space. Also, some of the general climate change-adaptation principles inherently require a scape-scale approach, such as maintaining and enhancing connectivity, protecting refugia, and expanding protected area networks. Similarly, some practices will benefit from coordinated implementation across a scape scale. For example, reducing nonclimate stressors in a particular locale may not have as useful an impact as reducing that stressor across larger areas. Alternately, there may be a strategic reason to implement an action in one part of a larger landscape than another, but knowing this requires examination at the larger scale.

The bedrock theme of this volume is the juxtaposition of scapes with climate change, and we begin with the straightforward premise that a primary conservation goal in response to climate change will be to ensure *connectivity* across a matrix of land/ocean types and uses to enable species to relocate as needed. We define connectivity as a measure of the ability of an organism to move among separated patches of suitable habitat at various spatial scales (Hilty et al. 2006). Increasing attention to the need for connectivity has grown in tandem with increasingly empirical evidence that (1) the relatively small portion of protected areas scattered across the globe will be insufficient for the task of protecting anything close to the full extent of the planet's biodiversity, and that (2) effective conservation of biodiversity will depend on conservation measures taken across broader scapes around and between formally protected areas (or what is often described as "the matrix"; see Franklin and Lindenmayer 2009). As Willis and Bhagwat (2009, 807) put it: "Although every measure should be put in place to reduce further fragmentation of reserves, we must determine what represents a 'good' intervening matrix in these human-modified landscapes."

Opportunities

At the same time that this volume focuses on scape-specific responses to the challenge of climate change, climate change also presents a number of *opportunities* for the cause of biodiversity conservation. Since climate change constitutes one of the most serious threats to global biodiversity, it is counterintuitive—perhaps mercenary—to consider it as an opportunity for conservation. However, the reality of a rapidly changing climate offers at least three potential opportunities at the scape scale for conservation planning that should not be squandered.

First, public awareness of the threat of climate change has dramatically increased the potential for public engagement in issues related to natural resource management. Numerous influences, notably the campaigns by Al Gore (*An Inconvenient Truth*) and Bill McKibben (350.org), and the attention received by the IPCC through their reports and receipt of the Nobel Peace Prize in 2007, have raised public awareness of the vulnerable conditions upon which society depends. Although in some countries such as the United States there remains substantial doubt and denial over whether climate change is anthropogenic (or is even occurring), public perception at a global level has come to the understanding that societal planning at local, regional, and national scales is now critical for sustained human welfare. Current societal concern is focused on sea-level rise, flooding, access to drinking water, maintenance of resource-based economies (e.g., forestry and fisheries), food production, energy sustainability, and public health. Yet conserving biodiversity and maintaining ecosystem services provided by conservation lands (including carbon sequestration) are being increasingly recognized as endeavors critical to human welfare. Conservation planning is thus moving more fully into the public arena as a result of climate change, which in turn increases the potential for broader public engagement.

Second, the damages caused by climate change, such as increased flooding, heightened storm surges, and increased tree mortality through disease and physical disturbances, will—if our economies allow—result in an accelerated pace of social investment in redesigning and reconstructing our built environment. For example, the expense of rebuilding after flood damage to human infrastructure in riparian flood zones may ultimately lead insurance companies and land-use planning agencies to bar or eliminate incentives for building in designated flood zones.

Third, the broad impacts of climate change across multiple social sectors, especially transportation, public works, and public health, increase opportunities for collaboration and decrease the perception that conservation issues are irrelevant compared to more immediate social concerns. This is simply because nature provides many societal benefits including human health and well-being. For instance, riparian vegetation and wetlands are increasingly valued for their role in protecting biodiversity and civil infrastructure in a future that may involve dramatic changes in water availability (Hassan et al. 2005; Kellert 2005). In addition, habitat conversion and fragmentation not only decrease the potential for biological adaptation to climate change and impair ecosystem services (water filtration, flood control, air filtration, and other functions that healthy ecosystems provide to humans), but also facilitate the spread of zoonotic diseases.

To no small degree, this third opportunity has given rise to the increasingly widespread concept of *ecosystem-based adaptation*. The term was apparently coined around 2008, at which time a well-attended workshop in Costa Rica proposed defining ecosystem-based adaptation as "adaptation policies and measures that take into account the role of ecosystem services in reducing the vulnerability of society to climate change, in a multi-sectoral and multiscale approach" (Vignola et al. 2009). Notably, that same year the IUCN was using the term in a position paper submitted to negotiators under the UN Framework Convention on Climate Change, calling on member countries "to mainstream ecosystem-based adaptation as an integral element of overall climate change adaptation in poverty reduction strategies and development planning" (IUCN 2008). In the years since, the term has been rapidly and widely adopted by myriad conservation-related organizations, agencies, international organizations, and research institutions. Indeed, it seems to have become a central, organizing concept in the broader international arena of climate change adaptation and is probably the closest the conservation community has moved biodiversity conservation to the "mainstream" of international climate policy.

Despite our use of the term *opportunity,* none of this is to say that natural resource manager and conservation planners should welcome climate change. The climate change challenges of the coming years will mostly exacerbate the challenges that began with rapid loss of natural habitat and extirpation of species, and that continue on with hyperacceleration of the roads network, air and water pollution, spread of invasive species, and globalization. Climate change will thus make an already daunting task even more difficult, and we have no choice but to face the prospect of even greater biological losses due to climate change before we finally succeed in developing a society that can live in peace with the rest of nature (Hannah 2011). Nonetheless, we would be foolish to ignore the potential for the climate crisis to provide conservation planners with an expanded circle of allies and increased public engagement.

Overview of the Book

Readers will find a wide variety of approaches being taken across the case studies of large scapes in this volume, and different considerations as well as prioritizations of particular issues. At the same time, readers will also find a string of commonalities woven throughout the different stories that must be addressed regardless of the scape. After the next chapter, which takes a systematic look at planning for climate change adaptation, the following nineteen chapters examine particular scape case studies from around the globe.

The case studies begin with three equatorial and tropical landscapes: the Albertine Rift in Africa, the Brazilian Amazon, and the Mesoamerican Biological Corridor. These share the common challenges of (1) a pressing number of urgent conservation priorities even without considering current and future climate change impacts and of (2) taking place in a context of highly impoverished communities. We then explore five temperate and Mediterranean regions including Canada's Boreal Forest, the Cape Floristic region of South Africa, Eastern Mongolia's Grassland Steppe, the Northern Great Plains of North America, and Washington State, USA. Relatively speaking, these temperate regions have strong climate science capacity but, with the exception of the boreal, are heavily impacted by human activities. The next section focuses on water scapes, with one freshwater case study of the Alps and three marine seascapes including the Sundarbans Mangrove Forest in Asia, the Vatu-i-Ra Seascape in Fiji, and the Wider Caribbean. These emphasize the essential connections between terrestrial systems and aquatic and marine health that will be particularly affected by climate change. This is followed by five montane landscapes—the Altai-Sayan in Eurasia, the Great Eastern Ranges in Australia, the Madrean Sky Islands across the US–Mexico border, the Northern Appalachian/Acadian ecoregion in northeast North America, and the Yellowstone to Yukon region—which all share a strong connectivity theme and showcase the potential for strong climate change-adaptation planning and action across topographically diverse areas. We close with the two polar regions, which describe intense climate impacts already occurring and the strong need to (1) identify key areas for refugia and (2) plan for increasing human activities as a new variable affecting conservation.

Our goal in selecting these case studies was to compare how a multiplicity of conservation actors in various scapes around the globe were responding to the threat of climate change. Each of these was written by an author or set of authors who have extensive experience in their respective region, and who are actively engaged in working on climate-related issues in their region. We asked authors to introduce the region, provide a historical overview of conservation in the region, discuss what is known about the impacts of climate change, and review what is currently being done and priorities for the future to incorporate climate change into conservation and natural resource management. The result is a collection of geographical case studies that allows for effective cross comparison but does not attempt to frame the challenge of climate change with a cookie cutter. The final chapter offers a synthesis of what scientists, managers, and practitioners have not only discovered but also created in these varied scapes, and offers recommendations for others working on similar issues around the world.

Chapter 2

Landscape and Seascape Climate Change Planning and Action

MOLLY S. CROSS, ANNE M. SCHRAG, EVAN H. GIRVETZ, AND CAROLYN A. F. ENQUIST

As discussed in chapter 1, revisiting our conservation goals and actions in light of climate change does not mean that we must discard what we have learned about critical conservation tools and strategies. We do, however, need to explicitly consider future climate change as we make conservation investments. Although climate change will not always be the most urgent problem facing biodiversity conservation, it will increasingly constitute a lens through which we must examine our actions. By anticipating the direct and indirect consequences of different scenarios of climate change for the particular species, ecosystems, and natural processes we aim to protect—and then determining appropriate and necessary conservation actions in light of those changes—we increase the likelihood of making informed decisions about both near- and long-term threats that will improve our ability to achieve conservation success.

This chapter expands on the opportunities and challenges involved in undertaking structured and participatory adaptation planning and action at large landscape and seascape (hereafter referred to as "scape") scales. Such climate change-adaptation planning and implementation processes are likely to include several steps (fig. 2.1; Cross et al., in review; Glick et al. 2009; Heller and Zavaleta 2009):

- Identify focal conservation features (e.g., species, ecological processes, ecosystems) and associated conservation objectives.

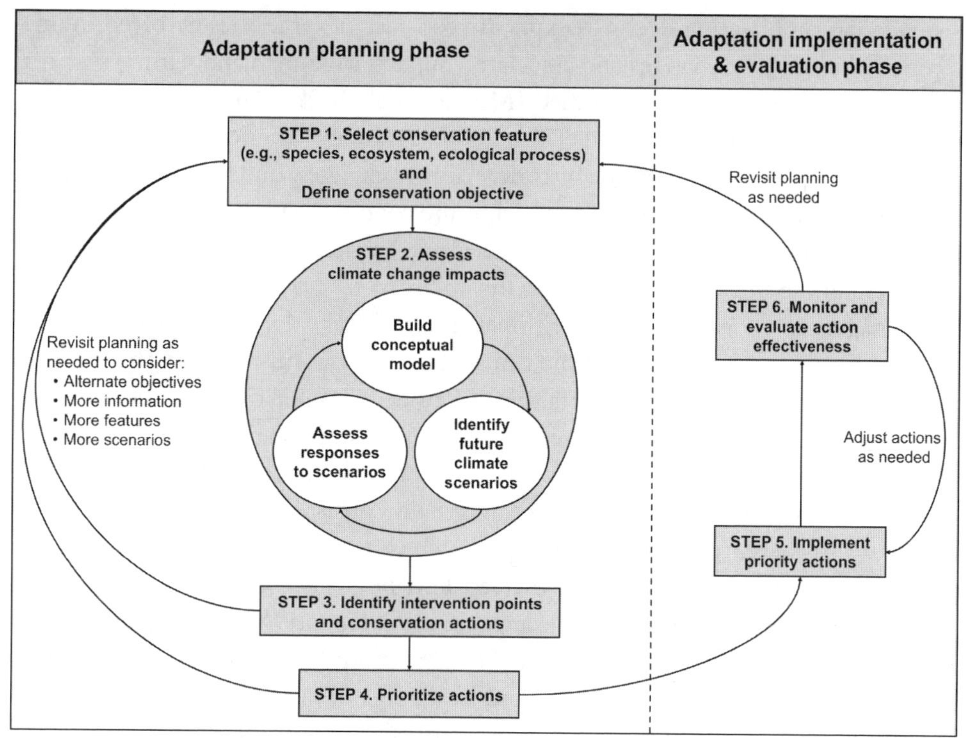

FIGURE 2.1 Example framework for climate change-adaptation planning and action showing the steps in a structured and iterative process for integrating the best available climate science into conservation decision making (modified from Cross et al., in review).

- Assess climate change impacts and vulnerabilities for a range of plausible climate change scenarios.
- Identify and prioritize climate change-adaptation actions needed to achieve stated objectives.
- Implement adaptation actions.
- Monitor action effectiveness to identify when actions need to be adjusted or planning revisited.

While not always taken in exactly this order, these steps should feel familiar to conservation practitioners since they mirror general approaches to conservation planning (e.g., Groves 2003; Wilson et al. 2009). Yet they are designed specifically to address the threat of climate change. While targeted climate change planning may be necessary in the near term as people become more familiar with the threats and strategies associated with climate change, the ultimate goal is to integrate climate change into comprehensive decision-making processes that simultaneously address multiple stressors (IPCC-WG2 2007).

Adaptation planning needs to explicitly consider uncertainties involved in projecting future climate changes and associated impacts to human and natural systems, only some of which can be quantified (Morgan et al. 2009). An additional concern is the lack of baseline climate and ecological data in many locations, which makes it difficult or impossible to examine historical climate changes, understand how climate affects species and ecosystems, and model future conditions with a high level of confidence. As with other institutional and scientific barriers to effective decision making, these uncertainties can undermine the ability to take conservation actions. And yet managers and conservation practitioners need to make decisions even when they do not have perfect knowledge of the future. Planners can turn to a number of tools to acknowledge and embrace uncertainty in their decision making. In cases with high and uncontrollable uncertainties, scenario-based planning can be a useful tool for guiding conservation decisions (Peterson et al. 2003). In this context, the IPCC-WG2 (2007) defines a scenario as "a coherent, internally consistent, and plausible description of a possible future state of the world," which is not meant to be a forecast or prediction of the future, but rather is used to describe an alternate, plausible trajectory for the future whose probability of occurrence is not known (Mahmoud et al. 2009). By looking across outputs from multiple climate models and greenhouse gas emissions scenarios (box 2.1), we can characterize a range of plausible futures for a scape in terms of the rate, magnitude, and seasonality of changes in climate variables such as temperature and precipitation. We can then identify actions that will be necessary to achieve conservation goals in light of each future scenario. Scenarios can be used within a variety of existing and emerging approaches to integrate climate change into conservation and management decision making, including risk assessment and management (e.g., Willows and Connell 2003), structured decision making (e.g., Ohlson et al. 2005), and adaptive management (e.g., Peterson et al. 1997).

Undertaking climate change-adaptation planning across large scapes that cross jurisdictional boundaries offers both challenges and opportunities. Private landowners often have different conservation and development incentives than do government landholders, and in some cases it can be difficult to discern who (or which agency) has land tenure. Within a single government, multiple agencies tasked with managing public lands across the same scape often have dramatically different and even conflicting missions. Regardless of the presence or absence of climate change impacts, it is difficult to establish a common goal for a species or an ecosystem across jurisdictional and public-private boundaries. Under altered climate conditions, the task will only be that much more challenging, especially since each agency, organization, or individual involved may have different perceptions of, and tolerance for, the risks associated with climate change. Despite these challenges, climate change can be a unifying issue for conservation, providing an opportunity to align different efforts under one umbrella (Grantham et al., in press). Moreover, when applied in a participatory manner that engages multiple decision-making entities, climate change-adaptation planning may

BOX 2.1 CLIMATE MODELING AND ISSUES OF SCALE

Projecting the direction, rate, and magnitude of climate change that species, ecological processes, or ecosystems may be exposed to in the future is essential to understanding ecological effects and vulnerabilities. Future climate conditions are usually assessed using outputs from three-dimensional global climate models that capture the key physical, chemical, and biological drivers affecting the Earth's climate, and which are driven by various scenarios of future atmospheric greenhouse gas concentrations (IPCC-WG1 2007). The greenhouse gas emissions scenarios developed for the IPCC Third Assessment Report represent plausible storylines of future economic, demographic, social, technological and environmental trajectories. The A1 and A2 family of scenarios tend to represent more economic-focused futures with relatively higher greenhouse gas emissions, while the B1 and B2 scenarios are more focused on environmental sustainability and therefore relatively lower greenhouse gas emissions (IPCC-SRES 2000). The IPCC Fifth Assessment Report (due in 2013–14) will rely on climate model projections that consider new atmospheric greenhouse gas concentration scenarios called "representative concentration pathways" (Moss et al. 2010).

The resolution of global climate models is fairly coarse, with grid cells typically ranging from about 200 to 300 kilometers. Global climate model outputs can be "downscaled" to finer resolutions using dynamical approaches that resolve physical climate processes to resolutions of roughly 10 to 50 kilometer-grid cells. Downscaling can also be conducted using statistical approaches that establish relationships between global model outputs and locally observed climate data from the recent past, which allow for downscaling to resolutions less than 10 kilometers. While fine-resolution climate projections are often desired for conservation planning, both downscaling techniques are ultimately subject to biases or inaccuracies that are embedded in the global climate models. These errors are related to the current inability of global climate models to account for finer-scale physical processes (e.g., cloud formation), various feedbacks (e.g., land-atmosphere interactions), and topographic effects on climate (e.g., orographic precipitation) (IPCC-WG2 2007; Hansen and Hoffman 2011; Wiens and Bachelet 2010). There are also uncertainties in the downscaling techniques themselves, related to (1) shortcomings in the quantity and quality of weather station data used for spatial interpolations, (2) assumptions about whether relationships between large- and small-scale climate processes might be altered by climate change, and (3) limitations in the ability to dynamically model fine-resolution climate processes, especially in topographically heterogeneous scapes (Daly et al. 2007; Hayhoe et al. 2011).

There are benefits and pitfalls to examining future climate projections at a large-scape scale. For sufficiently large scapes, the relatively coarse grid cells of global and regional climate models are less of a problem when the entire spatial extent of the scape is considered all at once. In particular, regional climate models, which are better able to account for some topographic effects on climate, can produce projections at a scale relevant for summarizing changes across an entire scape. However, if the goal is to differentiate the direction and rate of climate changes—and therefore potential ecological effects—within a scape, then finer-resolution information will be necessary. Climate projections that have been downscaled to sub-scape resolutions can be used for these types of analyses, if the associated caveats mentioned here are properly considered.

provide a bridge across jurisdictions by fostering increased communication and coordination. This type of coordination facilitates implementation of recommended actions and is critical to long-term biodiversity conservation in the face of climate and other environmental changes (Heller and Zavaleta 2009).

An example of efforts to coordinate climate change science, planning, and action across jurisdictional boundaries is the formation of twenty-one Landscape Conservation Cooperatives (LCCs) across the United States that involve multiple federal, state, and local agencies; tribes; universities; and nongovernmental organizations (USFWS 2011). Working in conjunction with eight regional Climate Science Centers (USDOI 2010), LCCs are intended to address landscape-scale stressors such as climate change by fostering a more networked approach to conservation, leveraging resources, and strategically targeting science to inform conservation decisions and actions across public and private jurisdictional boundaries. Efforts like these, to improve cross jurisdictional coordination across public and private boundaries on climate change conservation, can be found throughout the globe, as evidenced by the chapters included in this volume.

Selecting Focal Species, Processes, and Ecosystems

Conservation scientists have long recognized both the need for, and the difficulties of, conservation planning across large spatial extents (e.g., Noss 1983). With the emergence of the discipline of landscape ecology and tools such as GIS (geographic information systems, a spatioanalytical tool) and GPS (global positioning system) and satellite collars (units that can remotely track animals with high detail), the ability to discern patterns and processes across larger scales has contributed to expanding the focus of conservation beyond individual protected areas to large scapes. In some ways this makes natural resource decision making more complex, since it is unrealistic to plan conservation actions for every species, ecological process, and ecosystem across a large area (Groves 2003). The same holds true for climate change-adaptation planning for large scapes. Scape-scale adaptation planning will likely require the identification of a suite of focal conservation features (e.g., species, ecological processes, ecosystems) based not only on traditional prioritization criteria, such as the level of conservation concern, socioeconomic value, provision of ecosystem services, and/or usefulness as an umbrella for other species, but also on specific information about the conservation feature's vulnerability to climate change.

Climate change vulnerability is defined as a function of a feature's (1) exposure to climate changes, (2) sensitivity to those changes, and (3) adaptive capacity to cope with those changes (Glick et al. 2011; IPCC-WG2 2007). Assessing exposure to changes in climatic variables such as temperature and precipitation can be done retrospectively through analysis of recently observed climate trends, or by projecting future climate conditions using global and regional climate models (box 2.1). Exposure to other physical and biological responses to climate change can be evaluated through a number of experimental, observational, and modeling techniques (more on this in the section on assessing climate change impacts). Species and ecosystem sensitivities to climate change are often determined through correlative or mechanistic research that identifies species and ecosystem tolerances for specific climate, physical (e.g., snow, ice, hydrology, fire, etc.), and biological (e.g., physiology, phenology, species interactions, etc.) conditions. Adaptive capacity is comparatively more difficult to quantify, but generally relates to the ability of a species or ecosystem to persist in place (e.g., through evolutionary changes or phenotypic plasticity), move to local microhabitats that are more suitable, or migrate longer distances to more suitable regions (Dawson et al. 2011).

Assessments of these three components of climate change vulnerability—exposure, sensitivity, and adaptive capacity—can help indicate which species, ecological processes, ecosystems, and places within a scape are most likely to be affected by climate change. This information can be used to prioritize species, processes, and ecosystems for adaptation planning, as well as to identify places within a larger region where adaptation efforts are most urgently needed.

It may be useful, or even necessary, to assess climate change vulnerabilities at larger scape scales. In some instances, spatial variations in climate change effects could lead to a species being deemed "highly vulnerable" within a particular area of high exposure, but "less vulnerable" across the majority of a larger scape. Taking a larger perspective on that species' vulnerability would suggest that dramatic interventions may not be necessary since the species as whole may continue to persist in many parts of the scape, even though a vulnerable population might disappear. On the other hand, if an ecosystem is highly vulnerable across most parts of the scape, but less vulnerable in one specific location, it may be desirable to invest heavily in the protection of the less vulnerable occurrence since it may be the only remaining stronghold for that ecosystem type. Assessing vulnerabilities at multiple spatial scales (i.e., across larger areas in combination with smaller spatial extents) allows us to consider a species' or ecosystem's vulnerability in multiple contexts, which may influence adaptation decisions. While vulnerability analyses can be useful to decision making, they do not in and of themselves dictate whether to focus on the most or least vulnerable, or somewhere in between. That choice is determined by the availability of options for achieving conservation goals and is often a value-laden decision.

Framing Climate-informed Conservation Goals

In the context of climate change, we can think about framing—and, importantly, reframing—conservation goals and objectives in three principal ways: resisting change, increasing resilience to change, and facilitating transformation (Millar et al. 2007). Resisting change entails forestalling the undesired effects of climate change. Most commonly, resilience refers to the amount of change or perturbation a system can absorb before it shifts to a fundamentally different state (Holling 1973; Gunderson 2000; Millar et al. 2007; West et al. 2009). Although the concept of ecological resilience to climate change has been contemplated for at least three decades (see Hare 1980), the term's exact meaning in the context of climate change is not always explicitly defined. Here we interpret both resisting change and increasing resilience as implying a goal of maintaining a system's present structure and function as climate changes. Facilitating ecological transformation, on the other hand, acknowledges that ecological transitions from one state to another are possible or even inevitable and focuses on passively or actively enabling those transitions to occur (Galatowitsch et al. 2009).

For example, consider a valued wetland ecosystem that is vulnerable to projected increases in drought frequency and severity. Managers may decide to take extreme measures to resist climate-related changes in this system by reengineering local hydrology to channel dwindling water resources into the wetland, with the hope of maintaining current conditions. A resilience-promoting approach for this same system might be to retain the natural hydrology but reduce withdrawals (e.g., for human consumption or agricultural irrigation) so that the water supply to the wetland is less stressed and the system is better able to naturally cope with droughts. An approach of facilitating transformation for the wetland ecosystem could involve acknowledging that the site may not be suitable for wetland vegetation in the future, and seeding the area with alternate plant species that are better suited to warmer and drier conditions and thus more likely to flourish as drought conditions become more frequent.

Much of the recent dialogue on preparing for the impacts of climate change has focused on efforts to promote the resistance and resilience of species and ecosystems (Hansen et al. 2003; Heller and Zavaleta 2009; Poiani et al. 2011). However, these approaches may simply not be achievable as the magnitude of climate change increases and species assemblages are rearranged, ecological processes are significantly altered, and ecosystems transition from one state to another. In situations where we expect an ecosystem to undergo major changes in ecological function and structure, we will need to shift toward actively managing that ecosystem through a period of transformation (West et al. 2009). It is imperative that whichever goal is adopted, agencies and organizations take time to set clear objectives for the future desired state of the ecosystem or species given the potential trajectory of climate change impacts. In general, the more transparent and explicit these goals and objectives are, the easier it will be to identify necessary and appropriate conservation actions that will be required as climate changes, and to determine the success of those actions toward achieving our goals.

For some species, ecological processes, and ecosystems that are widely distributed across large scapes, the rate and magnitude of climate changes they are exposed to may vary across that area. If this variation is great enough, it may be possible to identify places where conservation goals of resistance or resilience are an option because the magnitude of expected changes is lower and it may be possible to reduce that conservation feature's vulnerability to climate change. In other places, the magnitude of expected changes may be sufficiently large to demand a focus on facilitating transformation. Where such variability exists, it may be possible to make decisions about where to invest resources in the long-term conservation of species and ecosystems that currently occupy a scape, and where to instead focus on managing for transformative change (Hobbs et al. 2009).

Assessing Climate Change Impacts

As mentioned earlier, future climate change exposure is typically assessed using output from global or regional climate models (box 2.1). Historic and future climate data is becoming increasingly available to conservation practitioners and planners through tools and websites such as Climate Wizard (climatewizard.org), Data Basin (databasin.org), WorldClim (worldclim.org), PRISM (prism.oregonstate.edu), World Bank Climate Portal (sdwebx.worldbank.org/climateportal), Consultative Group for International Agricultural Research (ccafs-climate.org), and EcoClim (ecoclim.org), among others. Climate model outputs can provide information about the range of future changes expected for a given region or locality, as well as spatially explicit climate data layers that can be used as inputs to biological and physical models (as described later and in box 2.2).

Approaches for anticipating the biological and ecological impacts of climate changes include experimental manipulations, analyses of historical and paleoecological observational records, and modeling (Dawson et al. 2011). Experiments in the field and laboratory allow us to examine direct responses to manipulated climate conditions (e.g., Shaw et al. 2002), but because they are usually conducted at a relatively small scale, it is difficult to extrapolate results to larger scape scales. Historic observational records over decades or sometimes centuries allow researchers to examine correlations between climate and ecological conditions for more mobile species and for more numerous points across larger scapes (e.g., Parmesan 2006), but are dependent on the existence of sufficiently long records. Paleoecological proxies of ecological and climate variables can reveal correlations over centuries or longer, but both recent historical and paleoecological observations are limited in their ability to illuminate causal relationships. They are also unable to predict the effect of conditions that may be outside the range of variability experienced during the time frame of the study.

Models provide an alternate approach for projecting future impacts of climate change at larger spatial scales. A range of modeling approaches exists that can provide input to climate change-adaptation planning, including species bioclimatic envelope

BOX 2.2 INCORPORATING CLIMATE CHANGE INTO SPATIAL CONSERVATION PLANNING

Several general adaptation principles described in chapter 1 relate to spatial conservation planning at the scape scale, such as the design of protected areas and connectivity, and the identification and protection of climate refugia. Some efforts to incorporate climate change into reserve design and connectivity planning take a relatively qualitative approach by calling for expanding and aligning protected areas and connectivity along north-south latitudes and low-high elevation gradients to capture poleward and upward range shifts (Cross et al. in press). Other efforts rely on quantitative analyses that integrate data on current and future distributions of climate, species, and ecological conditions into spatially explicit planning software such as Worldmap, SITES, Marxan, and Zonation (e.g., Hannah et al. 2007; Carroll et al. 2010).

Planners have also used climate model outputs to identify areas of future climate refugia where particular climate conditions continue to exist either in situ (overlapping with the current distribution of those conditions) or ex situ (outside of the current distribution) (Ashcroft 2010). Spatial climate data can also be used to examine questions related to how the optimal climate envelope for particular species (e.g., Pearson and Dawson 2003) or plant functional types and biomes (e.g., Gonzalez et al. 2010) may shift as climate changes. However, caveats related to the use of fine-resolution climate data (box 2.1) need to be accounted for in these analyses. In cases where modeled climate data is less reliable, or there are concerns about uncertainties associated with downscaled data, expert opinion on where to focus spatial conservation priorities can offer a valid alternative. However, biases can also confound expert opinion-based assessments.

An alternative to using climate and species distribution data to drive spatial conservation priorities is to take a relatively coarse-filter approach by identifying geophysical settings such as geology and elevation (Anderson and Ferree 2010) or areas of high diversity and interspersion of land facets, described by Beir and Brost (2010, 307) as "recurring landscape units with uniform topographic and soil attributes." The assumption is that these areas represent important biological arenas for supporting biodiversity, and yet they are composed of enduring features that are less likely to be altered by climate change (Anderson and Ferree 2010; Beier and Brost 2010). This approach of "conserving the stage rather than the actors" does not depend on uncertain climate-model projections, but it is data intensive and may be inappropriate for areas that do not have good baseline data on land facets. Since this type of coarse filter approach may not meet the needs of every species or taxa of conservation concern, it is likely to be an important complement to other approaches for setting spatial conservation priorities, rather than a substitute.

models, also called species distribution models (e.g., Pearson and Dawson 2003); terrestrial and aquatic ecosystem models (e.g., Cramer et al. 2001; Field et al. 2006); hydrological models (e.g., Gray and McCabe 2010); sea-level rise models (e.g., Park et al. 1989); phenology-based process models (e.g., Chuine and Beaubien 2001); and models that simulate changes in disturbance regimes such as wildfires, coral bleaching events, disease outbreaks, etc. (e.g., Hoegh-Guldberg 1999; Logan and Powell 2001; McKenzie et al. 2004). Also, various types of climate change risk assessments rely on these and other modeling approaches to estimate the probability of particular impacts (e.g., Schrag et al. 2011; Williams et al. 2009). Efforts to model human responses to climate change, such as regional shifts in agricultural patterns (e.g., Bradley et al. 2012; IPCC-WG2 2007) and land-use change related to renewable energy development (e.g., McDonald et al. 2009), can also be relevant for biodiversity conservation. Each of these approaches to modeling future climate change impacts has trade-offs in terms of the resolution, scale, and dynamic nature of projections, and whether they are able to include information about complex species and ecological interactions. For additional information on climate change modeling for conservation applications, see Lovejoy and Hannah (2005) and Hansen and Hoffman (2011).

Rather than relying on one single approach, it is best to combine these complementary methods for assessing climate change impacts and vulnerabilities (Dawson et al. 2011). Efforts to integrate experiments, observations, and modeling may be especially valuable for large scape adaptation planning since the scale limitations of experimental approaches can be balanced by the ability of modeling and observational approaches to consider larger areas. The limitations of models to incorporate the full complexity of real ecosystems can in turn be balanced out by inclusion of research methods that better account for those complexities.

For scape-level adaptation planning, it is useful to examine whether and how impacts may vary within a scape. This especially may be the case in regions with diverse topographic, microclimate, or microhabitat variation, such as mountain ranges or other areas of complex terrain. Even in more homogeneous scapes, climate change could negatively affect conservation features in some places while benefiting them in other places. For example, in regions that span substantial north-south latitudinal gradients, some widely distributed species may experience negative effects of warming and drying in the southern portion of the landscape, but positive effects farther north due to a reduction in cold-temperature limitations to survival. This variability in climate change impacts across a scape can inform efforts to select focal conservation features and determine whether conservation goals might vary across a larger area (as described earlier), and assist in the translation of general adaptation principles into concrete actions (see next section).

At any scale, incorporating the best available climate change science into adaptation planning will be challenging. One particularly difficult challenge stems from the fact that when extensive research exists to draw upon, not all of it can be directly compared. For example, not all ecological modeling studies are driven by the same climate data inputs since they may have chosen to use different global models or apply different downscaling techniques. Or the spatial and temporal scale at which climate change research was conducted may not match up with the geographic boundaries or time frames being considered during planning. A more basic challenge is that in many parts of the world, little or even no targeted climate change research exists to inform planning. For both of these reasons, adaptation planning often not only draws on quantitative science research, but also integrates a more qualitative component such as expert-driven syntheses of a breadth of climate change research that has both direct and indirect applicability to the focal scape. While targeted research aimed at informing adaptation planning and action is desirable, the absence of such information should not inhibit adaptation planning using more qualitative and expert opinion-driven approaches.

Identifying and Prioritizing Adaptation Actions

Chapter 1 reviewed the general adaptation principles that are important to shaping broad discussions about whether and how to adjust conservation in light of climate change. Conservation practitioners need ways to translate these general principles into practical actions appropriate for the features of interest in a particular scape. Identifying scape- and feature-specific strategies and actions for adaptation is challenging due to (1) the range of conservation goals held by various agencies and organizations at different spatial and temporal scales, (2) varying social, political, and economic circumstances (Adger et al. 2005; Smit and Wandel 2006), and (3) the inherent uncertainties in projecting future climate changes and ecological responses (Lawler et al. 2010).

Climate change science can inform which adaptation principles are necessary for achieving conservation goals for a given species or ecosystem. For example, if a species' habitat is projected to become fragmented, or existing habitat patches are likely to disappear or shift to new locations, enhancing or protecting connectivity between those isolated patches may be an important tool for increasing that species' ability to cope with climate change (Cross et al. in press). If the effects of climate change on a target cannot be directly ameliorated, then one possible adaptation strategy may be to reduce nonclimate stressors. For example, we can do little to prevent a decrease in summer precipitation from reducing summer streamflows in a rain-fed river system. Given this, conservation efforts may focus on increasing natural water storage and slowing water runoff, such as through restoration of riparian vegetation or restoration of beaver populations and their dam complexes. If the science suggests that the magnitude of negative impacts is likely to be so significant that there are no options for mitigating those negative effects, then it might be necessary to focus conservation efforts elsewhere or on other features.

Once the menu of adaptation options is filtered for those that are most applicable to a particular species or ecosystem, planners can focus on translating those general strategies into concrete actions, and prioritizing among those actions. Adaptation planning might initially concentrate on identifying and prioritizing those actions that are recommended under current climate conditions as well as multiple future scenarios. These are considered beneficial regardless of how climate changes—sometimes called "no regret" actions (Willows and Connell 2003). While no regret actions may be relatively easy to identify, they are not likely to be the only measures needed to achieve conservation goals as climate changes. In addition to the potential utility of actions across multiple climate scenarios, planners might consider the following criteria for prioritizing among potential actions (modified from USAID 2007):

- relative contribution to achieving a particular conservation goal
- economic, social, and political feasibility
- potential for positive synergistic effects on other conservation targets
- risk of unintended consequences
- ability to be modified or undone (i.e., if subsequent planning indicates that actions need to be altered or abandoned)
- opportunities for implementation including whether actions are similar to current practices or involve significant modification of current practices

Having identified key adaptation actions, it may be desirable to prioritize where within a scape to implement those actions. Several approaches exist to use qualitative and quantitative information on climate and other abiotic factors to direct the design of protected areas and connectivity, and the protection of climate refugia (box 2.2). Climate change information can also direct where within a scape to focus efforts to ameliorate climate-related impacts and reduce nonclimate stressors. For example, conservationists aiming to protect cold-water fish species may choose to restore riparian vegetation and limit stream water warming in those areas that are most likely to stay sufficiently cold in the future because they are at a higher elevation, are fed by cold-water underground springs, or are expected to undergo less warming according to climate model projections. Similarly, practitioners may concentrate conservation activities such as reducing agricultural runoff near those coral reefs that are likely to be less exposed to future climate-related stresses (Maina et al. 2011).

Even with these criteria and tools to assist in prioritizing adaptation actions in particular places within a scape, prioritizing among actions is often one of the most difficult adapation planning steps to accomplish (e.g., Seimon et al. 2011). This may partly be related to the unwillingness of decision makers to take risks in the face of uncertainties associated with both projecting climate change impacts and knowing which actions will be necessary to accomplish a given conservation goal. It also may stem from the fact that risk tolerance and conservation goals and priorities can vary widely across a diverse set of stakeholders involved in collaborative scape scale-adaptation planning.

Implementing Actions at a Scape Scale

Once priority adaptation actions are identified, an action plan can be developed and implemented (fig. 2.1). As with other steps in the adaptation planning and action process, implementing large scape-adaptation actions offers both challenges and opportunities. It is rare that we can implement conservation actions across a large area all at once. For example, while occasional opportunities exist to increase the protected status of very large areas, as recently occurred in Nahanni National Park in Canada (Ford 2011), it is more common that smaller areas are protected piece by piece through conservation easements, agreements with local inhabitants, or incentive-based payments for stewardship management. Scaling up the application of conservation actions at individual sites is a challenge to accomplishing larger scape-scale conservation goals. The need for cross jurisdictional coordination across scapes also presents a hurdle when attempting to scale up conservation actions. However, when adaptation planning is undertaken collaboratively across the public and private jurisdictions that make up a large scape, that collaboration can provide the groundwork for implementation at that scale. Another opportunity for implementing adaptation actions for conservation is to focus on those actions that benefit human as well as natural systems. Ecosystem-based adaptation (discussed in chap. 1) explicitly draws on the connection between conservation of biodiversity and human well-being to increase the likelihood of acceptance of adaptation actions.

Monitoring and Revisiting Plans

It is important that structured climate change-adaptation planning processes be revisited periodically to accommodate updated ecological information, revised climate projections, and shifts in conservation goals or sociopolitical priorities. With limited resources, it may be sufficient to do a relatively simple review of a plan to ensure that new scientific findings or monitoring results do not undermine previous assumptions. More formal plan revisions might be appropriate when a new generation of climate modeling and associated impacts assessments are completed (e.g., the Intergovernmental Panel on Climate Change [IPCC] provides updated syntheses of climate model results and impacts research roughly every five years). By committing to a structured and iterative process for considering the consequences of climate change, practitioners will better understand when it will be necessary to focus on climate change as a new and distinct threat, to revisit conservation goals and priorities, or to reprioritize conservation actions.

Because of the uncertainties in projecting future climate changes, ecological responses, and the effectiveness of conservation actions, it is often recommended that adaptation actions be implemented in an adaptive management framework (Hansen et al. 2010; Hansen and Hoffman 2011). In other words, the implementation of adaptation actions should be supported by efforts to monitor appropriate climate-sensitive indicators of environmental change and resultant impacts. For example, scientists, conservation practitioners, and trained volunteers (such as citizen scientists) could track key variables related to climate (e.g., precipitation and temperature, including growing degree days), physical system responses (e.g., snow, ice, and hydrology), and the response of species and ecological systems (e.g., metrics related to demography, phenology, ecological mismatches, seasonality, and range shifts). Moreover, carefully designed monitoring can be used to measure the efficacy of implemented conservation actions in light of climate and environmental changes. Based on this information, adaptation actions or management goals can be adjusted—or adaptation planning can be revisited altogether (fig. 2.1). One advantage to coordinating monitoring efforts across large scapes is the ability to leverage funding to support spatially distributed monitoring networks over sufficiently long time periods. In the United States, examples include the Long Term Ecological Research (LTER) network, the National Ecological Observatory Network (NEON), and the USA National Phenology Network. Developing consistent, standardized monitoring protocols across jurisdictions will also be necessary to maximize the comparability of resulting data.

Conclusions

Because of the myriad ways that climate change stands to alter physical, biological, and human systems, it is imperative that we explicitly address the challenge of climate change when making biodiversity conservation decisions. Looking at biodiversity conservation through the lens of climate change requires us to reconsider how we frame our conservation goals, integrate the best available climate change science into decision making, and identify appropriate and necessary conservation actions. Looking at conservation through the lens of climate change at large-scape scales has challenges, such as the ability to set priorities and goals across jurisdictional boundaries and uncertainties related to modeling climate variables at fine resolutions. Taking a larger perspective on conservation also has advantages, for example the ability to implement important adaptation principles such as designing, enlarging, and establishing protected areas, enhancing connectivity, and using ecosystem-based adaptation for people. While there has been a recent surge in attempts to address the threat of climate change in conservation planning and action, our understanding of how to do so is constantly evolving. Given this, iterative and flexible approaches to conservation that allow for the incorporation of new information, and changing societal values and political realities are needed.

The case studies included in this volume illustrate conservation efforts at varying points in the process of looking at conservation through the lens of climate change and implementing scape scale-adaptation actions. To various degrees they follow the planning steps outlined in this chapter and illustrate the struggles and prospects that conservation practitioners are facing. They serve as models to inform and inspire other practitioners aiming to integrate climate change into conservation goals, planning, and action.

Acknowledgments

We are grateful to James Watson and Erika Rowland for their helpful comments on earlier versions of this chapter.

PART 2

Equatorial and Tropical Landscapes

The three case studies that follow indicate that incorporating climate change adaptation into these landscapes is challenging for two central reasons. First, all three landscapes face inadequate investment in systems for climate science monitoring and data generation. Second, human communities in all three landscapes are beset with high levels of poverty. Consequently, successful climate adaptation for biodiversity in these regions must give considerable attention to livelihood issues.

Chapter 3

Albertine Rift, Africa

ANTON SEIMON AND ANDREW PLUMPTRE

Despite high human population density and extreme levels of poverty, Albertine Rift remains one of Africa's most important conservation priorities. Incorporating what are likely to be major impacts of climate change into conservation in this landscape requires the development of basic information on transboundary cooperation relationships between climate and biodiversity and addressing short-term conservation needs as well as longer-term planning. Given uncertainties in how climate change will unfold in the Albertine Rift landscape, conservation priorities are focused on safeguarding high-elevation and mountainous habitats, and maintaining or reestablishing connectivity between those areas.

Introduction to the Region

Africa's Albertine Rift runs from the northern end of Lake Albert to the southern end of Lake Tanganyika, encompassing watersheds that straddle several countries: the Democratic Republic of Congo (DRC), Uganda, Rwanda, Burundi, Tanzania, and Zambia (fig. 3.1). It forms Africa's western rift valley, and links with the Eastern or Gregory Rift Valley at Lake Malawi. Rich in diversity, it is home to more than 50 percent of Africa's birds, 39 percent of mammals, 19 percent of amphibians, and 14 percent of reptiles (Plumptre, Davenport, Behangana, et al. 2007). It contains more endemic and threatened vertebrate species than any other region on mainland Africa, including the mountain gorilla (*Gorilla beringei*), the golden monkey (*Cercopithecus*

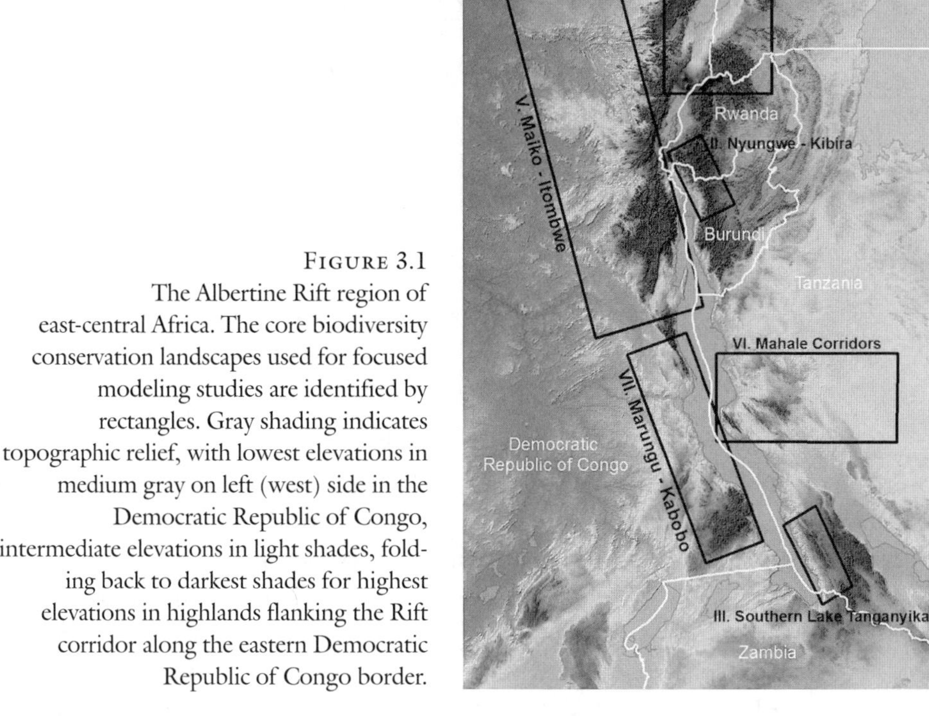

FIGURE 3.1
The Albertine Rift region of east-central Africa. The core biodiversity conservation landscapes used for focused modeling studies are identified by rectangles. Gray shading indicates topographic relief, with lowest elevations in medium gray on left (west) side in the Democratic Republic of Congo, intermediate elevations in light shades, folding back to darkest shades for highest elevations in highlands flanking the Rift corridor along the eastern Democratic Republic of Congo border.

kandti), forty-two bird species, and numerous reptiles, amphibians, fish, and invertebrates. Many of the endemic and threatened species—including plants—are confined to mountaintops or highland ranges/massifs that are separated from each other by lowland areas that either consist of different ecosystems or have been extensively cultivated for hundreds of years.

Across Africa there is a significant relationship between high biodiversity and human population density and infrastructure (Burgess et al. 2007). The Albertine Rift is home to some forty to fifty million people, the majority of whom are subsistence farmers and hunters. With some of the highest population densities of rural people on the continent—up to 800–1,000 people/km^2 at some sites—the region also suffers from some of the highest levels of poverty on the continent. As a result, people rely on access to the forests for their livelihoods, with income generated from forest products contributing to between 8 and 30 percent of annual household income where surveyed (Bush et al. 2004; A. Plumptre, unpublished data). This is of particular importance in supplementing seasonal "hungry gaps" when crops are not available in fields, as well as in providing funds for education and health treatment. Over 90 percent of the region's population consists of subsistence farmers, a low percentage of whom receive adequate education and health treatment (Plumptre et al. 2004). Between 55 and 65 percent of the population is under the age of twenty, and human population growth is between 2 and 3 percent per year; as a result, land is scarce and young people are migrating to cities or elsewhere to find work or land to cultivate.

Historical Overview of Conservation and Science Initiatives

Formal conservation of the fauna and flora of the Albertine Rift dates back to the early 1900s, although prior to then certain tribal groups practiced traditional forms of conservation. The European countries that colonized the Albertine Rift—notably Britain, Germany, and Belgium—applied the nascent ideas of parks and reserves in the region. Africa's oldest park, the Virunga Park in DRC, established in 1925, is in the Albertine Rift. Protected area authorities were initially created to manage hunting concessions for the most part, as it became clear that large mammals were becoming locally extinct in some areas. Each of the countries in the Albertine Rift has its own national parks authority that is legally mandated to conserve the fauna and flora in its protected areas and throughout the country as a whole. Many nongovernmental organizations (NGOs), both national and international, also promote conservation in these countries by supporting these protected area authorities or by working with local authorities and community groups.

In 2001, a meeting supported by the MacArthur Foundation and coordinated by the Wildlife Conservation Society (WCS) brought all of the protected area authorities and NGOs together in the Albertine Rift region to identify the threats to, and needs for, conservation in the Albertine Rift. Participants recommended (1) the drafting of a strategic plan for the conservation of the Albertine Rift, and (2) the establishment of a regional monitoring program. A core group of NGOs was elected to move these two objectives forward. These included the Albertine Rift Conservation Society, Dian Fossey Gorilla Fund International, Institute for Tropical Forest Conservation, International Gorilla Conservation Program, Makerere University Institute of Environment and Natural Resources, WCS, and the World Wide Fund for Nature (WWF). This led to a compilation of the information that existed on species distributions in the Albertine Rift (Plumptre et al. 2003; Plumptre, Davenport, Behangana, et al. 2007) and the development of a strategic plan for the Albertine Rift in 2004 (ARCOS 2004). This plan identified six key landscapes of protected areas for conservation in the Albertine Rift (fig. 3.1). Since 2004, protected area authorities and NGOs have been working to conserve these landscapes, where necessary, surveying the landscapes to identify new target areas for protection and to refine the boundaries of existing protected areas, and supporting transboundary and landscape conservation within them. Work on the regional monitoring program has also progressed with the ten principal Albertine Rift research stations working to share their information and results; this effort has culminated in the publication of a book on long-term changes at the station sites (Plumptre 2012). A regional database was also established by Makerere University to house data on species and survey results from the region.

Current Conservation in the Region

In most of the region, conservation areas were initially created to conserve large mammals. The priority landscapes identified in figure 3.1 contain several protected areas that in some cases constitute most of the landscape (e.g., 100 percent in landscape 4) while in other landscapes large areas remain unprotected (e.g., only 11 percent of landscape 5). Overall, 50 percent of the six landscapes are protected as some form of reserve or national park (Plumptre et al. 2009), and efforts are under way to increase this percentage, particularly in landscapes 3, 5, and 6. Transboundary conservation has been established at a formal level in landscapes 2 and 4, both of which straddle international borders and where studies have shown the importance of trans-boundary connectivity for the long-term conservation of some large mammal species (Plumptre, Kujirakwinja, Treves, et al. 2007; Treves et al. 2009). Elsewhere, conservation programs are focusing on conserving corridors within landscapes where they occur within the same country. For instance, within the Murchison-Semliki landscape in Uganda, WCS has identified key corridors for species such as chimpanzees (*Pan troglodytes schweinfurthii*), golden cats (*Profelis aurata*), and forest raptors, and is looking at the options for Reduced Emissions from Deforestation and Degradation (REDD) funding to conserve corridors where there is existing natural habitat between protected areas.

Efforts to conserve biodiversity in Albertine Rift are at present debilitated by minimal environmental monitoring and other data resources, weak institutions, and economic underdevelopment that characterize much of the region. The high human population density and its increasing growth have precluded attempts to establish corridors in cultivated land because of the difficulties involved. However, some interventions are being designed in the Murchison-Semliki landscape to find incentives for farmers to grow trees in key corridor areas where the forest has been destroyed. Mining concessions in eastern DRC have also been allocated on lands located between protected areas where natural habitat could form corridors, resulting in the need to engage with mining companies to minimize impacts such as bushmeat hunting that follows the opening up of the forest in concession areas. While of little consequence at present, climate change constitutes a long-term impediment to conservation that is likely to grow inexorably in significance and become a major concern, both as a direct force for change and as a factor influencing other landscape drivers and threats.

Regional Effects of Climate Change

The Albertine Rift straddles the equator in east central Africa, and thus falls within a belt of projected rainfall increases encircling most of the globe over the inner tropics. General circulation model (GCM) predictions for the twenty-first century, such as those performed for the IPCC Fourth Assessment Report, provide relatively consistent projections for the Albertine Rift of significantly warmer and wetter climates as time progresses (IPCC 2007). The individual models under various prescribed greenhouse gas emissions scenarios vary in degrees of warming and precipitation, but by the latter part of the century the signals are consistently positive for both parameters across almost the entire Albertine Rift.

Although such consistent trend signals entail a relatively advantageous situation for adaptive planning that incorporates climate change, the baseline states of Albertine Rift climates are only known at a very general and broad-scale level. The capacity to anticipate the impacts of predicted climatic changes upon biodiversity will require a holistic understanding of the relationship of the present-day climate system to ecological systems and their spatial distributions. Yet climate change studies focused on the Albertine Rift are hindered by a lack of prior investigation on regional climatology. This contrasts markedly with the Lake Victoria basin to the east, where various academic studies have addressed climate and climatic variability of the past, present, and future. Moreover, the relative absence of quality-controlled, multidecadal, climatological data means that baseline knowledge on climatic conditions is poorly developed, making comparison of climate model output to on-the-ground conditions especially problematic. There are no formally registered climatological observation sites (i.e., by the Global Historical Climatological Network) inside any of the Albertine Rift protected areas. Consequently, baseline climatologies utilized in modeling studies are interpolated surfaces from a sparse network of often distant observation sites in dissimilar climatological environments.

The study and integration of climate change into Albertine Rift conservation management and planning are still at an early stage. Recent funding initiatives, particularly from the MacArthur Foundation, have catalyzed multiple research efforts by international and regional groups that include governmental, NGO, and academic scientists (Seimon and Picton Phillipps 2011). Initial results are becoming available, mainly in the form of vulnerability assessments utilizing geospatial modeling and/or downscaled climate model output. Some tentative first steps toward actual adaptive actions are being considered, though as of yet have not been widely applied.

The most comprehensive project to apply climate change to conservation planning at a regional scale to date has been the Albertine Rift Climate Change Assessment, a study conducted by WCS to develop understanding of potential impacts of anthropogenic climate change on wildlife conservation and protected area management. In its first phase (2007–09), the project examined baseline climatological conditions within protected areas, used downscaled IPCC model output to quantify predictions of regional climate change across the Albertine Rift, assessed possible future impacts, and developed products that aid in estimating future distributions of biodiversity in the Albertine Rift. The project has since been developing and applying these findings through outreach to the wider biodiversity conservation community. Through the utilization of dynamic vegetation and crop models, the modeling approach was designed to generate a suite of products that now offers a first look at the potential impacts of anthropogenic climate change on wildlife habitat, key cultivars, and carbon budgets throughout the Albertine Rift region (Picton Phillipps and Seimon 2009). An additional output has been detailed climatological analysis within Albertine Rift protected areas, shedding light on previously unrecognized phenomena such as high-amplitude rainfall fluctuations within wet seasons, and helping to ascertain baseline conditions for assessing climatic changes within protected areas (Seimon and Picton Phillipps 2011). The second phase of the climate assessment project is focusing on implementing long-term monitoring for climate change principally through climatological observations and vegetation and species monitoring within protected areas. The assessment has identified near unanimity in climate model projections for the region, showing a strongly warmer and wetter climate evolving by the end of the century that will likely promote biome transitions, large displacements in ecotones, species range changes along with socioeconomic implications through impacts on cultivation and pastoralism, among many others.

Other groups have been conducting environmental modeling studies under projected future climatic conditions within the Albertine Rift domain and adjacent regions. These include BirdLife International, University of Durham, Albertine Rift Conservation Society, and WCS for avifauna in an ongoing study; the University of Edinburgh for vegetation (Doherty et al. 2009); the African Wildlife Foundation and International Gorilla Conservation Program on mountain gorilla vulnerability (AWF et al. 2011); and the International Livestock Research Institute for cultivation (Thornton et al. 2009). Although these research efforts have considerable complementarity, they have been largely performed in isolation from one another, and to date have only reached relatively limited subsets of potentially relevant stakeholders in biodiversity conservation. There is as yet no common repository for model output and related data for the Albertine Rift.

Approaches to Conservation under Climate Change

Recent scientific evidence gathered in the IPCC Fourth Assessment Report indicates that the Albertine Rift, similar to many parts of Africa, will be both particularly vulnerable to climate change-related stresses and subject to a very low adaptive capacity (IPCC-WG2 2007). Specifically, the predicted increases in temperature and concurrent changes in rainfall will exacerbate existing environmental degradation in many parts of East Africa, and threaten large populations of subsistence farmers and hunters with low financial, technical, and institutional capacity who are struggling to survive and are employing unsustainable land-use practices. Nevertheless, human subsistence pressures are causing the replacement of suitable habitat for wildlife and biodiversity with crops and pastureland, leading to ever smaller and more isolated populations of the diverse fauna and flora found here. This is particularly true for the nonnomadic and nonvolant strict endemic species prevalent in the Albertine Rift that may have limited options for adapting to climate change as their habitats continue to shrink.

The dire context of this situation is amplified in the Albertine Rift, a region that also suffers environmental change drivers such as rapid human population growth, challenges of human and wildlife diseases, poverty, and, in some areas, political instability. It is therefore important that potential climate change impacts on biodiversity in the Albertine Rift be more fully explored and quantified to facilitate timely planning of priority research, and measures applied correspondingly to mitigate the threats.

While conservation agendas across the Albertine Rift region are beginning to incorporate climate change, this response appears to have been prompted more by external influences than by concerns from conservation interests within the region. Specifically, there are two major drivers prompting the conservation community to focus on climate change: one related to mitigation, the other to adaptation. On mitigation, interest in greenhouse gas emissions mitigation has been spurred by the development of international funding mechanisms that value standing forests for their carbon content and other ecosystem functions, specifically, the Reducing Emissions from Deforestation and Degradation (REDD and REDD+) process. Projects are under way at many sites through the Rift region to map and assess the carbon content for REDD project applications, motivated by win-win prospects for habitat preservation and revenue generation.

In terms of climate change adaptation, new project-funding opportunities—particularly those of the US-based MacArthur Foundation—have likewise catalyzed considerable activity. This occurred as an expansion of the foundation's decade-long funding initiative of the Albertine Rift as a priority "hotspot" for biodiversity conservation. To date, the research groups receiving support for climate change-adaptation initiatives have not coordinated their efforts amongst each other, resulting in largely independent studies with only limited informational interchange. In an effort to address this, WCS hosted a meeting in Rwanda in 2011 that brought researchers and stakeholders together to develop a series of recommendations that would improve the linkages between science and action in terms of climate change adaptation for conservation in the Albertine Rift region. Participants included representatives from major conservation NGOs; scientists and policy makers from the governments of Rwanda, Democratic Republic of Congo, Uganda, Tanzania, Burundi, and Kenya; as well as educators and scientists from academic institutions and major funders for conservation and research. With key scientists, decision makers, and stakeholders all participating, the joint recommendations thus developed will represent a valuable consensus perspective on climate change threats and opportunities for the region.

Landscape Conservation Needs Assessment in Response to Climate Change

The conservation landscape of the Albertine Rift covers a large and heterogeneous spatial domain, so a region-wide assessment on climate change would be a large undertaking and perhaps less informative than more focused subregional studies scaled to individual landscapes and protected areas. Identification of the priority conservation landscapes (fig. 3.1) were based on natural habitat cover and presence of endemic and threatened species, and not on climate change impacts. We have been undertaking an assessment of how climate change might affect these individual landscapes with a view to providing recommendations about their shape and size so that they will be more adaptable to climate change. Additionally, there have been some efforts at needs assessment for individual protected areas. For example, by drawing upon the combination of extensive field data and model output generated by the climate assessment study, we have been working to integrate climate change adaptation and monitoring into strategic planning for the Nyungwe National Park in Rwanda. Particular threats identified include increasing fire occurrence, changes to hydrological regimes, and possible destabilization of an effective buffer zone along park margins currently provided by tea plantations, which may lose their agricultural viability if climate warming crosses a thermal tolerance threshold for this cultivar. Also of relevance are national-level assessments for the Albertine Rift countries for the National Adaptation Programs of Action (NAPA) process of the United Nations Framework Convention on Climate Change; these are not conservation focused, however, and the attention paid to environment and conservation issues varies among the published reports (UNFCCC N.d.). With relatively little guidance currently available, it is not surprising that climate change-adaptation planning for conservation is still for the most part undeveloped throughout the Albertine Rift region. Efforts to date have mostly been focused on vulnerability assessments and modeling of future states, which are among necessary building blocks for deliberate actions but which also generally fall short of implementation objectives. There are some activities that are currently under way. One example is the conservation corridor mapping effort described earlier, and there is a comprehensive fire management strategy being developed for the Nyungwe National Park in Rwanda where increasing dry season severity is expected to lead to corresponding increases in fire occurrence and extent.

Roadblocks and Opportunities

As suggested earlier, there are several roadblocks to the incorporation of climate change into conservation planning and action in the Albertine Rift. These are related to the availability of climate information, land-use changes around protected areas, and the low sense of urgency assigned to the threat of climate change.

Severe observational data deficits handicap efforts to model and assess the impacts of climate change on biodiversity in the Albertine Rift. These deficits come in both the form of poor climate-monitoring data, in terms of both quality and quantity, and in the absence of biodiversity monitoring tailored to the detection of climate change impacts. Critically needed are sustained, research-grade, *in situ* observations that can serve as baselines for assessing change and trend behavior, and that can be incorporated into increasingly sophisticated climate and landscape models (Seimon and Picton Phillips 2011). The absence of baseline knowledge on how local climatology within protected areas controls ecological systems makes it difficult to assess how changes in a set of modeled parameters in a given park's domain will drive ecological responses. Furthermore, the resolution of the current suite of climate models is wholly insufficient to capture the topographic and ecological complexity of the Albertine Rift, a region of complex terrain characterized by large relief. While techniques such as statistical down-scaling can be applied to GCM output, as was done for the WCS Climate Assessment project, the resulting model grid scale (~50 km) still falls far short of that necessary to capture the landscape complexity (fig. 3.2).

Over time, the traditional fixed-boundary model for protected areas in the Albertine Rift has resulted in some extremely abrupt ecological discontinuities along many park margins, rather than buffered transition zones. As a result, protected areas are becoming clearly delineated islands of biodiversity increasingly isolated from one another except where they are contiguous. Human settlement and agricultural activity bordering protected areas are often intensive, with land transformed for agriculture and settlement extending to the park limits. In the case of Volcanoes National Park in Rwanda, a recent survey assessed the human population within 5 km of the park boundary at more than 1,000 per square kilometer in some parishes. These settlement patterns run contrary to many of the recognized characteristics of effective protected areas, as well as the capacity of resident species to adapt to climate change. Many species, particularly larger mammals, are effectively confined within park borders, and the only options for movement as an adaptive response to changing climatic conditions and attendant habitat shifts are either upward, where topography within park boundaries allows, or to other localities within the restrictive domain of the protected area.

Finally, the intense pressure placed on protected areas from present-day human land-use practices and exploitation of resources—such as timber and protein from

FIGURE 3.2 Three-dimensional digital elevation model of the Greater Virunga Landscape in the Albertine Rift viewed to the north and illustrating the scale of climate model grid cells, after downscaling to half-degree (~50 km) resolution, relative to landscape features. Notable features are the Virunga volcanoes in the south of the scene (bottom), the Rwenzori Mountains to the north (top center), and Lake Edward (center) and Lake George (upper right). The major part of the Rwenzori, an entire mountain range with an elevation range of more the 3,500 meters, falls within a single grid cell.

bushmeat—renders the threat from coming climatic changes a comparatively low, esoteric priority. It is therefore difficult to make the case that conservation planning should be changed in the near term to account for climatic change expected in the relatively distant future.

Despite these challenges, there are opportunities to think about conservation in the Albertine Rift through a climate change lens. From a macroscale perspective, the mountainous highland regions of the Albertine Rift flanking the region's great lakes collectively stand as a best-hope refuge for sustaining biodiverse ecosystems with a strong semblance to present form over the long term. This is readily apparent inasmuch as the mountainous uplands of the Albertine Rift contain the greatest topographic relief and elevated terrain, although environmental modeling amply reaffirms this inference and allows it to be quantified. The Albertine Rift's rich geographic endowment, which is without parallel in scale and extent throughout tropical Africa, means that this region has long-term potential to serve as a haven for biodiversity in the face of anthropogenic climate change—with the critical proviso that other, more imminent conservation threats such as habitat conversion and overexploitation are adequately constrained and controlled. On the other hand, the upward taper of topography also means that range contractions of biota will occur for species that track their preferred thermal envelopes as they rise upward.

The magnitude of overall climatic and ecological change by century's end appears to be greatest in the southern parts of the rift domain, from Mahale through southern Lake Tanganyika. In this region, modeling results from the WCS Climate Assessment for plant functional types under significantly increased rainfall regimes indicate a shift from the current dominance of deciduous toward evergreen forests by the close of the twenty-first century. At more focused scales, the adjacent Nyungwe and Kibira transfrontier parks of Rwanda and Burundi, respectively, exemplify prospects for long-term viability of highland protected areas. From a biodiversity conservation perspective, environmental modeling shows an overall picture of the future for this subregion to be relatively favorable due to the high mean elevation with large local relief of the landscape and extensive native forest cover in protected areas. In the face of climate change, such biodiversity strongholds could be anticipated to have particularly enduring sustainability compared to many other Albertine Rift protected areas, especially those at lesser elevations.

Therefore, among highest priorities for conservation from a climate change perspective in the Albertine Rift should be to safeguard as many high-elevation and mountainous habitats as possible. Existing protected highland forests such as at Nyungwe, Kahuzi Biega in DRC, and Rwenzori National Park in Uganda offer high potential to serve as biodiversity refugia over the long term. As a corollary, priority would be to maintain or otherwise reestablish ecological corridor linkages interconnecting the various ranges, massifs, and volcanoes. Protected areas where the major part of the habitat occupies low elevations along the Albertine Rift valley floor appear to face a particularly dire predicament over the long term from excessive heat and desiccation; this includes some very important parks such as Uganda's Semuliki and Queen Elizabeth National Parks.

Conclusion and Recommendations

The Albertine Rift, one of the world's most important priority regions for biodiversity conservation, is now confronted by climate change as a key threat to the long-term persistence of plant and animal species as well as the human livelihoods that depend upon them. Research and conservation groups have begun to conduct environmental modeling studies that are bringing forth a wealth of new insights on the potential impacts of climate change within the Rift. The absence of precedent on how to proactively engage climate change is a major hindrance. At the same time, chronic threats related to human population pressures, natural resource extraction, and landscape conversion result in conservation attention being occupied by short-term needs and, frequently, crisis management.

Despite such serious limitations, the ecological future of the Albertine Rift appears somewhat more favorable than for many other tropical regions: although the climate will warm inexorably, the Rift is expected be the beneficiary of rainfall increases that may offset some of the more deleterious aspects of hotter climates. Characterized by high basal elevations and landforms with large vertical relief, the region hosts a considerable protected areas network that encompasses a strong representation of the region's species and habitats. Critically needed are (1) improved information on how anthropogenic climate change is likely to drive ecological responses at local scales, (2) comprehensive monitoring to detect such changes and understand their dynamic causation, (3) strengthened institutions more able to develop and apply knowledge of climate change into conservation planning and management, and (4) more refined and accurate models.

Chapter 4

The Brazilian Amazon

ENEAS SALATI, MARC DOUROJEANNI, AGENOR MUNDIM, GILVAN SAMPAIO, AND THOMAS LOVEJOY

The Amazon represents one of the most biodiverse regions in the world, much of which is poorly known and highly threatened by conversion to agriculture. Because of the low capacity in the region to conduct science, conduct conservation planning, and enforce laws and regulations, it is also at high risk. Global climate change is exacerbating local land-use impacts that are affecting the region's climate on a landscape scale. For the unique climate and functioning system of the Amazon to persist in the future, halting forest loss is imperative, or the thermal and water balances largely creating the unique landscape will collapse.

Introduction to the Region

Covering 670 million hectares, the Amazon basin includes portions of nine South American countries: Brazil, Peru, Bolivia, Colombia, Ecuador, Venezuela, Guyana, Suriname, and French Guyana. Covered mostly by dense forests that start as cloud forests at around 3,800 meters in the Andean Amazon region, this region holds an important place in the international environmental arena due to its vast territory and diversity.

The Brazilian Amazon is the largest and flattest part of the Amazon basin with an area of 410 million hectares, half of which is below an altitude of 100 meters. However, its northeastern portion has a few major very old massifs, including Neblina Peak (2,994 meters) and Mount Roraima (2,810 meters). Some other isolated lower massifs are located in its southern and southwestern portion.

The whole Amazon biome comprises fifty-three major ecosystems and over 600 different types of land and freshwater habitats, resulting in an extremely rich biodiversity. In total, the fifty-three major Amazonian ecosystems are classified as thirty-four forest areas (78 percent of the total biome), six typically Andean environments (1.5 perçent), five floodplains (5.83 percent), five savanna areas (12.75 percent), and two tropical steppes (1.89 percent). The diversity within and across these ecosystems is legendary if little known, including 1,400 fish species, 163 amphibians, 387 reptiles, 1,300 birds, and over 500 mammals, including 90 species of primates. Despite the territory being relatively level, endemism is common: 87 percent of amphibians, 62 percent of reptiles, 20 percent of birds, and 25 percent of mammals found nowhere else in the world. There is also enormous plant diversity estimated at 30,000 to 50,000 species.

Out of the broader Amazon basin's fifty-three ecosystems, thirty can be found in the Brazilian Amazon, nineteen of which are forests (77.5 percent of the area), followed in order of extent by savannas, floodplains, and steppes. The average temperature of the Brazilian Amazon is 24–26 degrees Celsius, and its average annual rainfall is around 2,300 millimeters, but with wide regional variations.

In some areas, the Brazilian Amazon may have been occupied by a far more developed civilization than usually considered, one that disappeared before the arrival of the Europeans. Its current indigenous population is not precisely known, with estimates varying from 250,000 to 700,000 persons belonging to some 180 ethnic groups. Overall, more than 16 million people inhabit the Brazilian Amazon today. The two largest cities are Manaus (1,739,000 inhabitants) and Belém do Pará (1,438,000 inhabitants). Another seven cities have a population of over 200,000, and eleven others have over 100,000 inhabitants.

In addition to federal oversight, the Brazilian Amazon territory is administered by seven states (the largest being Amazonas and Para; the smallest, Amapa and Acre) with a number of incorporated counties within each state. Each state is quite autonomous, with elected governors, state assemblies, a number of state secretariats (including ones for environment and sciences), and a varying number of public institutions to facilitate administration. Each state is served by several public and private universities.

The traditional economic base has been agriculture, increasingly focusing on exports (mainly soybeans, cotton, corn, and cattle). Other important sectors are mining, oil exploration, and hydropower generation. Extensive cattle ranches still occupy the largest expanses.

Historical Overview of Conservation and Science Initiatives

The major and best known scientific players in the Brazilian Amazon are the National Institute of Amazon Research based in Manaus, and the Emilio Goeldi Institute of

Pará based in Belém do Pará. The National Institute for Space Research has been involved in major projects in the region in cooperation with regional agencies. Additionally, today there are dozens of other public and private institutions, especially universities, which work across the region. Independent views come from a number of national research institutions (e.g., the Amazon Institute of People and the Environment and the Socio-Environmental Institute), international research institutions (e.g., Woods Hole Research Center), and well-known organizations such as World Wildlife Fund (WWF) and Greenpeace.

The federal government plays the leading role in conservation through its Brazilian Institute for the Environment and Renewable Natural Resources (IBAMA) and, more recently, the Chico Mendes Biodiversity Conservation Institute, which is responsible for establishing and managing all federally protected areas. No less important are the states' environmental agencies that are in charge of state public and private protected areas. Conservation issues, such as deforestation control, depend on IBAMA and corresponding state agencies. The National Foundation for the Indian (FUNAI), responsible for the overall administration of indigenous lands, is also important in natural resource decisions.

Amazon conservation and sustainable development is also furthered by a large number of international multilateral and bilateral agencies that provide a considerable amount of funding, one of the main externally funded efforts being the Amazon Region Protected Areas program. Also, many international NGOs are significant stakeholders, the WWF and Conservation International being two well-recognized examples. For many years, funding has been directed toward strong human resources training and scientific cooperation between the National Institute of Amazon Research and the Max Planck Institute of Limnology of Germany.

Over the past thirty years, some of the notable scientific research studies on the Brazilian Amazon have been (1) the Manaus-based Biological Dynamics of Forest Fragments Project conducted under the leadership of Thomas Lovejoy; (2) the studies on Amazon hydrology and the importance of the forest in maintaining the regional climate parameters, led by Eneas Salati; and (3) the multifaceted research of Philip Fearnside covering issues such as deforestation, climate change, biodiversity, and conservation policy, among others. Most of these works are coordinated with the National Institute of Amazon Research.

Current Conservation in the Region

A large number of detailed federal and state legislation and regulations cover all aspects of Brazilian Amazon development and conservation. Federal and state forest legislation provides strict regulations to preserve natural forests, such as, for example, the compulsory protection of 50 to 80 percent of rural properties. Legislation on protected areas is also comprehensive. However, institutional capacity to enforce such a voluminous legal body is quite limited, and there is a great concern that most of this legislation is not at all applied or respected.

There are two main types of protected areas defined by law in Brazil. First, conservation units are protected areas designated in *sensu stricto*, with the prime objective of conserving nature and its biodiversity, which may be established at the three levels of government: federal, state, and local. At the federal level, they are linked to the ministry for the environment through the Chico Mendes Institute for Biodiversity Conservation. Conservation units can be designated either for full protection or for sustainable use, each of which is further subdivided into several categories. Those with full protection include ecological stations, biological reserves, parks, wildlife refuges, natural monuments, and private natural heritage reserves. Those that allow sustainable use include extractive reserves, sustainable development reserves, forests, areas of relevant ecological interest, fauna reserves, and environmental protection areas. In Brazil, conservation units occupy around 17 percent of onshore territory and 1.5 percent offshore, totaling approximately 150 million hectares. The National Conservation Unit System includes more than 1,600 conservation units, of which slightly more than half are public. Conservation units under federal and state governments protect more than 100 million hectares of natural environment in all Brazilian biomes. In the Brazilian Amazon, 230 conservation units currently cover 98 million hectares or 23.8 percent of the entire biome.

Second, a number of other types of lands may also act as *de facto* protected areas. Indigenous lands contribute greatly to conserving biodiversity and reducing deforestation—especially in the Amazon, where these lands cover vast areas and have natural vegetation and are managed with the prime objective of preserving the culture and society of the traditional indigenous peoples living there. Indigenous lands cover almost 110 million hectares, 10 million of which have been recently sanctioned. In total, conservation units and indigenous lands cover 50.7 percent of the national territory (fig. 4.1).

Other major specific areas that, although not formally designated as protected, nonetheless contribute greatly to the protection of nature and conservation of biodiversity:

- Areas defined by the Forest Code, federal law that defines (*inter alia*) forests in the national territory as being assets of common interest of the Brazilian people and that preserve the natural vegetation and its associated biodiversity, even if not demarcated or without a special administration system.
- Military areas under the ministry of defense are designed to guarantee the country's security, and are demarcated and managed by a separate regime.

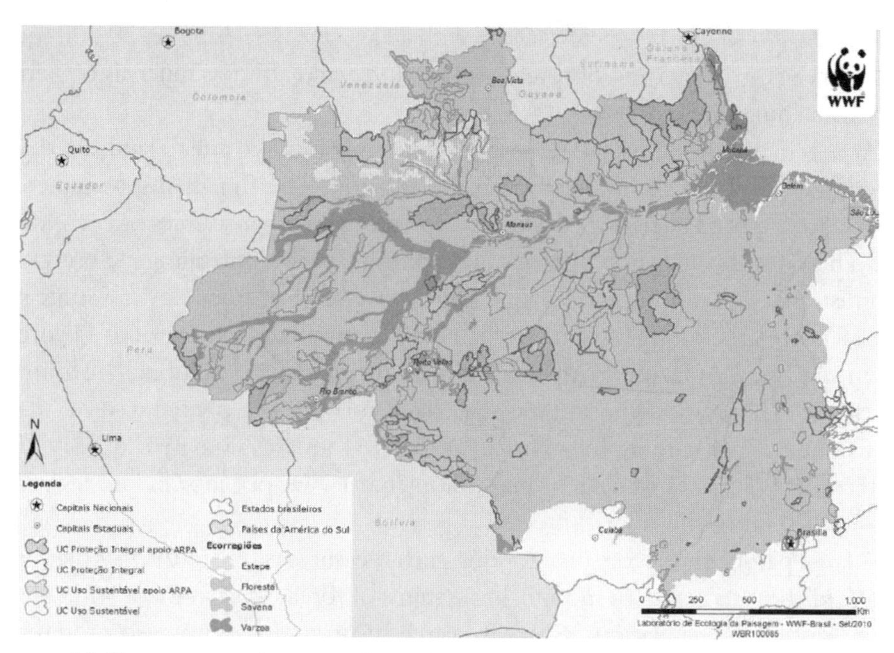

FIGURE 4.1 Ecosystems and conservation units in the Brazilian Legal Amazon.

In addition to these land categories, the federal Amazon Region Protected Areas Program is key to implementing the government's main policies and strategies for the conservation of the Amazon, including the Sustainable Amazon Plan, the Action Plan for Prevention and Control of Deforestation in the Legal Amazon, the National Protected Areas Plan, and the National Plan on Climate Change. As to the latter, the thirteen conservation units, created in the Amazon between 2003 and 2007 with support from the Amazon Region Protected Areas Program, will prevent the emission of 0.43 billion tons of carbon through avoided deforestation by 2050.

At the transboundary level, currently the only ecological corridor formally established in the Brazilian Amazon is the Binational Iténez-Guaporé Corridor with Bolivia (established 2001), an area known for having the region's largest diversity of fish. IBAMA is also studying the implementation of seven new corridors through its neighboring frontiers, including the vast "Central Corridor of the Amazon."

Deforestation in the Brazilian Amazon

Institutional roadblocks to enforcing legislation for natural forest preservation results in a high rate of deforestation and forest degradation. Total accumulated deforestation in the Brazilian Amazon is estimated at 756,842 square kilometers (18.9 percent). Annual deforestation varies from a minimum of nearly 6,000 square kilometers to as much as 29,000 square kilometers a year, with an average bordering on 20,000 square kilometers a year. The extent of forest degradation could affect as much as 60 percent of the remaining forest.

About 55 percent of Brazil's greenhouse gas emissions are due to land-use change and deforestation essentially in the Amazon Region. The Brazilian government is, however, committed to reducing deforestation and also other sources of emissions as well. There has been major progress in recent years in controlling deforestation. Administrative, economic, and legal measures have been adopted under a strategy of political action, including the Action Plan for Prevention and Control of Deforestation in the Legal Amazon. With this series of measures, the National Institute for Space Research (INPE) reported the deforested area was sharply reduced by 73 percent from 27,772 square kilometers in 2004 to 7,464 square kilometers in 2009. It fell to a record low in 2011 of 6,451 square kilometers (www.obt.inpe.br/prodes/prodes_1988_2010.htm).

A large part of the success in adopting such measures is due to the fact that Brazil has one of the most modern monitoring systems of forest areas in the world through remote sensing by the National Space Research Institute. Brazil was also a forerunner in the use of meteorological satellite data to monitor burnings in the country, culminating in the creation of the Forest Fires and Burnings Prevention and Control Program. Implemented through a partnership between IBAMA and INPE, the goal of the program was to prevent and control burnings in the country and thus prevent forest fires.

One important outcome in the effort to avoid further deforestation was the 2008 establishment of the *Fundo Amazonia*, under the responsibility of the National Bank of Economic and Social Development, for raising resources to be applied in projects aiming at forest conservation. In 2009, the government of Norway made the first donation to this fund of about US$100 million, out of a total pledge of US$1 billion.

Regional Effects of Climate Change

The Amazon region, which has been in dynamic equilibrium over the past few thousand years, is now undergoing two types of anthropogenic changes to climate that may alter this age-old balance. The first is regional deforestation that may alter localized thermal and water balances and, consequently, the characteristics of the forest and biodiversity of the associated fauna. This impact largely results from government policies of the countries in the Amazon Basin and could be controlled if recently approved policies are properly implemented.

Deforestation further contributes to altering the water regime, compounding changes in source water. Salati et al. (1979) showed that deforestation can alter the water balance in the Amazon region through measuring the isotopic compositions of rainfall in the Amazon region, both demonstrating that there is a strong recirculation of water in the region and disproving the notion of complete condensation of all inflowing moisture in the basin. Generally, replacing forests with pastures reduces evapotranspiration, thereby increasing the surface-sensible heat flux and, consequently, surface temperature. In particular, forest conversion increases the surface albedo roughness, reduces the leaf-area index and associated canopy interception, and reduces the available soil moisture since pasture plants generally have shallower roots than rainforest trees (Gash and Nobre 1997). As a consequence, tropical deforestation is expected to lower the ability of the land surface to maintain a high rate of evapotranspiration throughout the year, leading to changes in the latent heating of the atmospheric boundary layer and the strength of tropical convection. In general it is expected that these changes in the surface energy and water balance lead to a significant reduction in rainfall and an increase in surface temperature (Sud et al. 1993; Costa and Foley 2000; Sampaio et al. 2007).

The second anthropogenic factor that may affect the regional balance of the Amazon is global climate change. A 2007 World Bank study on Andean countries that contribute to Amazon River formation concluded that increased temperatures in that region could directly affect the ice stored in the peaks of the Andes (Orlove 2009). Such alterations to water supply in these regions could also affect the production of hydropower plants, agriculture, and natural ecosystems. For example, Bolivia's Chacaltaya Glacier has lost almost 82 percent of its surface area since 1982 and may completely melt by 2013 (Francou et al. 2003). This rapid shrinkage has resulted in a temporary net increase in hydrological runoff (Pouyaud et al. 2005).

Altered river flows are already being observed and predicted to further change. Marengo et al. (2009) used global models to predict river discharges in the Amazon Basin for present climate and double CO_2 future scenarios, while Milly et al. (2005) assessed changes in streamflow in various rivers worldwide. Both study results suggested a possible 10 to15 percent reduction in streamflow in the Amazon basin. Studies at a large scale in the Tocantins and Araguaia Rivers, in the eastern Brazilian Amazon, concluded that rapid land cover changes since 1960 are associated with about a 25 percent increase in the annual mean discharge despite no significant change in precipitation (Coe et al. 2007; Costa et al. 2003). Coe et al. (2009) illustrate the influence of historical and potential future deforestation on local evapotranspiration and discharge of the Amazon River system with and without atmospheric feedbacks and clarify a few important points about the impact of deforestation on the Amazon River. With extensive deforestation (viz., greater than 30 percent of the Amazon basin), atmospheric feedbacks brought about by differences in the physical structure of the crops and pasture replacing natural vegetation can cause water balance changes and a decrease of about 25 percent in annual discharge.

The possible impacts of global climate change on the flow of the Amazon River within the Brazilian territory were also studied using a model of the impact of future climate on rainfall and temperature and its consequences on water resources. Using the data from the HADRM3P model on the 50 × 50 kilometer scale, the impacts on surplus water were estimated for scenarios A2 and B2 (see chap. 2). It was found that the Amazon River flow may be affected by global climate change. Based on the HADRM3P model, it was estimated that the variation in contributions from the river within Brazilian territory is currently 131,947 cubic meters per second. Under scenario B2 using the 1961–90 period as a baseline, the study showed a 7 percent drop in the flow between 2011 and 2040, 15 percent in the 2041–70 period, and 25 percent in the 2071–2100 period. For scenario A2, the figures obtained indicated a 7 percent drop for the 2011–40 period, 26 percent for the 2041–70 period, and 30 percent for 2071–2100.

An additional World Bank report, using a model on a 20 × 20 kilometer scale, showed a possible variation in the flows of the River Amazon in Obidos, Brazil, with an increase from 200,000 to 230,000 cubic meters per second in the high-water season, and a drop from 80,000 to 60,000 cubic meters per second in the low-water season for the years 2075–79 compared to the 1979–2003 period (Vergara and Scholz 2010).

The stability of the Amazon forest-climate equilibrium is being perturbed by a number of human drivers of change, including deforestation, climate change, forest fires, higher atmospheric CO_2 concentrations, and increased frequency of droughts and floods (Nobre and Borma 2009; Malhi et al. 2008; Betts et al. 2008). For the Amazon forest, if climate warming reaches above 3.5 to 4 degrees Celsius, there is a risk of passing a "tipping point" leading to "savannization." (The term *savannization* has been defined as changes in the regional climate caused by either land cover change [Nobre et al. 1991; Oyama and Nobre 2003; Sampaio et al. 2007] or climate change in such a way as to increase the length of the dry season and to turn the regional climate into the typical climate envelope of savannas.) (See page 53.) The IPCC (2007) put the tipping point at 2.5 degrees, and subsequent refined Hadley Centre models put it at 2.0 degrees (Vergara and Scholz 2010). Quantitative assessments for the maintenance of the tropical forest indicate that such tipping points may be passed if the deforested area exceeds 40 percent of Amazonia (Sampaio et al. 2007) or if climate change results in a temperature increase of > 3 to 4 degrees Celsius (Cox et al. 2000) later revised downward to 2.0 by the Hadley Centre. The likelihood of crossing a tipping point can be greatly exacerbated by increases in forest fires and droughts (Vergara and Scholz 2010).

Roadblocks and Opportunities

Outside protected areas, the main issue for conservation is the rampant continuous deforestation and degradation of natural forests. There is no indication of any ongoing action that can effectively tackle illegal land appropriation, limit deforestation to authorized sites, and stop anarchical logging and fires. Illegal gold is also a growing threat. Ambitious government plans to expand infrastructure promote environmental protection, but do not invest enough in effective law enforcement.

Ecological representation in the protected area system is good, although some gaps are still evident. The main issue concerns management deficiencies and, with regard to state protected areas, the permanent risk of being downgraded or even eliminated by state assemblies to accommodate development claims. The State of Rondonia recently eliminated twenty state protected areas.

Despite lack of funding and staff, conservation is relatively well accomplished in federal full-protection areas and larger indigenous reserves. The federal biological reserves, national parks, and ecological stations are the best preserved patches of the natural ecosystems. However, few are fully implemented and management deficiencies are substantial. Lack of staff, training, funding, and equipment are indicators of problems that are increasing every year as a consequence of new development pressure. Even less protected are the areas "conserved" by individual states, and conservation is almost nonexistent in the so-called Areas of Environmental Protection (a lesser federal category). Other categories of protected areas for sustainable use offer different levels of protection, but never at the necessary level.

In conclusion, approximately 20 percent of the Brazilian Amazon today is actually protected from destructive development, although 50.7 percent is nominally protected by the national and state governments collectively, including indigenous areas (27 percent). This includes federal full-protection areas, a number of state full-protection areas, a few protected areas for "sustainable use," and a large portion of the major indigenous lands. Yet the studies mentioned in this chapter show that as a result of the changes in rainfall and temperature, the entire Amazon region could be affected by global climate change and deforestation with regard to available water resources and biodiversity. Extensive study will be necessary with various scenarios of future climate and models to provide more precise forecasts. Most importantly, real "experiments" of large-scale deforestation or climate change must not be conducted, since the return to current conditions would be very costly, if not impossible, because a large part of the biodiversity of the fauna and flora will have disappeared with no chance of rehabilitation.

Approaches to Conservation under Climate Change

Unfortunately, climate change has not up to now been considered a major organizing factor in conservation, infrastructure, and soil-occupation planning in the Brazilian Amazon Region. Climate change impacts on conservation in this large landscape have been more a concern of a few scientists and of the local and international communities rather than a government concern. Almost no landscape conservation-needs assessment has been taken in response to climate change; for instance, hydroelectric power plants are being planned and constructed in the region without any consideration of possible impacts of future climate change on water availability. Current priorities and activities related to climate change in the Brazilian Amazon are limited to scientific studies. In such a context, the government pledge and action for the reduction of deforestation relates to climate change mitigation but not to adaptation. Major government concern has been with biodiversity conservation, but even here there has been no planning or implementation regarding forest protection as related to climate change. Moreover, no adaptation planning for climate change is occurring in the region. Ultimately, there could be several opportunities for conservation under climate change, mainly related to adaptation and enforcement under current environmental legislation.

On the other hand, according to the recent Law 12,187 (2009), Brazil will adopt, as a voluntary national commitment, actions to reduce its emissions of greenhouse gas up to the year 2020 by 36.1 percent to 38.9 percent. The emissions forecast for 2020 and details of the actions to achieve this objective will be provided by decree, based on the second Brazilian Inventory of Anthropogenic Emissions and Removals of Greenhouse Gases not controlled by the Montreal Protocol concluded in 2010. As formally declared by president Luiz Inacio Lula da Silva at COP-15, in Copenhagen, Brazil is "hereby committed to reducing the deforestation rate of the Amazon by 80% in 2020." Some conservation scientists have suggested priorities for conservation that consider climate change for the Amazon. Some specific priorities recommended include conservation corridors that cover altitudinal gradients and ecotonal gradients, which would provide opportunities for species to move as climate change takes place; and maintenance of riparian vegetation particularly for the southern tributaries, which flow from drier savannas into transition forest and then the rain forest itself and provide a form of corridors themselves (Malhi et al. 2008; Killeen and Solorzano 2008). Likewise, Killeen et al. (2007) and Killeen and Solorzano (2008) indicated additional priority of places in the western Amazon beyond Brazil where land formations might provide some climatic stability.

In the end, however, the non-Brazilian Amazon depends on the westward movement of moisture from the central and eastern Amazon, so the Amazon needs to be conserved as a whole system. The climate change models suggest dieback or savannization in the southern and eastern Amazon, and the extreme droughts of 2005 and 2010 may be previews of what could lie ahead. Deforestation and fire also affect the hydrological cycle, so the work reported by Vergara and Scholz (2010) and conducted by Carlos Nobre (Nobre et al. 1991), among others, is particularly important. This study was the only time deforestation, fire, and climate change have so far been modeled together. This suggests that a tipping point to Amazon dieback could occur at 20 percent deforestation. Current deforestation is at 18 percent, so an obvious policy response is to build a margin of safety through aggressive reforestation (which could be a mix of natural forest and plantation forests for economic return).

Conclusion and Recommendations

The carbon content of the Amazon forest is of such a magnitude that conserving the Amazon as a whole is important from climate change mitigation considerations alone. It also is vulnerable to climate change in particular in the south and the east from savannization. The droughts of 2005 and 2010 can be considered a preview of future climate change impacts. Some ecosystem restoration is critical to avoid Amazon dieback and ensure that the region does not reach a tipping point. Finally, climate change planning needs to be integrated into conservation planning and management of the Amazon as a whole.

Chapter 5

Mesoamerican Biological Corridor

MARGARET BUCK HOLLAND

When the Mesoamerican Biological Corridor (MBC) was established in 1997, many heralded it as the world's largest and most ambitious transboundary conservation and development project. Within three years, over US$280 million in donor financing was directed to MBC projects, which heavily emphasized regional cooperation and integration. Just over a decade later, and with more than US$500 million invested, many have judged the MBC a failure. And yet the Central American Commission for Environment and Development has retained MBC as a component in its most recent regional environmental strategy, and portions of the original MBC persist. After tracing the rise and perceived fall of the MBC, this chapter focuses on the impacts of climate change in Central America and Mexico, and then explores how the experience of the MBC can guide regional climate change mitigation and adaptation projects so as to optimize benefits for biodiversity and ecosystem health.

Introduction to the Region

The seven nations of Central America lie nestled within the Mesoamerican isthmus, a small strip of land that has bridged the continents of North and South America for nearly three million years. Constituting less than 2 percent of the Earth's land surface, this bridge is home to 12 percent of the world's known species, a result of the Great American Biotic Interchange (Woodburne 2010; Programa Estado de la Nación 2008). Yet today the Central American landscape is a patchwork of land uses shaped by a more recent history of tumultuous sociopolitical development, rampant deforestation, and rapid population growth. The post–World War II era in this region was witness to a boom in population growth of more than 3 percent per year in the 1960–80 period (Rosero-Bixby and Palloni 1998). Although the majority of the region's 42.5 million people live in urban centers, more than 40 percent persist in rural areas (UNPD 2009). According to global standard measures of poverty, between 12 and18 percent of the population in Guatemala, Nicaragua, and Honduras live on less than $1.25/day, while El Salvador (6.4 percent) and Panama (9.5 percent) have slightly lower poverty counts. Costa Rica (2 percent) and Mexico (4 percent), both considered upper-middle income countries, are considerably better off in terms of this overall measure (World Bank 2011).

Similar to population trends, regional deforestation rates were among the highest in the world for the years 1970–90, with the UN Food and Agriculture Organization estimating an annual deforestation rate of 1.5 percent during the 1980s (Kaimowitz 2008). Population growth, livestock development, and conversion to other agricultural uses were principal drivers of forest loss, and some have posited a direct demand-driven relationship between US imports of beef and deforestation due to the conversion of forest to pasture for ranching (Kaimowitz 1996; Rosero-Bixby and Palloni 1998). This "hamburger connection" (Myers 1981) has since garnered criticism as it fails to account for the other factors influencing pasture increase and forest loss during that time, such as government incentives, subsidies, and a cultural ideal embodied by the cattle farmer ("la cultura del potrero," or culture of the pasture) (Edelman 1995). By the 1980s, the need for conservation to stem the tide of deforestation and biodiversity loss became increasingly dire, and it was at this time that the regional protected areas network began to develop in earnest (Sader and Joyce 1988).

Historical Overview of Conservation and Science Initiatives

Two decades ago, the combined protected area system within the region consisted of 233 protected areas covering more than 16 percent of land area (Coates 1997). Since 1992, the protected area networks from each country have been formally integrated into what is known as the Central American Protected Area System (CCAD 2003). Today, the Central American Protected Area System includes over 670 parks, covering close to 124,000 square kilometers (~24 percent of land area). The majority of these protected areas are very small in size, with 83 percent covering less than 150 square kilometers, and only 4 percent extending over 1,000 square kilometers (Estado de la Nación 2008). This trend is consistent with that of the global protected area network, where more than 75 percent of protected areas measures less than 100 square kilometers in extent (West et al. 2009). Most of the larger protected areas in the region were created in the 1980s and 1990s as biosphere reserves and transboundary protected areas. More recently, establishing parks of this size and scope has become an increasingly elusive goal both due to the fragmented landscape and to the political challenges presented by cross boundary collaboration within the region.

Paseo Pantera: Path of the Jaguar

In the early 1990s, Dr. Archie Carr III of the Wildlife Conservation Society and David Carr of the Caribbean Conservation Corporation conceived of a series of corridors providing connectivity between the existing protected area networks. Dubbed *Paseo Pantera*, focusing as it did on the charismatic jaguar, the initiative drew conceptually from the system of wildlife corridors that had been proposed between Florida's parks (Noss and Harris 1986). Around the same time, steps were taken to create two international peace parks, one between Nicaragua and Costa Rica under the International System of Protected Areas for Peace (SIAPAZ) initiative, and the other between Costa Rica and Panama, named La Amistad (Coates 1997). Both of these areas would be key components in the regional proposal for *Paseo Pantera*.

The plan behind *Paseo Pantera* was to promote landscape linkages between parks through Mexico and Central America to allow for the movement of larger mammals, thereby triggering a cascade of conservation benefits (Kaiser 2001). USAID (United States Agency for International Development) provided a matching grant to both the Wildlife Conservation Society and the Caribbean Conservation Corporation for a five-year pilot project of *Paseo Pantera* in June 1990. Around this same time, the conservation community in Costa Rica was also circulating the concept of making land-use connections between reserves (Grandia 2007). By the conclusion of the funding period, *Paseo Pantera* was embraced widely by the conservation community and seen as a key opportunity for financing conservation in the region. A regional map vision for *Paseo Pantera* proposed a route for corridor establishment based on a combined suitability and feasibility analysis using GIS (Lambert 1997). The concept quickly met with resistance, however, from indigenous groups and advocates for the rural poor who were concerned with land-use restrictions without adequate compensation or alternative opportunities. This vocal opposition resonated with the growing regional emphasis on sustainable development, and subsequently the original mission behind the *Paseo Pantera* shifted to include both conservation and development goals.

Formation of the Mesoamerican Biological Corridor (MBC)

In 1997, an agreement was signed by all governments in Central America, as well as five states of southern Mexico, marking the creation of the MBC. The official mission called for the corridor to:

> offer a set of environmental goods and services to the Mesoamerican and global society, through the sustainable use of natural resources and thus contributing to the betterment of life of the inhabitants of the region. (CCAD 2002, 18)

Overall management and coordination of the MBC was placed under the purview of the Central American Commission for Environment and Development. In October of 1998, Hurricane Mitch devastated the region, and governments turned their attention to recovery and reconstruction efforts. However, the tragedy of Mitch ultimately drove home the message to governments and communities alike that ecosystems and human livelihoods are inextricably linked.

Despite a windfall of funding from international donors, the MBC hovered in limbo between concept and reality, trying to straddle the dual regional objectives for conservation and development (Miller et al. 2001). The map of proposed elements of the MBC, published in 2002, is symbolic of this struggle (fig. 5.1). There was little consistency or ecological basis provided—including from the *Paseo Pantera* efforts—in the proposed corridor designations between countries, with braided networks planned in Guatemala and Costa Rica, while seemingly larger swaths of land were proposed as corridor areas in Mexico and Panama. This inconsistency suggests that

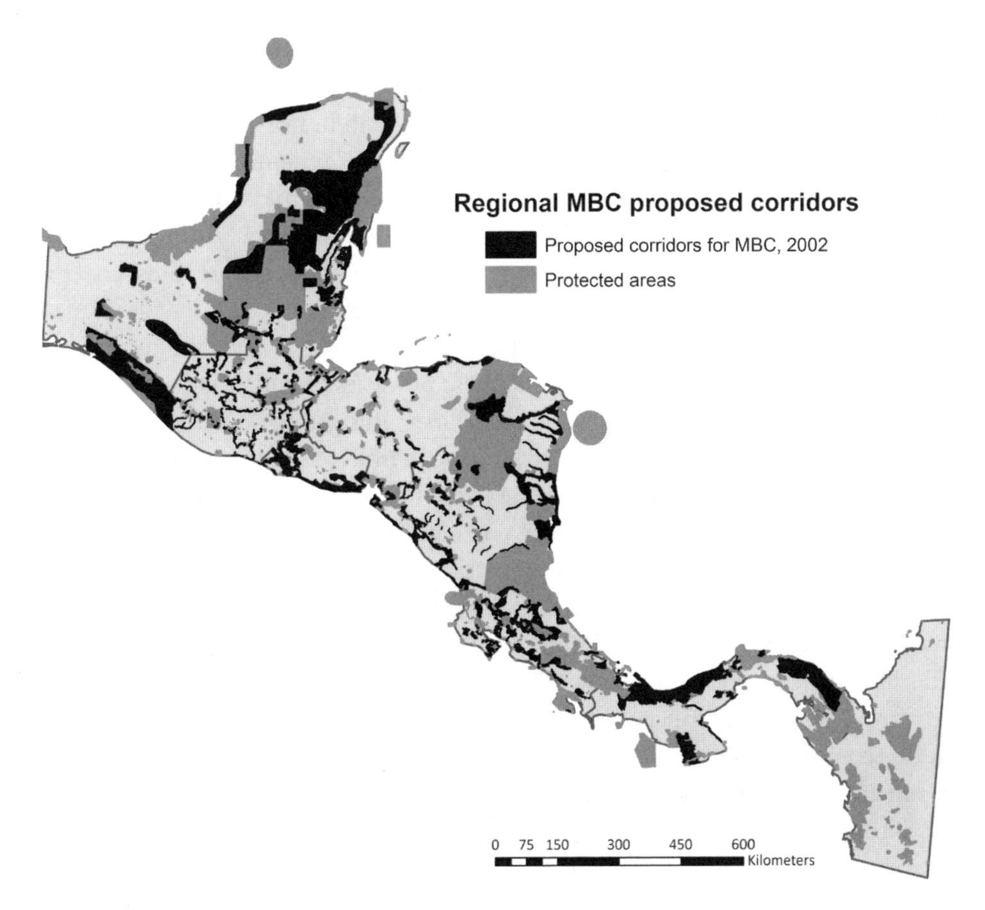

FIGURE 5.1 Conceptual map of the proposed elements of the Mesoamerican Biological Corridor (2002). Map produced by author. Source data: World Database on Protected Areas (protected areas) and CCAD (corridors).

transboundary cooperation was politically challenging and elusive even at this level of planning.

After reviews of the first phase of the MBC exposed the complex challenges of implementing a transboundary project with such broad scope, a second phase commenced with a focus on regional consolidation of the MBC. In 2003, the first official Mesoamerican Parks Congress focused on the MBC, with at least half of the symposia centered on connecting corridors, poverty alleviation, regional integration, social participation, and indigenous issues. Dr. Carr, one of the visionaries behind *Paseo Pantera*, found the theme of the convention to be symbolic of the shift to a "donor-driven" agenda for the MBC, dominated by a focus on regional economic integration and poverty reduction. Carr lamented that this represented a change in the vision for protected areas in the region, from wildlife refuges to entities approaching "welfare nuclei," or "utopian bubbles of peace and tranquility, each bubble centered on a protected area" (Carr 2004, 36). Others have suggested that the MBC fell victim to "green neoliberalism," and that it was co-opted by elites both within the region and internationally who emphasized public-private partnerships over community-driven projects (Grandia 2007; Finley-Brook 2007).

Other critics point to an overall lack of baseline indicators for tracking progress and evaluating the impact of the MBC. Few national or regional analyses of land use exist that would permit an evaluation of the investments in the MBC. Using MODIS satellite imagery from 2000–2001, a team of NASA scientists analyzed forest cover in the region based on a subset of sites, which indicated at the time that forest cover was higher and conversion lower inside the MBC corridor areas than outside (Sader et al. 2001). The World Bank also financed the production of a regional ecosystem map assessment from 1999–2001 data. While these products provided some indication of the effectiveness of protected areas and corridors in mitigating deforestation in the region, they also pointed to the potential increase in isolation of key habitat patches.

In 2006, nearly one decade after the formal creation of the MBC, a new phase of emphasis on transboundary priority corridor areas and projects began (fig. 5.2) with the goal of encouraging increased binational and trinational cooperation among government agencies, indigenous and community groups, and other civil society groups. However, by the end of 2006 the regional coordination office for the MBC officially closed its doors, with few of the ten proposed transboundary corridors having moved beyond the concept phase. Support for the implementation of the MBC in Mexico continues, as does a focused effort on the transboundary corridor areas and engagement with indigenous communities, all with funding support primarily through the European Union and the Inter-American Development Bank. And yet, in candid conversations with regional practitioners and researchers, the MBC is widely viewed as an initiative that failed under the combined weight of its own promises and the burden of donor-driven agendas.

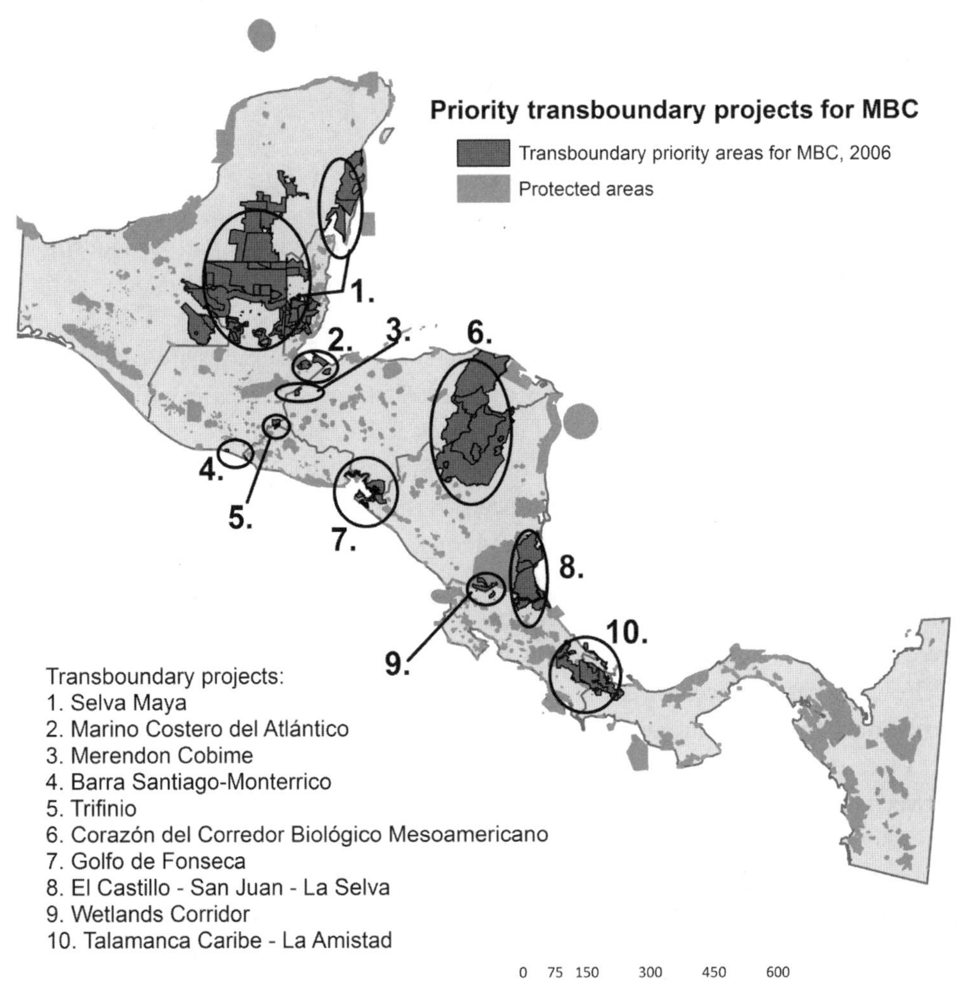

FIGURE 5.2 Proposed transboundary corridors for the Mesoamerican Biological Corridor (2006) Map produced by author. Source data: World Database on Protected Areas (protected areas).

Nevertheless, there are indicators of effective conservation in the region over the past decade, forcing the question of how influential the MBC has been in those successes. Most notably, recent estimates of forest cover at the regional scale indicate that the rate of deforestation has greatly slowed since the 1990s, with some areas of overall forest regeneration evident in parts of Costa Rica, El Salvador, and the Pacific portions of Panama and Nicaragua (Kaimowitz 2008). While some deforestation hotspots still exist within the region, mainly focused at the forest-farm interface and in areas that were previously under conflict (the Darien in Panama and the Petén in Guatemala), the current regional-scale trend appears to be one of increasing forest regrowth, indicative of a forest transition (Rudel et al. 2005; Kaimowitz

2008). This is a positive development for Mesoamerica, as a new stage is set for conservation throughout the region with a strong emphasis on climate change mitigation and adaptation.

Regional Effects of Climate Change

In Central America, the threats of climate change are real and tangible. The site of less than 0.5 percent of global CO_2 emissions, the region is already experiencing a documented increase in the frequency and intensity of extreme events, both precipitation and drought related, with future impacts predicted to exacerbate current conditions (Anderson et al. 2008; Ramírez et al. 2009). One global climate risk index for individual countries, calculated based on trends across the last decade, places Honduras and Nicaragua at number 1 and number 3, respectively (Harmeling 2009). Extreme hydrometeorological events have been on the rise for the past fifteen years, with floods and windstorms affecting over six million Central Americans. While the impacts of large-scale disasters factor significantly in the region (with Hurricane Mitch alone impacting more than one million people), more than 85 percent of the population were affected by small-scale, localized events between 1990 and 2005. The occurrence of droughts has also been on the rise, affecting more than two million people over this same time period (Ramirez et al. 2009).

Over the next century, average temperatures across the region could increase between 3.6 and 4.7 degrees Celsius, based on predictions across three climate models and assuming a business-as-usual scenario. An average reduction of 28 percent in precipitation is predicted by this same modeling protocol (Ramírez et al. 2009). This also translates into increased risk of drought, particularly for El Salvador and Guatemala. Some analysts suggest that the expected reduction in average precipitation will be compensated for by an increase during hurricane and storm events (Vergara 2009). The intensity of such extreme events is also predicted to increase in this century by 5 to10 percent, causing the greatest impact on Belize and Honduras, followed by Costa Rica and Panama. In terms of the economic costs of these events, Mexico is expected to be hardest hit, with a tenfold increase in losses over the next ten to fifteen years alone (Ramírez et al. 2009).

Due to the effects of rising temperatures, the accumulated economic impacts of climate change over the course of this century for the Central American region are valued at close to US$103 billion (current), representing more than 70 percent of regional gross domestic product at 2008 values (Ramírez et al. 2009). This is based on an assessment of measurable costs across four thematic areas: agriculture (19 percent), water resources (15 percent), extreme weather events (18 percent), and biodiversity (19 percent). This same economic analysis concludes that, for the year 2080, the region's index of potential biodiversity could be reduced by 38 percent, with Nicaragua, Honduras, and El Salvador experiencing the greatest impact (Ramírez et al. 2009).

With support from USAID and NASA, the Water Center for the Humid Tropics of Latin America and the Caribbean (Cathalac) produced a regional-scale analysis of potential impacts of climate change on biodiversity (measured as species richness) and ecosystems. An index of climate change severity was derived from baseline climate data on temperature and precipitation, along with monthly anomaly data. The researchers used the WorldClim datasets, downscaled to a spatial resolution of one square kilometer, and focused on the IPCC A2 (worst case or business-as-usual) and B2 (best case) greenhouse gas emissions scenarios (chap. 2) from the Hadley Centre Coupled Model (HADCM3). The methodology was adapted from a prototype index developed by Tremblay-Boyer and Anderson in 2007 (Anderson et al. 2008). The climate change severity index indicates the degree to which a particular location will be pushed beyond its natural variation, or climatic "comfort zone." With the index spatially defined across the region, the research team identified critical areas for conservation action, based on the intersection of highest species richness per country and the most severe shifts predicted in the climate change index (Anderson et al. 2008).

The resulting overlay predicts that by 2020, based on A2 scenarios, critical conservation areas will be concentrated within the Caribbean lowlands of Nicaragua and Costa Rica, central Panama, and the Darien peninsula (Panama). By 2050, this same definition of critical areas is predicted to intensify and expand northward, to include the Caribbean lowlands of Guatemala and Honduras, as well as the Pacific coastal region of El Salvador, an upland region in the Yucatan, and the Osa peninsula in Costa Rica (Anderson et al. 2008).

This analysis is representative of Cathalac's emerging role as a regional climate change center for research and action. Since 2005, Cathalac has been the institutional home to SERVIR, a regional visualization and monitoring system. SERVIR acts as an Earth observation platform, disaster risk analysis, and geospatial data portal as well. Apart from SERVIR, Cathalac has developed a monitoring tool for estimating carbon stocks for tropical ecosystems across various scales, and created a climate one-stop web platform for information exchange on the latest developments in climate change research most relevant to the region. Products developed by Cathalac also support the Regional Program for Reducing Vulnerability and Environmental Degradation.

The main challenges presented in regional modeling efforts are related to transnational coordination, data and information politics, sustainable financing, and consistency in methodological and research design. Most recently, a Cathalac team has managed a region-wide effort to map current land-use and forest cover change as a product (due to be released in 2011). This map is the first regional product of its kind. The only previous regional mapping effort was the Central American ecosystems map, developed in 2003, and intended as a planning product for the MBC.

Cathalac's climate change research products are generally regional in scale and based on large-scale modeling efforts. It also produces some site-level and national-scale research, but typically within the context of Panama where Cathalac is based. The

Costa Rica–based Center for Education and Research in Tropical Agronomy (CATIE) is also a key regional node for climate change science. CATIE's research spans from regional modeling to local-scale forest carbon investigations at the plot level, focusing on projects related to climate change adaptation and mitigation (e.g., forest conservation and payments for ecosystem services), with particular emphasis on vulnerable forest and agricultural systems.

Research by institutions such as Cathalac and CATIE is essential to understanding the potential impacts of climate change throughout the region. Moreover, their role in analyzing forest carbon is critical for the current stage of conservation planning in Mesoamerica, much of which is focused on developing incentive-based programs and other forest conservation policies related to the recently approved global climate change policy of Reducing Emissions from Deforestation and Degradation (REDD) (chap. 1). Such programs offer an opportunity and incentive to achieve both mitigation and adaptation goals.

Approaches to Conservation under Climate Change

The political and landscape realities in the Mesoamerican region have made it increasingly difficult to create new protected areas over the past decade. As many conservation practitioners shifted their focus to corridors, often under the umbrella of the MBC, they have looked to incentive-based programs as a way to achieve forest conservation objectives. The majority of these have been structured as payment for ecosystem services (PES) schemes, which are considered the main vehicle through which national-level REDD strategies would function in Central America and Mexico (Kaimowitz 2008).

The first (and still only) national PES programs in the region were developed in Costa Rica and Mexico, with links to the MBC. The Costa Rican program was first launched in 1997 through a national fund for forest financing that established three types of contracts for conserving standing forests, reforestation, and sustainable forest management. After this first generation of PES contracts, the World Bank and the GEF helped to cofinance the second phase (2001–06) of the program for a total of US$49 million. Known as the EcoMarkets project, the purpose of this cofinancing was to target the Costa Rican PES program in priority corridor areas of the MBC, and to join climate change mitigation efforts with biodiversity conservation (Sills et al. 2005).

According to an external evaluation, the EcoMarkets project achieved all of its original objectives, including those related to promoting and amplifying the number of PES contracts focused on forest conservation within high-priority corridor areas (Sills et al. 2005). While the MBC is named as the organizing framework for spatially targeting the PES agreements, the map of Costa Rica's priority corridors developed under the EcoMarkets project does not reflect those proposed for the MBC. Rather, in 2006, the Costa Rican conservation community updated its own spatially refined GAP analysis to define priority corridors (SINAC 2007). It is also difficult to determine whether Costa Rica's fund for forest financing continued to prioritize payments in priority corridor areas after the conclusion of the EcoMarkets project. Despite these discrepancies and the consensus that PES contracts have had negligible impacts on deforestation thus far, the degree to which Costa Rica has been successful at integrating corridors with programs related to climate change mitigation is notable.

The Mexican national PES program began in 2003 and originally focused on linking forest conservation and hydrological services. This quickly expanded to include carbon sequestration, and is also heavily financed by the World Bank and the GEF. The GEF recently cofinanced a US$31 million project to amplify and refine the Mexican PES program to include prioritizing contracts in priority ecological corridors related to the Mexican portion of the MBC. Similar to Costa Rica, the current Mexican map of corridors differs from those proposed for the MBC in 2001 moving to a finer-scale set of analyses.

Other countries in the region have also had experience with implementing PES schemes, but primarily on a watershed or municipal scale. PES has proliferated in Mesoamerica over the past decade, and many believe that the main way to channel REDD funding to the region will be through national and subnational PES schemes (Kaimowitz 2008). To be eligible and compliant with REDD policy, however, the PES programs will require refinement in terms of planning, implementation, and monitoring. Institutions such as Cathalac and CATIE are well placed to provide the resources and expertise needed to guide the planning and monitoring efforts. Along with support from regional institutions, these institutions could facilitate REDD implementation through a regional vision that draws from the MBC.

While most efforts focused on integrating forest conservation and corridors with PES/REDD relate primarily to climate change mitigation, there is an increasing push to identify other opportunities for ecosystem-based adaptation to climate change. Given the type of climate change impacts in Mesoamerica, much attention has targeted response to large-scale storm events and natural disasters. One of the biggest challenges in approaching climate change adaptation is that so much remains unknown about how different tropical ecosystems respond to impacts. To this end, current research focus is to understand how current ecosystems and species are responding to existing change. For example, CATIE is both the coleader and regional focal point for the Tropical Forests and Climate Change Adaptation research initiative, which focuses on understanding the impacts of climate change of tropical forest ecosystems and developing robust methods for assessing vulnerability of forest-dependent communities (Nkem et al. 2009).

Within the region, some recognize the potential role of corridors in climate change adaptation. For example, Mexico's 2007 national climate change strategy calls for the establishment of corridors and priority conservation regions as an adaptation strategy to "improve the adaptive capacities of ecosystems and species" (GoM 2007). There is also the recognition that previously established corridors need to be analyzed, evaluated, and potentially redefined to account for climate change. To this end, CATIE researchers developed a model of existing protected areas and priority corridors in Costa Rica to assess the role of different corridors in reducing the vulnerability of protected areas to future climate change. More focus is needed, however, on how to prioritize conservation given the potential role of corridors in both regional mitigation and adaptation efforts. Until there is better understanding of how different ecosystems might be affected, implementing adaptation measures at the local scale remains elusive.

Conclusion and Recommendations

Although the MBC is still regarded as one of the most ambitious transboundary conservation and development projects in the world, its current status could hardly be more ambiguous or tenuous. The widespread belief is that the program has quietly dissolved, eclipsed by more powerful, regional, sustainable-development projects. However, the pieces and footprint of the MBC continue to persist in many current efforts, with MBC serving as an institutional framework for conservation management and planning in the region. Climate change now stands to force a new paradigm for conservation in the region, and along with the MBC presents an opportunity in terms of future corridor financing and implementation. This was evident in a recent report on the implementation of REDD in Mesoamerica, where a main recommendation was to create a "Mesoamerican Community Carbon Reservoir," and use that regional vision to influence the future design of the MBC (Fundación PRISMA 2011). Considering the strict requirements for planning and monitoring to be required for REDD+, there is a significant need for spatial planning and research, as well as a conservation vision that strongly emphasizes both core areas and corridors as tools to help systems adapt and be resilient to change. For both champions and critics of the MBC, as its definition and focus shifted over time to respond to the changing sustainable development vision of the region, now is the time to once again redefine and redesign the MBC: learning from its failures and its legacies, and demonstrating the MBC's relevance for conservation in a region already confronted with a rapidly changing climate.

PART 3

Temperate and Mediterranean Landscapes

These five temperate landscapes encompass relatively high degrees of human impacts, with the exception of the boreal landscape where the conservation focus is on projected future impacts. Most of these landscapes have sophisticated climate science and conservation planning relative to other case studies. Given this longer experience in targeted examination into climate change impacts, they offer some of the most advanced examples of how to incorporate climate change into conservation planning and action.

Chapter 6

Boreal Forest, Canada

MEG KRAWCHUK, KIM LISGO, SHAWN LEROUX, PIERRE VERNIER, STEVE CUMMING, AND FIONA SCHMIEGELOW

The vast expanse of the boreal forest in Canada is home to a diversity of wide-ranging animals, from migratory land birds and waterfowl to the largest caribou herds in the world. Boreal ecosystems are likely to experience dramatic changes in this century, particularly through anticipated alteration in vegetation and wildfire regimes as a result of the greater-than-average rate of warming predicted for more northerly latitudes. Conservation efforts aimed at addressing the challenge of climate change are focused on finding win-win strategies that accomplish both mitigation and adaptation by protecting the carbon storage potential of boreal ecosystems and developing innovative tools for integrating the effects of economic land uses, natural ecosystem dynamics, and climate change into a unified approach to conservation planning in a multiuse, yet largely intact, landscape.

Introduction to the Region

As the most extensive terrestrial ecosystem on Earth, the circumpolar boreal forest represents one quarter of the world's remaining frontier forests (Bryant et al. 1997). One third of it, roughly 600 million hectares, is found in Canada, where vast and intact boreal landscapes persist. Canada's boreal ecosystems are dominated by coniferous forests in northern regions, with mixed wood forests more prominent in the south. Major tree species include coniferous black and white spruce, jack and

FIGURE 6.1 Wood Buffalo National Park spans areas of Alberta and the Northwest Territories. Wood Buffalo is the largest national park in Canada, a place where fires can be left to burn across the landscape. Ecological processes such as fire and hydrologic flow contribute to structural diversity that is astonishingly complex and beautiful, especially when viewed from the air. Photo courtesy of Marc-André Parisien.

lodgepole pine, balsam fir, and tamarack larch, along with broad-leaved trembling aspen, balsam poplar, and white birch. Canada's boreal contains an estimated one to two million lakes and ponds (NRTEE 2005), and wetlands cover 20 percent of the area (NRC 2009). Natural disturbances, most prominently fire and insect outbreaks, but also flooding, ice storms, landslides, and windstorms, contribute to the structural heterogeneity of these systems (fig. 6.1).

The region supports over one-third of the breeding populations of North American migratory land birds (Blancher 2003), a significant proportion of the continental breeding grounds for migratory waterfowl, and predator-prey assemblages that include the largest caribou herds in the world. Range contractions of many North American carnivores and ungulates (Laliberte and Ripple 2004) and the latent extinction risk of mammals in northern Canada (Cardillo et al. 2006) highlight the importance of Canada's boreal regions for the persistence of many mammalian species, especially those with large home ranges.

Ownership of Canada's boreal region is almost entirely public, intersecting eight provincial and three territorial jurisdictions wherein governments have primary responsibility for natural resource management. More than five hundred First Nations communities in the region maintain strong cultural ties to the lands and waters, generally relying on natural resources for subsistence and economic development. Along with federal and regional governments and First Nations, other stakeholders in the boreal region include resource extraction industries, conservation organizations, local communities, and the public at large. Effective conservation requires resolution of many conflicting interests between these stakeholders. The specter of rapid climate change adds further complexity to this challenge.

Historical Overview of Conservation and Science Initiatives

There are approximately 1,931 either legally designated or interim/proposed protected areas in Canada's boreal (fig. 6.2; table 6.1), ranging in size from < 1 hectare to 4.5 million hectares (CEC 2010; CCEA 2011; Lee and Cheng 2011) and protecting roughly 10 percent of the region. While there are no formal corridors designated between protected areas, over three quarters of boreal landscapes remain intact based

TABLE 6.1 The percentage of area protected and intact of Canada's boreal ecozones

		Percentage protected		
Ecozone	*Area (ha)*	*Legally designated*	*Interim/proposed*	*% intact*
Arctic Cordillera	43, 141	66.7	0.0	100.0
Atlantic Maritime	1, 544, 888	0.0	0.0	22.2
Boreal Cordillera	43, 995, 911	10.5	1.2	94.4
Boreal Plain	70, 538, 454	8.9	0.2	38.6
Boreal Shield	160, 645, 537	4.2	3.2	72.7
Hudson Plain	37, 509, 896	10.3	1.0	98.3
Montane Cordillera	6, 111, 367	35.2	0.0	83.3
Pacific Maritime	78, 283	0.6	0.0	99.4
Prairie	1, 286, 678	1.3	0.0	0.3
Southern Arctic	9, 566, 028	9.9	0.5	98.7
Taiga Cordillera	24, 790, 728	9.3	3.8	97.7
Taiga Plain	65, 735, 815	8.1	6.0	73.1
Taiga Shield	130, 121, 089	2.5	6.9	95.9
Total	551, 967, 815	6.5	3.6	78.7

Note: Intactness is based on Global Forest Watch Canada's Intact Forest Landscapes (P. G. Lee et al. 2010).

FIGURE 6.2 The distribution of legally designated (dark gray) and government sponsored interim/proposed (mid-gray) protected areas within Canada's boreal region. A black bounding outline identifies the extent of the boreal, and within it, light gray shading illustrates Global Forest Watch Canada's Intact Forest Landscapes, areas that are > 5,000 hectares and are devoid of detectable human disturbance (P. G. Lee et al. 2010). Ecozones are outlined and those contributing substantial area to the boreal are labeled. Canadian provinces and territories are outlined for reference.

on Global Forest Watch Canada's intact forest-landscapes analysis (fig. 6.2), suggesting current high levels of natural landscape connectivity. The majority of anthropogenic disturbance falls along the southern boundary of the boreal where agriculture and forestry are dominant activities. However, development interests are now turning further north to rich mineral deposits, oil and gas reserves, and untapped waterways, focusing on hydroelectricity in the eastern boreal, and oil and gas exploration and extraction in the west. In particular, oil sands operations in the west-central boreal are expected to affect an estimated 13.8 million hectares (Schneider and Dyer 2006). Meanwhile, economic research estimates that ecosystem services in the boreal—including carbon storage, flood control, and water filtration—are almost fourteen times greater than the market value of resources extracted from timber, mining, oil and gas, and hydroelectric generation (Schindler and Lee 2010).

Accelerating development of natural resources, gaps in knowledge, and poor management of competing demands are substantial challenges for conservation in Canada's boreal region. Allocation of natural resources to the demands of varied industries such as oil, gas, and forestry typically occurs without consideration of cumulative effects, resulting in significant potential to exacerbate human influences in these landscapes. The rate of industrial development is outpacing the acquisition of knowledge about the effects of human activities on ecological patterns and processes, and this imbalance introduces significant uncertainty into conservation planning, especially in the context of climate change. Even basic data, such as a detailed forest inventory for commercial forests, are frequently either outdated or proprietary; beyond this commercial zone, vegetation inventories rely primarily on coarsely interpreted satellite imagery. Moreover, there is no comprehensive national-scale monitoring system to track and measure changes in Canada's boreal ecosystems (FPTGC 2010), limiting an objective evaluation of the effectiveness of existing conservation strategies.

Current Conservation in the Region

Until the early 1990s, the scientific focus in boreal regions of Canada was on commercial forests, and largely the domain of government- and university-based researchers. Both the community and spatial extent of interests have since expanded. In 1995, the Sustainable Forest Management Network partnered universities, government, industry, First Nations, and nongovernmental organizations in a broad array of research on commercial forests. An extensive amount of knowledge was generated by this twelve-year effort (SFMN 2011). In 2003, the Canadian Boreal Ecosystems Analysis for Conservation Networks Project, described in greater detail in this chapter, began developing a scientific framework for proactive and comprehensive conservation planning across all boreal regions of Canada.

Environmental organizations and First Nations have catalyzed most recent boreal conservation initiatives. More than twenty-five million hectares of the boreal forest are now managed according to Forest Stewardship Council certification standards, the most stringent now in use, which consider "social, economic, ecological, cultural and spiritual needs of present and future generations" (FSC 2010). The efforts of Ducks Unlimited Canada to conserve and manage waterfowl breeding habitats now encompass wetland and upland sites throughout boreal Canada. In 2003, the Boreal Forest Conservation Framework united environmental organizations, First Nations, and forest and energy companies toward conservation of large interconnected areas of the boreal and in support of sustainable communities (BLC 2003). The Provinces of Ontario and Quebec have since committed to protect large portions of the boreal within their jurisdictions (Government of Ontario 2010; Gouvernement du Québec 2009). The Gwich'in, Sahtu, Deh Cho, and Akaitcho Dene First Nations are also leading initiatives to protect regions of the boreal. Their efforts, in cooperation with the federal government, have significantly furthered the permanent protection of boreal ecosystems, for example by the expansion of Nahanni National Park in Canada's Northwest Territories. In 2010, the Canadian Boreal Forest Agreement allied nine environmental organizations and twenty-one forest companies in a commitment to conservation and best practices over more than 72 million hectares of land, including substantial deferment of logging in woodland caribou habitat while conservation planning takes place (CBFA 2010). Unfortunately, these promising developments may be seriously compromised if conservation planning and implementation does not take into account the anticipated effects of global climate change.

Regional Effects of Climate Change

Climate change poses a high risk to Canada's boreal ecosystems because of projected rapid future warming at northern latitudes and the sensitivity of these systems to environmental change (Ruckstuhl et al. 2008; Schindler and Lee 2010). Global climate models vary in their simulation methods and prediction of future conditions, resulting in projected future climates with wide bands of uncertainty. However, the Intergovernmental Panel on Climate Change (IPCC) identified a number of consistent trends that apply to northern latitudes of North America, including boreal ecosystems: annual mean warming is likely to exceed the global mean; warming is likely to be largest in winter; minimum winter temperatures are likely to increase more than average; mean annual precipitation is very likely to increase; and snow season length and snow depth are very likely to decrease (IPCC-WG1 2007). On the ground, one outcome of these changes is that the pronounced seasonality, with long cold winters maintaining extensive areas of permafrost, is expected to diminish (IPCC-WG2 2007). Melting permafrost and drying peatlands will release globally significant quantities of greenhouse gases (Schuur et al. 2009) that could contribute to further global temperature rise. Altered forest fire regimes and patterns of insect outbreaks; changes to regeneration, growth, and mortality of boreal vegetation; altered phenology of flora and faunal cycles; earlier snow/ice melt; and changes in species ranges are just some of the expected consequences of climate warming in the boreal. Some of these changes have already been detected (Williamson et al. 2009; Lemieux et al. 2010; Schindler and Lee 2010). These alterations will affect the entire boreal forest ecosystem, including the people who live and work on the land (Chapin et al. 2006).

Fire is a dominant disturbance process in much of Canada's boreal. Because the size, frequency, seasonality, intensity, and severity of fires are tightly coupled with climate and weather, rapid changes in fire regimes are expected in the future, acting as a catalyst to alterations in plant and animal communities (Weber and Flannigan 1997).

Extensive retrospective and prognostic research on fire-climate coupling in the boreal has been based on national datasets of fire occurrences dating back to the 1950s, tree ring research, and other indirect historical reconstructions (Flannigan et al. 2009). Recent (1959–99) increases in summer temperature have been correlated with increased area burned in Canada's boreal (Gillett et al. 2004), and variance in annual area burned has increased over the last 150 years (Girardin and Sauchyn 2008). The degree and nature of changes in fire risk has varied from place to place, for example, over the last 100 years, fire risk has diminished in Canada's southeastern and southwestern boreal, apparently due to increased summer moisture; in the northern and northwestern boreal, small increases in fire risk are apparent, with other areas showing negligible change (Girardin et al. 2009).

In the future, fire activity is projected to change in response to interactions in altered patterns of precipitation and temperature (Balshi et al. 2009; Krawchuk et al. 2009). General expectations are that fire activity in Canada's eastern boreal may change little or even decline, whereas increased activity is expected in the north and west. Fire is an ecological process for which we have a relatively sound mechanistic understanding and abundant empirical information on past events, but even so the magnitude and polarity of predicted changes still vary with temporal and spatial scales of study and the modeling procedure used (Flannigan et al. 2005; Krawchuk and Cumming 2011).

With climatic controls on spatial patterns of vegetation being a key focus of many biogeographic studies, researchers have increasingly focused on potential effects of climate change. Bioclimatic or niche models of species ranges used with projected future climate are a common method for understanding potential climate change effects on the boreal biome and constituent flora (Hogg and Bernier 2005; Schneider et al. 2009). For instance, future projections of a climate moisture index forecast a substantial increase in the area of boreal forest stressed by drought by the end of the century, marking a reduction in boreal extent over the provinces of Alberta, Saskatchewan, and Manitoba (Hogg and Bernier 2005). Dynamic vegetation models provide an alternative method to project the future of boreal ecosystems, and show changes in terrestrial biomes across Canada under a suite of climate change-emissions scenarios (Lemieux and Scott 2005). Under climate conditions projected for a doubling of preindustrial atmospheric CO_2 concentrations, these vegetation models predict a northwards expansion of the boreal. This would be coupled with a contraction from the current southern limits of boreal forest due to conversion to temperate mixed-species forest, woodland, or grassland.

Global projects mapping potential changes in climate provide important information for ecosystem conservation in the boreal. The velocity of climate change quantifies the speed at which a species would need to move across the Earth's surface to maintain constant temperature space (Loarie et al. 2009). Based on this concept, most protected areas currently within the circumpolar boreal will maintain their present temperatures for less than 100 years. Present and future combinations of climate conditions mapped by Williams et al. (2007) predict that some combinations of climate conditions will disappear from the boreal region altogether, while others with no current analogue will appear. It is difficult to predict how Canada's boreal flora and fauna may respond to these novel conditions.

Approaches to Conservation under Climate Change

Boreal ecosystems have neither the high species richness of tropical biomes nor notable hotspots of endemism, characteristics that help explain why, historically, global conservation efforts have largely neglected the boreal. Yet recently, researchers have highlighted the boreal as a globally significant ecosystem where, compared to other terrestrial biomes, very large intact wilderness landscapes shaped by natural disturbances, such as fire, still exist (Bradshaw et al. 2009). Intact boreal systems provide a suite of ecosystem services, including water filtration and carbon sequestration. The boreal is a large carbon sink due to low temperatures suppressing below-ground decomposition rates, resulting in vast carbon stocks in peatlands and forest soils (Kurz and Apps 1999; CBI 2009; Schuur et al. 2009; Tarnocai et al. 2009). This carbon has generated global attention to conservation of the boreal forest as a measure to curb or mitigate the effects of climate change. Fortunately, the spatial overlap of carbon storage and biodiversity in terrestrial ecosystems (Strassburg et al. 2010) suggests that safeguarding carbon stocks may benefit biodiversity conservation. A comprehensive strategy to accomplish carbon and biodiversity conservation across the boreal represents a fundamental win-win for mitigation and adaptation to climate change.

Methodologies for incorporating climate change into conservation planning are in their infancy for the boreal, as elsewhere in the world (see chaps. 1 and 2). However, protecting large, intact ecosystems is one method widely proposed for both climate change adaptation and mitigation (Hannah et al. 2002; Bradshaw et al. 2009; Turner 2010). In Canada's boreal, "large" and "intact" still exist (fig. 6.2) and there is also some political will to protect them. Accordingly, the negative interactions between climate change, habitat loss, and fragmentation may be less dire in the boreal than in other regions where the human footprint is more developed (Noss 2001).

Roadblocks and Opportunities

As in many regions of the world, scientific studies in the boreal have focused on the effects of changing climate through paleoecology, observation and monitoring of recent trends, experimental warming, and model-based projections. The vastness of the boreal and variability across its extent in Canada compound the difficulty of finding adequate data to advance understanding of the potential future structure and function of boreal ecosystems. For instance, fine-scaled regional models that include hydrologic components are critical to identifying climate microrefugia (Dobrowski 2011) but still need to be developed across the Canadian boreal region. The availability of downscaled global climate datasets and remotely sensed imagery is growing (e.g., Coops et al. 2008), providing data on biophysical properties of the boreal. In particular, the BioSpace project (Duro et al. 2007) and the Boreal Avian Modelling Project (http://www.borealbirds.ca; Cumming et al. 2010) are notable in generating cohesive data that can inform conservation planning for large landscapes under a range of climate scenarios. These research programs are developing the empirical basis to model species distributions at boreal-wide scales, now and into the future.

In the past, uncoordinated efforts among government agencies, economic sectors, communities, and political jurisdictions have failed to develop a systematic and sustainable solution to the challenges of boreal conservation. Moreover, the conventional focus on protected areas has discouraged the development of a comprehensive vision for boreal ecosystems, despite the abundant opportunities such a vision would present (fig. 6.2). Although this is not surprising given the complexity of the conservation and management issues at hand, new "full-landscape" approaches are needed if the vast conservation potential of the boreal is to be realized.

Conservation Matrix Model

In response to the need for conservation planning methods tuned to the unique qualities and potential of Canada's boreal, the Canadian Boreal Ecosystems Analysis for Conservation Networks (BEACONs) Project has advanced the Conservation Matrix Model as a scientific framework for systematic conservation planning across Canada's boreal (http://www.beaconsproject.ca; Schmiegelow et al., in prep.). In this section, we present key features of the Conservation Matrix Model and illustrate how they might accommodate the challenges of conservation in a changing climate. The Conservation Matrix Model has four components:

1. *Ecological benchmarks* are defined as "ecologically intact areas that are representative of natural environmental variation, including vegetation communities and productivity gradients, and are sufficiently large to maintain key ecological processes and support natural ecosystem dynamics." Large, intact, and representative benchmarks are the anchors of a boreal conservation network.
2. Additional *site-level protected areas* conserve special elements and capture specific values that may be poorly represented in benchmarks.
3. *Managed forests* allow for active resource utilization.
4. The larger *conservation matrix* encompasses the first three components and provides ecological support for the landscape as a whole.

These four components are meant to contain the entire ecosystem and its constituent processes and to include substantial areas under active adaptive management (Walters 1986). Accordingly, implementation of the Conservation Matrix Model requires sound science and systematic planning that integrates ecological, social, and economic interests. New spatial simulation technology designed to support these requirements is being developed by companion initiatives (Hauer et al. 2010).

Ecological benchmark areas can be thought of as theatres large enough for a vast number of ecological and evolutionary processes to play out. For much of the boreal, fire is one of the largest-scale ecological processes. In the Conservation Matrix Model, fire statistics are used as spatially variable criteria to determine the minimum size of benchmark areas (Leroux et al. 2007; Schmiegelow et al., in prep.). Because of fire-climate coupling, fire size statistics can be projected under a range of climate scenarios and so the Conservation Matrix Model can be adapted to accommodate potential futures as well as present climate and fire regimes. Once a suite of appropriately sized and intact potential benchmarks is identified, the model uses ecological representation criteria to design networks of benchmarks. Some of these criteria are sensitive to climate, and efforts are under way to develop predictive models that support assessment of representation under changing climate. Representation of enduring features (Anderson and Ferree 2010) is a complementary approach that would capture the geophysical underpinnings of ecological diversity among benchmark areas. Benchmarks, with little or no trace of human development, will also allow carefully designed monitoring programs implemented as part of active, adaptive management to discern the independent effects of climate change and land management on ecological systems.

The Conservation Matrix Model offers an intuitively appealing approach to systematic conservation planning under climate change. Benchmark areas covering environmental gradients, nested within a conservation matrix, are more likely to allow ecological systems to move and adapt to changing conditions. Broad-scale connectivity under the Conservation Matrix Model results from permeability of landscapes throughout the matrix, sustained by careful planning and management of human activities. Accordingly, the Conservation Matrix Model encompasses strategies recommended for conserving ecosystems under the threat of climate change, specifically the protection of large areas and maintenance of landscape connectivity (Heller and Zavaleta 2009). The model has limitations, of course, and it acknowledges many by specifically invoking active, adaptive management to maximize the rate of learning about boreal ecosystems and their response to change. Given the known, unknown, and unknowable uncertainties in current and future ecosystem dynamics, acting on conservation now, while instituting procedures to learn from these decisions, will almost certainly provide better outcomes than no action at all.

To date, the concepts associated with the Conservation Matrix Model have informed discussions in Saskatchewan, the Yukon, Ontario, the Northwest Territories, and Quebec, with an emphasis on identifying representative networks of ecological benchmarks. Significant parts of the model are implemented in readily available software (CBP 2011; DataBasin 2011). However, some components of the model require enhancement. Most urgently, we require a better understanding of the conditions necessary to maintain the role of the matrix as a supporting environment for species range shifts under climate change. Predicting the future distribution of vegetation remains a major challenge and will rely on the efforts of the broader research community (e.g., Hamann and Wang 2006; Lemieux and Scott 2005).

Advances in large-scale planning initiatives, such as the Canadian Boreal Forest Agreement and requirements under the Far North Act in Ontario, provide opportunities to operationalize comprehensive conservation planning over vast areas. The Far North Act supports protection of at least 50 percent of the boreal Far North in Ontario, an area of approximately 45,000,000 hectares that contains valuable mineral deposits and renewable energy potential, as well as globally significant carbon stores. The act further requires development of a strategic plan to guide community-based, land-use planning throughout the region. First Nations will play a pivotal role in the development of local plans. Alignment of an extensive and relatively intact planning area, multiple stakeholder engagement, and a focus on proactive planning to address conservation and sustainable development under a changing climate, provides ideal conditions for application of a Conservation Matrix Model (FNSP 2010).

Conclusion and Recommendations

Canada's boreal forests provide one of the last places on Earth where intact ecosystems remain extensive enough that proactive conservation planning might sustain biodiversity and associated ecological and evolutionary processes while adapting to or mitigating the effects of climate change. Immediate efforts are needed to realize this potential. The Conservation Matrix Model provides a comprehensive and well-developed, science-based framework with which to conserve the global heritage of Canada's boreal forests, offering a method to include economic land uses, natural ecosystem dynamics, and climate change in a unified framework. The bold challenge that lies ahead is implementation; putting the science and concept to the test of application. It is only by getting the Conservation Matrix Model dirty, working with it on the ground, that the next steps will be made. These steps will include identification of gaps in understanding but also magnificent opportunities to conserve lands and waters, livelihoods, and beauty. We hope that coordinated efforts among local, national, and international stakeholders, in combination with scientific leadership, will emerge to make Canada's boreal region a global example for conservation and sustainable management in the context of climate change.

Chapter 7

Cape Floristic Region, South Africa

LEE HANNAH, DAVE PANITZ, AND GUY MIDGLEY

The highly diverse and unique Cape Floristic Region near the southern tip of South Africa has been a productive testing ground for evolving systematic conservation planning techniques. As protea plant species ranges contract, expand, and shift in response to climate change, new reserve selection methods are needed to account for this dynamism. Conservation planners are applying cutting-edge methods for integrating climate change species-distribution models into the design of connectivity and protected areas in the Cape Floristic Region. These modeling advances, when linked to specific land-use policies and opportunities for acquisition, provide a promising basis for implementing a set of actions to make conservation in the Cape robust to climate change.

Introduction to the Region

The Cape Floristic Region (CFR) is by far the most species rich of the five phytogeographic or botanically distinct regions in Africa and one of the most botanically diverse regions of the world. With an area of just 90,000 square kilometers, it comprises less than 5 percent of southern Africa yet contains an estimated 44 percent of Africa's 20,500 plant species. The Cape flora is so rich in species and distinct from elsewhere in southern Africa that the region is distinguished as one of the world's six floral kingdoms, and one of the few semiarid systems classified as a global biodiversity hotspot (Goldblatt and Manning 2002; Goldblatt 1978; Myers et al. 2000).

Fynbos heathlands cover the majority of the region (87 percent) and are dominated by fire-prone shrubs and small trees with sclerophyllous or ericoid leaves that grow primarily on sandstone soils (Goldblatt and Manning 2002; Richardson et al. 2001). Though mainly fynbos, the CFR contains four other vegetation classes: (1) renosterveld is dense shrubland, commonly populated by Asteraceae, which often grows near fynbos but is generally limited to richer soils; (2) succulent shrubland occurs in the drier valleys of the CFR and includes many species of Aizoaceae and Asteraceae (Cowling and Hilton-Taylor 1999; Cowling et al. 1998); (3) afromontane forest occurs in deeper soils and high-rainfall areas; and (4) forest transitions to shrubby or herbaceous vegetation such as subtropical thicket, consisting of dense, semisucculent, and spiny evergreen shrubland or low forest occur where soil qualities change and precipitation decreases or becomes more seasonal (Goldblatt and Manning 2002).

Overall, the CFR has 9,030 species of vascular plants in 988 genera (Cowling and Pressey 2001; Myers 1990; Goldblatt and Manning 2002). These are dominated by shrubs (53 percent), followed by geophytes (17 percent), other perennials (11 percent), graminoids (9 percent), annuals (7 percent), and trees (2 percent). The Cape hosts unusually high proportions of petaloid monocots, succulents, and sclerophyllous to microphyllous shrubs, and unusually small numbers of annuals for a largely semiarid region—7 percent compared to 30 percent for California and 16 percent for Chile (Kalin Arroyo et al. 1994; Goldblatt and Manning 2002). Asteraceae and Fabaceae are the region's largest families, together comprising roughly 20 percent of CFR species. Species diversity of Iridaceae, Aizoaceae, Ericaceae, Proteaceae, and Restionaceae in the CFR is virtually unmatched worldwide, and Rutaceae and Poaceae are also represented in significant numbers (Goldblatt and Manning 2002). Major fynbos families include Proteaceae (also called "proteas"), the largest and most charismatic floral family; Ericaceae, members of the heath family with dwarf-shrub forms and small tubular or bell-shaped flowers; and Restionaceae, reed-like plants resembling horsetails.

Notably, the CFR boasts one of the world's highest rates (69 percent) of local plant endemism, amounting to 1.9 percent of all endemic plant species globally. This high proportion of endemic plants is comparable to those in New Zealand (81 percent), Australia (85 percent), and Hawaii (92 percent) (Goldblatt 1978). In addition, the Cape is host to eight of the ten families endemic to southern Africa, six of which are found exclusively in the CFR (Goldblatt and Manning 2002; Myers et al. 2000; Cowling et al. 1996). Nearly 198 plant genera are endemic to the CFR; another 40 genera are strongly focused in the region. In the 18,000 square kilometer southwestern tip of the Cape, there are about 6,000 plant species, over 4,200 (70 percent) of which are endemic (Myers 1990). Endemism with the region tends to be highly localized; 80 percent of plant species have ranges smaller than 100 square kilometers (Cody 1986).

The recent and extreme diversification of a relatively small number of plant lineages is the most prominent evolutionary characteristic of the CFR (Cowling and

Holmes 1992). Twelve of its 172 plant families comprise 64 percent of the CFR's flora with over 200 species each. Similarly, 13 of 988 genera together account for 25 percent of the region's plants, each with over 100 species (Goldblatt and Manning 2000). This rich and diverse flora is thought to have evolved at the southern edge of the tropics, from a combination of southern African temperate flora and from a tropical African forest flora, as Africa became more arid (Goldblatt 1978). The bulk of this diversification is thought to have occurred relatively recently (post-Pliocene), after the development of seasonal Mediterranean-type climates in the late Pliocene allowed fire to assume its important ecological role (Cowling and Pressey 2001; Linder et al. 1992).

The Cape's mild, temperate, Mediterranean-type climate provides winter rain for most of the region, and the eastern Cape also receives a significant amount of summer rainfall (Goldblatt 1978). Most soils are nutrient poor, contributing to adaptation and specialization in plants that occur there, and the vegetation is largely adapted to the recurrence of fire every ten to thirty years (Cowling 1992). Strong winds support fire and plant dispersal, factors that are central to the region's rich biodiversity (Simmons and Cowling 1996).

Most of the natural habitat of the Cape has been converted to agriculture and other human uses. This high rate of conversion combines with high endemism to make the Cape a global biodiversity hotspot. Biodiversity hotspots have lost more than 70 percent of their original pristine habitat. In the Cape, this loss has been due to industrial agricultural crops such as wheat, as well as to specialty crops such as orchards and vineyards. Cape Town is the hub of a major urban area that coexists with major biodiversity on Table Mountain and south to the Cape of Good Hope. Population density elsewhere in the CFR, however, is relatively low. The loss of habitat over much of the Cape is due to extensive agriculture rather than to intensive residential development. The processes of urban and agricultural development are continuing, and in some cases intensification of agriculture is crowding out remnants of natural vegetation. The already high and ongoing loss of habitat means that conservation planning for the unique biodiversity of the Cape is a high priority.

Historical Overview of Conservation and Science Initiatives

Though the stunning biodiversity of the CFR has been prized and studied by scientists since the eighteenth century, concerted efforts to conserve CFR ecosystems did not begin until the first reserves were designated in the late 1800s, focusing primarily on protection of forests and game (Gelderblom et al. 2003). Conservation activity was high in the CFR through the mid-1980s, largely through the work of South Africa's Forestry Department, including expansion and legislative reinforcement of montane forest reserves, as well as programs for prescribed burning and invasive species control. Progress slowed dramatically in the late 1980s as apartheid-era economic sanctions and government cutbacks slowed conservation activity for the next decade (Gelderblom et al. 2003). Despite the prior century's success in placing over 20 percent of the CFR under conservation management, conservation implementation had not been directed by a coherent plan for representation of biological patterns and processes. Reliance on low opportunity-cost land acquisitions led to disproportionate representation of montane forest ecosystems, rendering the CFR reserve network of the 1990s unrepresentative of its major vegetation classes and the ecological processes required to maintain biodiversity (Rouget et al. 2003; Cowling and Pressey 2003).

Systematic conservation planning for the Cape (i.e., conservation planning that is data- and target-driven, efficient, explicit, and flexible) and planning for climate change have emerged only recently. Although reserve selection algorithms for identifying representative biodiversity patterns first appeared in the late 1970s with incremental progress through the following decade (e.g., Rebelo and Siegfried 1992), these efforts remained largely in the academic sphere. Later efforts incorporated measures of site irreplaceability (Pressey et al. 1994) and vulnerability (Pressey et al. 1996) to avoid losing critical features to destructive processes during the implementation phase (Cowling and Pressey 2003).

In the wake of South Africa's democratic elections in 1994, newfound international donor interest in conservation projects provided opportunity for professionals to apply systematic reserve design in South Africa's globally important ecosystems (Reyers et al. 2010). Lombard et al. (1999) and Cowling et al. (1999) identified alternate reserve systems for the Succulent Karoo biome adjacent to the CFR, both of which highlighted the importance of connecting conservation areas across climatic gradients to facilitate migration of low-dispersal plant species and protection of areas that will become suitable in future climates (note that the term *biome* in the South African context connotes a more geographically constricted usage than at the global level). At the same time, Rutherford et al. (1999) modeled major climate-driven vegetation shifts in both the Cape and Succulent Karoo by 2050, the latter losing more than 80 percent of its current range, due to the biome's southward collapse and, to a lesser extent, migration to areas disjunct from its present range.

Integrating Science into Biodiversity Conservation

In the late 1990s, conservation plans began to inform new land protections. Cowling and Mustart (1994) created a broad structure conservation plan for the Agulhas Plain that Lombard et al. (1997) followed to select a representative reserve network (fig. 7.1a) using a minimum-set approach (i.e., in which minimum thresholds are set for representation of targets such as vegetation types, populations, and/or proxies for ecological and evolutionary processes). The plan included some attention to biodiversity retention, but neither sufficiently accommodated the full range of critical ecosystem and evolutionary processes nor addressed the habitat loss and degradation

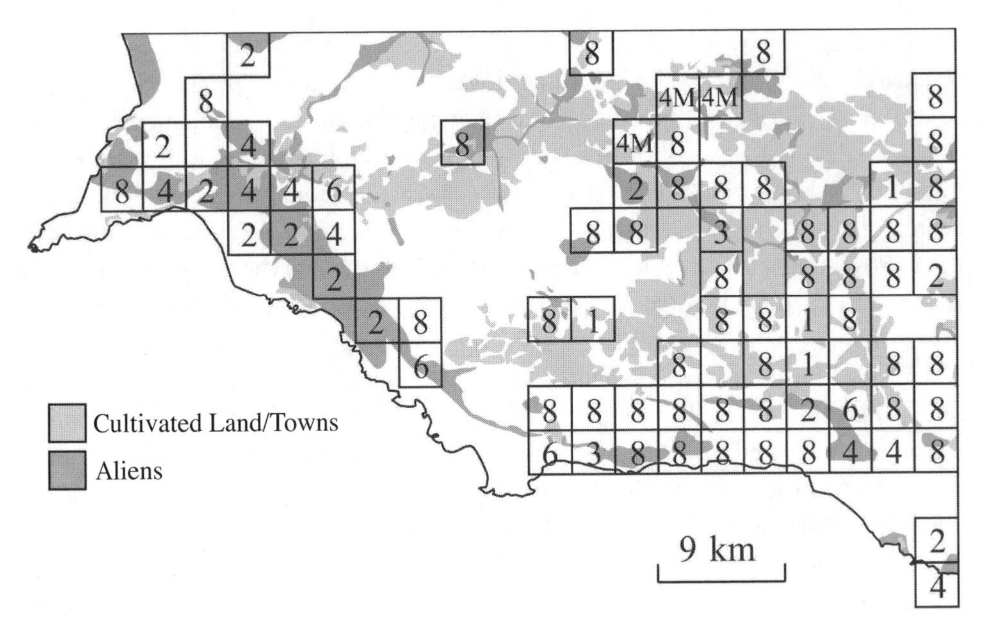

FIGURE 7.1a The systematic conservation plan for the Agulhas plain (from Lombard et al. 1997). Numbers indicate the number of reserve selection algorithm variations that selected each cell.

projected to continue during the plan's implementation (Heydenrych et al. 1999). The areas identified by Lombard et al. (1997) nevertheless formed the basis for the establishment of the 20,000-hectare Agulhas National Park in 1999 (fig. 7.1b) to conserve the highly fragmented and vulnerable lowland fynbos and westland ecosystems (Hanekom et al. 1995). This represented the first conservation application of systematic conservation planning, not only in the CFR, but in all of South Africa (Heydenrych et al. 1999).

In light of the Cape's high level of biodiversity and the substantial threats to its persistence, the Global Environmental Facility allocated funds to the development of a broad plan to support persistence of Cape biodiversity in the face of habitat loss and climate change (Cowling et al. 2003; Frazee et al. 2003; Lochner et al. 2003). The resulting Cape Action Plan for People and the Environment (CAPE) project was coordinated by the World Wide Fund for Nature South Africa in collaboration with government, private institutions, and local communities (Younge and Fowkes 2003). Project goals included identification of (1) large reserves (> 500,000 hectares) to support landscape-scale evolutionary and ecosystem processes, (2) lowland areas for establishment of a network of medium-sized reserves, and (3) financial incentives to encourage conservation on private lands. The approach was guided by systematic conservation planning to design a reserve network that set explicit targets for biodiversity (land cover types, populations of proteas and selected vertebrate species, population sizes for medium and large animals) and process (six spatial surrogates for ecological

FIGURE 7.1b The land units of the Agulhas National Park (shaded in gray). High-priority units from the conservation plan were captured in many instances by the final park configuration. Compare to figure 7.2, however, where many priority units for conservation under climate change fall in the hills above the Agulhas plain and outside the park, as shown here.

and evolutionary processes). The model and plan incorporated a virtually unprecedented number of biodiversity targets worldwide (Cowling et al. 2003; Pressey et al. 2003).

Achievements of the CAPE plan included integration of biodiversity conservation into land-use plans at the provincial scale, expansion and strengthening of reserves within the CFR (Tortell 2010), and development of off-reserve conservation mechanisms (e.g., Von Hase et al. 2010). The project facilitated novel institutional partnerships and cooperation between government agencies, including creation of a conservation planning unit within the Western Cape Province conservation management department and close collaboration between the Western Cape environmental policy and regional planning departments (Gelderblom et al. 2003). CAPE also paved the way for subsequent integration of conservation planning into government departments and legislation at multiple levels (Reyers et al. 2007). One notable successor was the National Spatial Biodiversity Assessment, whose products were used to develop policies and plans at the national (e.g., National Spatial Development Perspective; South African Environmental Outlook 2005; National Strategy for Sustainable Development; National Biodiversity Framework) and provincial (e.g., Western Cape Province Provincial Spatial Development Framework; Mpumalanga Provincial Conservation Plan) scales (Reyers et al. 2007).

Other plans worked at the scale of individual vegetation classes within the CFR. For example, the Subtropical Thicket Ecosystem Planning Project identified priority conservation areas for persistence of the Subtropical Thicket straddling the Western and Eastern Cape Provinces. The four-year (2000–2004) Global Environment Facility (GEF)-funded initiative was a systematic conservation assessment designed to effect long-term conservation via direct integration of outputs into policies and practices of governmental and private land-use planning and management organizations. The resulting conservation priority maps and guidelines identified large-scale corridors to reach biodiversity processes targets and conservation status categories for biodiversity features. Direct involvement of stakeholders from implementation organizations throughout development allowed the assessment to respond to real-world implementation opportunities and limitations (e.g., using cadastres, or actual land parcels, instead of arbitrary planning units), thereby increasing the plan's policy relevance and rapid adoption into land-use planning (Pierce et al. 2005).

Most of these representation-based algorithms assumed climate to remain static through time. This assumption was increasingly problematic as consensus mounted about the reality of climate change and associated risks to protected areas (Halpin 1997; Rutherford et al. 1999). Early efforts to systematically address the issue of climate resilience involved protection of landscape features supporting ecosystem and evolutionary processes, such as interspecies interactions, disturbance regimes, and migration (e.g., Cowling et al. 1999; Cowling and Pressey 2001). These strategies to protect climatic and elevational gradients for migration represented a step forward, but because they did not explicitly model climate changes and species responses, there was no concrete means to predict their effectiveness.

A key limitation of the initial CAPE plan was reliance on conservation of climatic/elevational gradients and costly ecosystem restoration as the primary mechanisms for climate change adaptation. Without incorporating explicit models of climate change impacts, the plan offered little ability to forecast species' capacity to migrate across those gradients and restored areas within the reserve network (Cowling et al. 2003). Nonetheless, the plan's broad policy integration provides a platform for incorporation of explicit climate change models into future versions of the plan.

None of the previously described conservation plans leveraged emerging capacity to predict climate change impacts on species. This means that the reserve selection methods are not fully robust to climate change, as future projections must now consider physiological tolerances, competition, and dispersal mechanisms to approximate species responses (Hannah et al. 2002). Nevertheless, the studies and applications above provide critical insight into the importance of close and early integration between systematic planning with stakeholder and policy frameworks. Even as more sophisticated climate and species models are essential to conservation planning for climate change, they must also be successfully integrated into extant planning processes and provide clear guidelines for implementation. It is perhaps due to these challenges that, despite a recent proliferation of bioclimatic niche and species models applied to the CFR, no reserve networks have yet been based on explicit climate and species models. The following sections discuss recent progress toward that goal.

Regional Effects of Climate Change

There have been several efforts to model climate change impacts on elements of the Cape flora, but the most extensively published is an effort conducted at the South Africa National Biodiversity Institute (SANBI). This work was carried out in collaboration with the SANBI Protea Atlas Project and modeled over 300 protea species (Midgley et al. 2002; Hannah et al. 2005). The species-distribution model used was a generalized additive model (GAM) driven by two midcentury climate model projections (using models from the Hadley Centre and National Center for Atmospheric Research, and the relatively high A2 emissions scenario; see chap. 2 for additional discussion of emissions scenarios and climate models). These models showed marked declines in range size across many species and served as one of the inputs to a pioneering work on extinction risk from climate change (Thomas et al. 2004).

This set of models was made possible by the extraordinary occurrence records available through the Protea Atlas Project. The Protea Atlas compiled known herbaria and scientific records of proteas with substantial contributions from amateur botanists; the development of the second of these datasets was defendable because the large, showy protea flowers are easily identifiable even by nonprofessionals. The Protea Atlas has GPS-mapped trails in the Cape, and provides amateur botanists with an identification toolkit that allows them to record and report occurrences of proteas (Rebelo 2001). Using the best georeferenced Protea Atlas data, 330 species had sufficient occurrence points to produce robust GAM species-distribution models.

The resulting models showed that by midcentury, under a no-dispersal assumption, approximately one-third of the species modeled lost all climatically suitable habitat or had severe range dislocations that eliminated all overlap between current range and future suitable climate. Range dislocation in many of these species was greater than estimated dispersal distances. Under this scenario, another third of the protea species lost most of their suitable range. Alternatively, in a full-dispersal assumption, about 10 percent of species lost all suitable climatic space and an additional ~10 percent lost most suitable range. Thus, the numbers of species losing most or all climatically suitable habitat in the future ranged from about 60 percent (no dispersal assumption) to about 20 percent (full dispersal assumption). Actual dispersal, and actual loss of suitable climate, would likely fall somewhere between these extremes. Species at low latitudes lost the most suitable area, presumably because of limited poleward landmass to accommodate range shifts.

Responding to criticism by Thuiller et al. (2004) of the original Cape Protea study, Thomas et al. (2004) acknowledged 33 percent variation due to choice of species distribution-modeling approach, but also highlighted the 100 percent variation from choice of dispersal assumption and climate change scenario, respectively. While acknowledging that the number of environmental variables used in models could also impact variation, they found no significant correlation with estimations of extinction risk.

The protea modeling effort used a second, biome approach to complement the species modeling effort (Midgely et al. 2002). The fynbos biome was sampled and projected into a future midcentury scenario using the same General Circulation Model (GCM) scenarios as used in the species modeling. The results of the biome model showed a pronounced southward collapse of the biome, consistent with the species modeling results.

Approaches to Conservation under Climate Change

The protea species models have been used to (1) assess protected area effectiveness in the face of climate change (Hannah et al. 2005), (2) identify protected areas and connectivity to compensate for the effects of climate change on species' ranges (Williams et al. 2005), (3) assess the relative contribution of climate change to species endangerment (Bomhard et al. 2005), and (4) estimate extinction risk from climate change (Thomas et al. 2004). These applications have helped define conservation priorities in the CFR, but more important, have helped define a set of tools that can be used in planning conservation for climate change anywhere in the world.

The implications of the protea species models for species representation in reserves were explored by Hannah et al. (2005). Araújo et al. (2004) had demonstrated that climate change might drive species out of reserves, and the protea species modeling confirmed this finding in a very different biogeographic setting. It is not surprising that climate change may reduce species representation in reserves, since reserve locations have not been selected to be robust to climate change. As ranges shift and often contract, representation in reserves changes, and in the majority of cases, declines.

To counteract these predicted declines in representation would require siting of connectivity areas and new protected areas specifically to address climate change. This topic was the subject of a reserve selection working group convened by Conservation International and SANBI in 2001. The group used multispecies modeling efforts from the Cape, Europe, and Mexico to design reserve selection algorithms for climate change (Williams et al. 2005; Hannah et al. 2005). The method developed used multiple timestep-species models to identify "chains" of suitable habitat within a species' dispersal ability, from the present to future. A reserve selection algorithm then located the areas that could conserve all species in all timesteps using the least area, starting

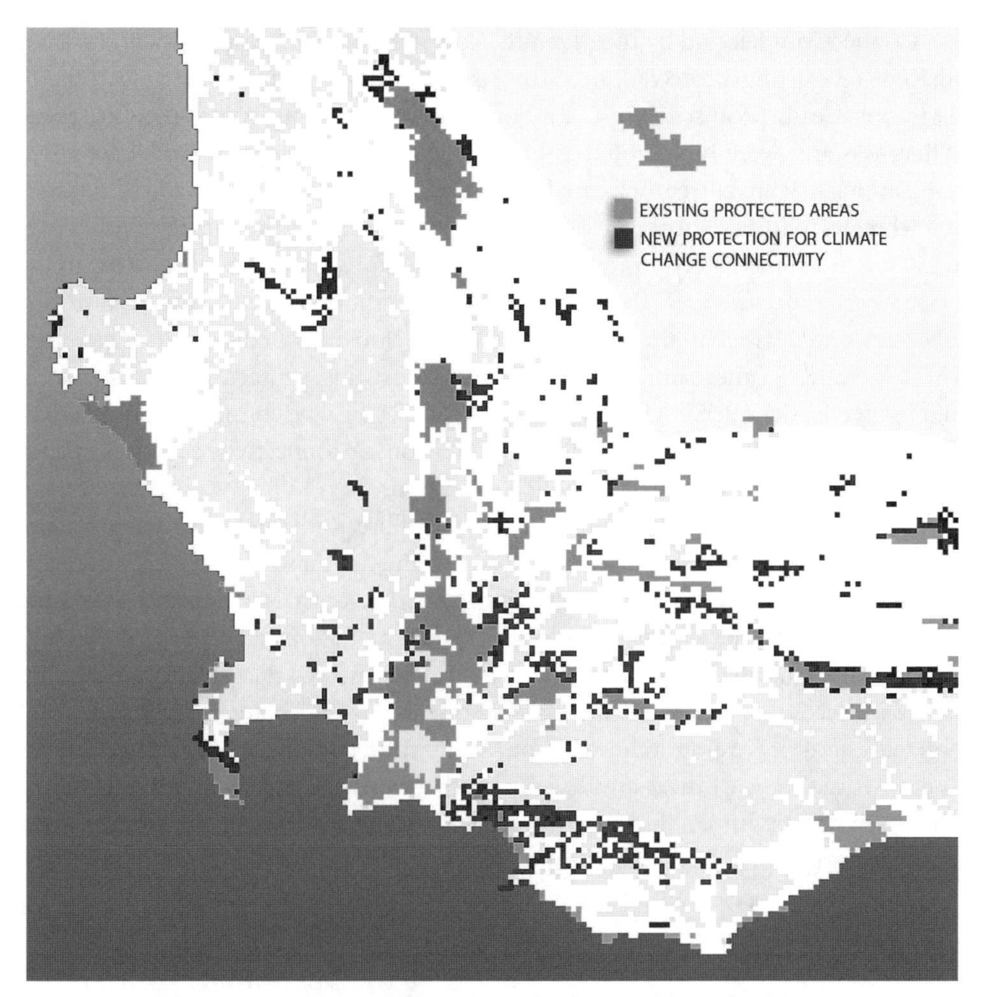

FIGURE 7.2 Climate change-connectivity analysis of Williams et al. 2005. Light gray areas represent existing protected areas. Dark gray areas are planning units selected by a conservation planning algorithm modified for climate change to select "chains" of climatically suitable habitat, from the present to the future, in ten-year timesteps. The shortest chain is a single planning unit in which climate remains suitable in all timesteps. Longer chains are selected when necessary, within dispersal distances appropriate for a given species. The Cederberg Wilderness area is the large protected area in the upper left; the Piketberg is the U-shaped group of cells recommended for new protection in the upper left; and the Agulhas plain is in the lower center (compare to fig. 7.1).

with existing protected areas and expanding as necessary (fig. 7.2). The result defines areas for future protection or connectivity to compensate for species' range shifts. Phillips et al. (2008) improved this method using commercial optimization software, increasing the efficiency of the system by about 20 percent. Where new connectivity or protection for climate change is not undertaken, an extinction risk is faced.

Connectivity selected by the planning algorithm in the chains analysis was often adjacent to existing protected areas, to take advantage of future suitable climates that were already protected (fig. 7.2). For instance, several units near the Cederberg Wilderness area were highlighted as important because multiple species have small populations that are currently located next to that protected area, and for which suitable climate conditions may be found inside the reserve in the future (Williams et al. 2005). By immediately protecting these populations and connecting them to the Cederberg Wilderness area, these protea species might only need to move short distances to find areas that are both protected and climatically suitable in the future. These areas, and other similar areas adjacent to existing protected areas that were highlighted by the reserve selection models (fig. 7.2), may be considered priority candidates for contractual conservation, new protection, or connectivity actions, depending on local land ownership and enabling conditions.

In addition to highlighting the importance of areas adjacent to protected areas, the reserve selection and connectivity analysis for the Cape also identified several areas that are largely unprotected at present, but which are crucial to conservation as the climate changes. These areas include the Piketberg region, and areas in the hills above and behind the Agulhas plain (fig. 7.2). The reserve selection working group visited a selection of the Agulhas areas with officials of CapeNature, a public institution with statutory responsibility for biodiversity conservation in the Western Cape of South Africa. Many of these areas were already prioritized for contractual conservation agreements with landholders; the climate change analyses underscored the urgency and priority of these agreements.

The protea species modeling effort and associated reserve selection working group results have underscored that the line between additional protected areas, expansion of existing areas, and connectivity for climate change adaptation is a spectrum. Areas adjacent to existing protected areas may often best be conserved by annexation to the existing reserve or emphasis in the buffer zone management of the reserve. So, one way to view these areas is as candidates for new levels of protection. At the same time, the protected areas are selected by the conservation algorithm because they harbor current populations of species expected to find suitable habitat in the adjacent protected areas in the future. The algorithm is therefore connecting present and future populations. In this light, conservation of these areas would provide connectivity. Whether they are viewed as additional protected areas or as landscape connectivity for climate change, these areas are of vital importance to conservation that can withstand the test of climate change.

Conclusion and Recommendations

The studies and conservation processes discussed earlier highlight significant progress over the last several decades in the application of systematic conservation planning and climate change projections to conserving Cape biodiversity. There is nonetheless additional progress to be made. In the research sphere, the critical role of uncertainty must be considered in the interpretation of results of fine-scale climate modeling efforts.

Variation is introduced into model results by choice of species distribution modeling approach, dispersal assumption, climate change scenario, and environmental variables. Such variation should be taken into account in study design, highlighted in findings, and minimized to the extent possible. Future modeling work should also leverage increased model sophistication and computing power to simulate not only time series of spatial habitat distribution (as in Williams et al. 2005 and others), but also population dynamics and interspecies interactions under climate change.

Such advances in simulating biological climate change responses can best be put to use if complemented with strong links to implementation and policy. Models should thus also incorporate existing constraints (e.g., economic, social, political) on planning in the Cape, and outputs should provide clear guidelines for integration into specific planning processes. Those processes must in turn be structured to incorporate model results, implying complete, early-stage integration of models into Cape planning processes. Policy mechanisms to standardize these links at regional and local scales would reduce the obstacles to integration.

The protea species modeling is now over ten years old and is in need of renewal. Using the same Protea Atlas database, improved species-distribution modeling techniques and more recent climate-model outputs would provide state-of-the-art biodiversity modeling for conservation planning in the Cape. Other taxa, notably the Restionaceae and Ericaceae of the Cape, could also be modeled, albeit with less rich distributional datasets than are available for the proteas. These modeling advances, combined with the recent upsurge in international funding for climate change adaptation, provide a promising basis for fully implementing a set of actions to make conservation in the Cape Floristic Region robust to climate change.

Chapter 8

Eastern Mongolian Grassland Steppe

EVAN H. GIRVETZ, ROBERT MCDONALD, MICHAEL HEINER, JOSEPH KIESECKER, GALBADRAKH DAVAA, CHRIS PAGUE, MATTHEW DURNIN, AND ENKHTUYA OIDOV

Nomadic livestock herders have coexisted with migratory wildlife in Mongolian grasslands for centuries, but both are now threatened by a combination of climate change, overgrazing, and increased extractive development (i.e., mining and petroleum). Both people and wildlife depend on grassland productivity for their survival, but increasing temperatures are projected to drive higher rates of evapotranspiration leading to moisture stress and decreased productivity. Combining ecoregional assessments with ecosystem-based adaptation strategies—such as the use of grass banks and sustainable grazing management—has the potential to help herders cope with climate change impacts and maintain sustainable livelihoods into the future. Accomplishing this will require coordinated regional planning that identifies where to implement specific on-the-ground conservation strategies to address climate change impacts such as desertification, soil loss, water scarcity, and overgrazing.

Introduction to the Region

Grasslands are globally some of the most converted and least protected of natural community types (Hoekstra et al. 2005). Today, Mongolia is home to the largest intact temperate grassland system left in the world, supporting endemic wildlife such as the endangered Mongolian gazelle (*Procapra gutturosa*) and pastoral human communities that have grazed livestock in these areas for centuries. However, these unique grassland ecosystems are threatened by a combination of climate change, habitat fragmentation, and land-use change from a growing mining and petroleum industry, as well as heavy grazing pressure from a pastoralist system that has only recently been introduced to the global market economy.

There is evidence that climate change is already leaving its fingerprint in Mongolia and these trends will likely become more prominent as global greenhouse gas emissions grow. During the period from 1951 to 2002, observed temperatures in Mongolia increased approximately 1.9 degrees Celsius, placing the country among the top ten fastest warming countries in the world. With precipitation rates having changed little (Girvetz et al. 2009), the warmer temperatures have likely contributed to both the expansion of the Gobi Desert northward (Yu et al. 2004; Zhang et al. 2011) as well as the drying of rivers and water bodies. A national inventory of water resources in 2007 found that 17 percent of surveyed streams, 24 percent of surveyed springs, and 32 percent of surveyed water bodies had dried up (MNET 2009).

In this chapter we provide background on the natural history and human dimensions of the Mongolian grasslands and then propose landscape-scale, climate-adaptation strategies to help both the Mongolian grasslands and the people who depend upon them to cope with projected climate change impacts. Due to the superior condition of its vast grasslands, Mongolia presents an important opportunity for integrating climate adaptation into conservation, but it will require strategies that meet the needs of the 40 percent of Mongolian people who make their living from the livestock trade, as well as ameliorate the increasing pressures from resource extraction. Overall, the climate adaptation strategies employed must consider the natural landscapes as well as the social, cultural, and political contexts in Mongolia.

Our particular focus is on the temperate grasslands and forest steppe of Central and Eastern Mongolia (fig. 8.1), which is delineated as the Mongolian portion of the following three terrestrial ecoregions: the Mongolia-Manchurian Grasslands, the Daurian Forest Steppe, and the Trans-Baikal Forests (Olson et al. 2001). This study area covers approximately 440,000 square kilometers. While these ecoregions extend into China and Russia, generally the grasslands in Mongolia are much less degraded than the grasslands in the two neighboring countries (Fernandez-Gimenez 2000; Angerer et al. 2008).

Over 800 years after Genghis Khan established the country as a unified kingdom and superpower, Mongolia is now rapidly recovering from the economic difficulties initiated by the collapse of the Soviet Union. With an economy centered on livestock husbandry and mining, the two greatest threats to these grasslands are related to grazing and natural resource extraction. With regard to the former, the problem results from a number of factors: changes in herder behavior, overgrazing due to the complex interaction of the shift to a market economy since 1990, and the increase in goat numbers following the global demand for cashmere wool. In terms of the latter, the threat arises from booming mining and petroleum extraction activities since the passage of a liberal mining law in 1997, as well as from the related infrastructure construction and changes in road traffic, population, and land use that emerged in the wake of the change to the new free market.

In short, the transition in Mongolia's political and economic system over the past two decades has brought the country to a crossroads where increased livestock grazing and rapid development of resource extraction now poses a substantial threat to the country's rare and remarkable natural landscapes. Although there is still a window of opportunity to ensure that these threats do not impoverish Mongolia's vast, nearly pristine grasslands, it will take a coordinated landscape-level conservation planning effort to find ways of balancing the protection of natural areas with the needs of pastoralists and the mining and petroleum sectors that are major parts of the Mongolian economy.

Historical Overview of Conservation and Science Initiatives

Mongolia is home to one of the world's first nature reserves, Bogd Khan Mountain Reserve, established in 1778 by the monarch Bogd Khan as a sacred area where hunting and livestock grazing were prohibited. In 1818, two additional protected areas, Otgontenger Mountain and Bulgan Mountain, were designated. Since the establishment of Mongolia's modern protected-area system in 1957, the system has expanded to include sixty-four protected areas, covering 14 percent of Mongolia's land area (approximately 21 million hectares) and preserving important examples of the nation's rich biological, ecological, and cultural heritage (Myagmarsuren 2008). In 1998, the Mongolian national government put forward an ambitious master plan for Mongolia's protected areas, which pledges to protect 30 percent of the entire country by 2020.

In the grasslands of the northeastern quadrant of the country, a significant area has received protected area designation: 6.6 percent of the Mongolia-Manchurian Grasslands (2,100,000 hectares), 5.3 percent of the Daurian Forest Steppe (500,000 hectares), and 43.7 percent of the Trans-Baikal Forests (1,600,000 hectares). To identify the most important areas to protect and guide the government toward its 30 percent protection goal, The Nature Conservancy led an Ecoregional Assessment for eastern Mongolian grassland steppe (fig. 8.1; Heiner et al. 2011) along with a wide range of partners from other international nongovernmental organizations (NGOs), Mongolian national and local governments, and international development organizations (Moore et al. 2010). The purpose of this assessment was twofold: first, to design a portfolio of priority conservation areas that optimizes for representation of habitat and ecological condition; second, to produce a decision support framework for bal-

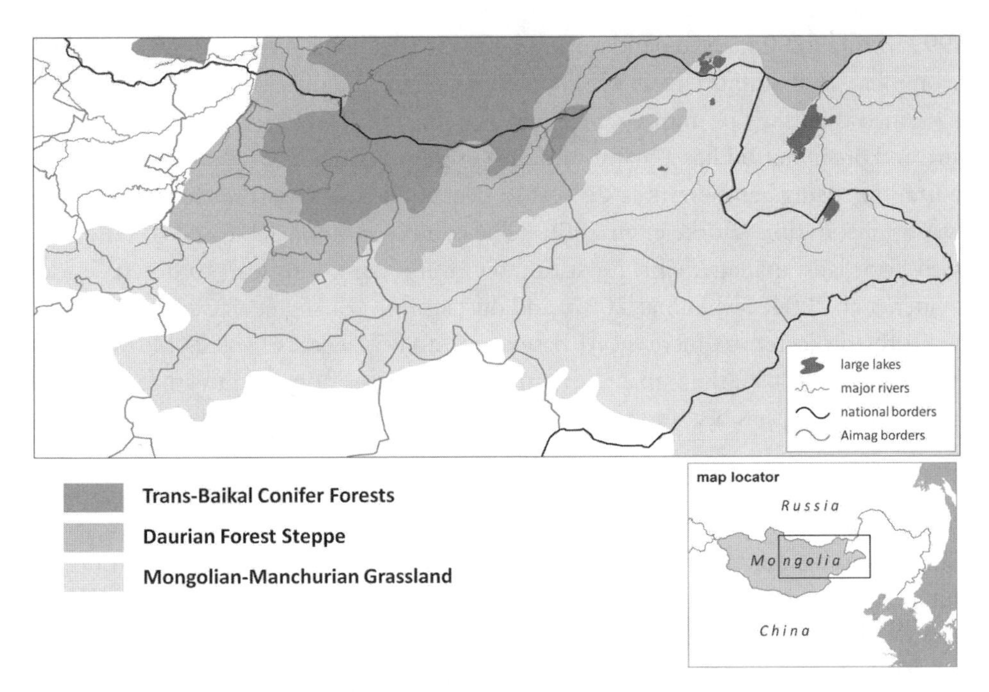

FIGURE 8.1 Map showing three ecoregions in eastern Mongolia considered in this analysis: Mongolian-Manchurian Grasslands, Daurian Forest Steppe, and Trans-Baikal Conifer Forest.

ancing priorities for biodiversity conservation, pastoral land use, and development of mineral and petroleum resources.

Regional Effects of Climate Change

The Eastern Mongolian grassland steppe is an arid system that relies heavily on spring and summer rainfall to drive grassland productivity. The spatial and temporal pattern of rainfall is irregular and unpredictable, as demonstrated by the nomadic movement patterns of Mongolian gazelle in search of forage (Mueller et al. 2008). The Mongolian gazelle is endemic to this region and plays a major ecological role in the grasslands. The large nomadic population, over one million individuals, is a prey base for predators and scavengers, and redistributes nutrients across its wide range. Mongolian gazelles are also an important food source for subsistence hunters (Olson 2008).

Pastoralist herders in Mongolia have adapted to climate variability over centuries. They are periodically plagued by the *dzud*, the Mongolian word for a natural disaster that occurs in some years when a summer drought is followed by a harsh snowy winter, resulting in massive die-offs of livestock and higher likelihood of overgrazing. The most severe dzuds in recorded history occurred during the winters of

1999 through 2002, and a dzud during 2009–2010 killed 8.4 million (20 percent) of Mongolia's forty million livestock.

Since this is a dry, moisture-limited system, the interaction between temperature and precipitation has a major impact on soil moisture. Even with increasing precipitation, rising temperatures can cause decreased soil moisture due to increased evapotranspiration (Burke et al. 2006). The combination of decreased soil moisture with poor land management, most notably overgrazing, can lead to desertification (Asner et al. 2004; Sivakumar 2007) and dust storms (Zhang et al. 2003). With temperatures projected to increase to between 2 and 5 degrees Celsius by the 2080s (Girvetz et al. 2009), the impact of temperature increases on soil moisture and grassland productivity has become a major concern.

Projecting Future Impacts

Understanding exactly how climate will impact grassland productivity is difficult, but we can gain insight into likely future impacts by modeling climate change impacts. Specifically, we related grassland productivity to moisture stress as calculated by a variant of the United Nations Environment Programme's Aridity Index (UNEP 1992). Used in many other studies (Gao and Giorgi 2008; Bannayan et al. 2010), the aridity index quantifies the supply (precipitation) versus demand (potential evapotranspiration) in moisture. We mapped the net primary productivity of the grasslands based on the Normalized Difference Vegetation Index, derived from satellite imagery, and then related these values to the aridity index, which shows a strong relationship ($p < 0.0001$) between increasing moisture availability and increased grassland productivity as measured by net primary productivity (NPP) (unpublished data).

Based on this relationship between NPP and the aridity index, combined with data from sixteen general-circulation climate models, we then estimated grassland productivity for the future projected period of 2070 to 2099 baseline and future projected time period as compared to the historic baseline of 1961 to 1990. We found that grassland productivity in the Mongolian grasslands is projected to decrease due to the exacerbation of moisture stress due to climate change. The decrease in productivity will be considerably greater with higher greenhouse gas emissions (IPCC-SRES 2000): a 6.9 percent decrease on average under the relatively high A2 emissions scenario, compared to a 1.9 percent decrease under the relatively low B2 emissions scenario (see chap. 2 for additional information on emission scenarios). The greatest proportional decreases in productivity are projected to occur in the most arid southwestern portions of the study area at the grassland's ecotone, with the Gobi Desert to the south. This area has already experienced a northward shift in the grassland-desert border (Yu et al. 2004).

Climate change may reduce water availability, both surface flows and groundwater. Mongolian herders depend on surface flows, springs, and wells for drinking water, both for themselves and for livestock. Therefore, droughts have the potential to compound the impacts discussed thus far on livestock and pastoralist livelihoods.

The combination of grazing pressure, decreasing productivity, water scarcity, and vulnerability to desertification has the potential to produce extensive grassland degradation and create challenges for herders to maintain livelihoods. Soil erosion and desertification is likely to increase—a problem that will likely be exacerbated by grazing pressure. This could trigger threshold events, such as rapid vegetation shifts, desertification, and dust storms (Angerer et al. 2008). Desertification and dust storms are already occurring and causing major problems in Inner Mongolia (a bordering autonomous region within China) and surrounding areas.

Climate change is also projected to increase winter precipitation across Mongolia (Girvetz et al. 2009). This combination of decreasing grassland productivity and increased winter precipitation would likely lead to increased frequency and magnitude of dzuds. Even with warming, the winters will still be cold and harsh. The massive livestock die-offs that occur during dzuds are largely the direct result of snow and wind, both of which could become more intense and frequent.

It is clear that climate change will impact both natural ecosystems and the people who depend on these ecosystems for their livelihood. In addition to the direct impacts of climate change, the human response to climate change has the potential to exacerbate impacts. Pressure to overgraze the grasslands may grow in order to compensate for reduced forage availability. Adaptation strategies must take into consideration not only the direct impacts to ecosystems and the people that rely on them, but also indirect effects via human responses to climate change.

Approaches to Conservation under Climate Change

Climate and weather have played a profound role in shaping ecosystem dynamics and human societies in Mongolia. Under conditions of anthropogenic climate change, effective conservation strategies must be organized around climate and weather in these grassland ecosystems. Driven by moisture conditions, the grassland-desert boundary naturally shifts back and forth across the landscape (Yu et al. 2004), and pastoralist herders have responded with a nomadic lifestyle of following shifting precipitation patterns in the spring and summer in order to provide their livestock with the best available forage. Conservation and nomadic herding have traditionally been conducted in a landscape that is dynamic. With climate change potentially driving these ecosystems toward a drier and less productive state, responses to this novel threat must become an integral part of grassland conservation planning in Mongolia.

It is clear that climate impacts to these natural grassland ecosystems cannot be separated from impacts to the people who live in them. People have lived sustainably as a part of this system for millennia, and the herding lifestyle will undoubtedly continue. Consequently, it would be futile to separate climate adaptation for natural systems from adaptation for people's pastoral livelihoods. Rather, what is needed is an ecosystem-based adaptation approach to climate change wherein nature conservation is the means to help both people and ecosystems adapt to future changes in the grassland systems. Ecosystem-based adaptation relies on nature conservation to protect, restore, and manage ecosystems for safeguarding both human societies and biodiversity from climate change impacts (SCBD 2009; Vignola et al. 2009). The benefits nature provides people—water, forage, storm protection—can be self-maintaining, and in the long run likely more cost effective than built infrastructure, particularly in responding to too much or too little water (Postel and Thompson 2005).

Ecoregional Assessments

Ecoregional conservation assessments provide an excellent opportunity for incorporating climate change considerations into large landscape planning (Groves 2003; Game et al. 2010). Climate change impacts will not be uniform at broad spatial scales. For example, the southwestern portion of the study area is already converting to desert (Yu et al. 2004) and is forecasted to lose grassland productivity more rapidly than the rest of the study area. This area is also among the most heavily grazed for livestock (unpublished data from the Policy Research Institute of Mongolia). In the face of such climate change impacts, a spatial planning framework is essential for effectively balancing priorities for conservation, pastoral land use, and resource development.

Here we provide an example of how regional patterns of grassland productivity and grazing pressure have been categorized for use in identifying climate adaptation strategies. This analysis was then overlaid with a set of priority conservation areas from the ecoregional assessment for the Eastern Mongolian Grassland steppe (Heiner et al. 2011) to identify conservation and adaptation strategies most appropriate for each conservation area.

We first identified areas that are potentially overgrazed by classifying the study area according to livestock density and steppe productivity based on (1) density of herder camps (unpublished data from the Policy Research Institute of Mongolia) as a rough proxy for livestock density, and (2) mean NPP from 2000 to 2006, described earlier (DAAC 2010). To define classes of livestock density, we divided herder camp density into three quantiles. To define classes of steppe productivity, we masked the analysis to the extent of steppe ecosystems, and classified NPP by sheep units (SUs) per hectare (Holechek et al. 1989; Winters 2001). One sheep unit is the amount of forage sufficient to feed one adult sheep for one year without damaging vegetation.

The resulting classification defines nine possible combinations of livestock density and productivity (fig. 8.2). To generate a conservative estimate of area overgrazed, we

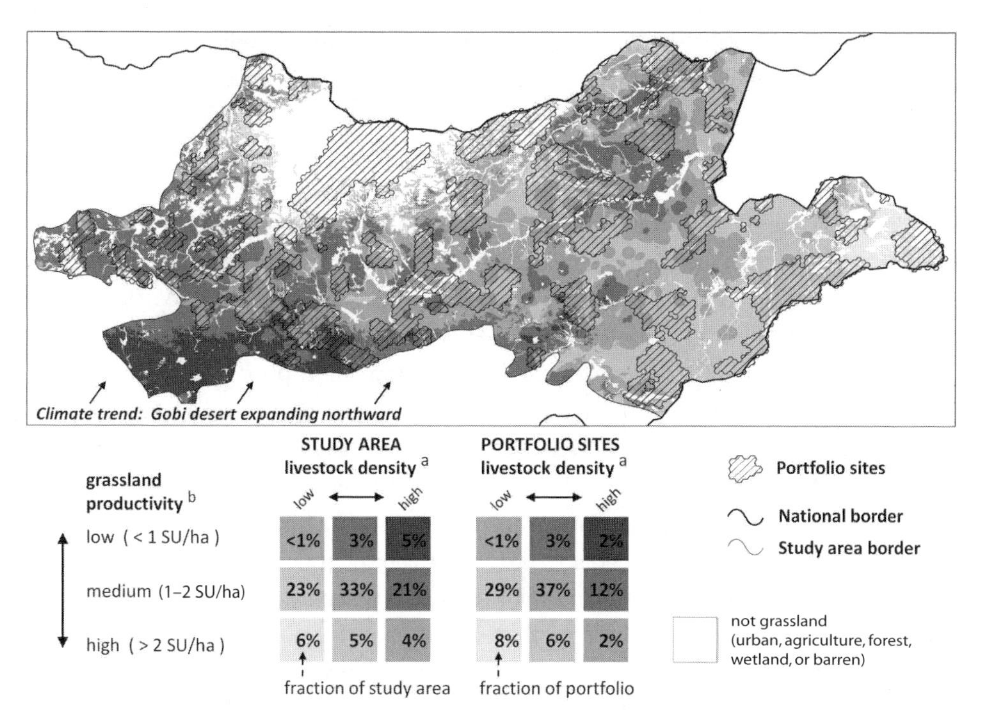

FIGURE 8.2 Classification of study area according to livestock density and grassland steppe productivity. [a] Density of herder camps (from Policy Reseach Institute of Mongolia), divided into three quantiles, is a rough proxy for livestock density. [b] Grassland productivity is represented by net primary productivity from 2000–2006 (MODIS Land Subsets 2010) and classified by sheep units per hectare (one sheep unit [SU] is the amount of forage sufficient to feed one adult sheep for one year without damaging vegetation). To generate a conservative estimate of area overgrazed, we defined potentially overgrazed areas as those with lowest productivity (less than 1 SU/hectare) and highest livestock density (highest quantile of herder-camp density). The resulting classification defines nine possible combinations of livestock density and productivity. Areas toward the upper right of the classification (high-livestock density and low-grassland productivity) are candidates for improved grazing management strategies, while areas toward the lower left of the classification are candidates for grass banks.

defined potentially overgrazed areas as those with lowest productivity (less than 1 SU per hectare) and highest livestock density (highest quantile of herder camp density). This definition classifies 5 percent of the study area (2,290,000 hectares) and 2 percent of the areas identified as priorities for conservation (280,000 hectares) as potentially overgrazed. To predict how this pattern might change in response to climate change, we applied the same classification schema using forecasted NPP (2070–99; SRES A2 emissions scenario) described earlier. Assuming that livestock density does not change, the forecasted NPP predicts that by 2070, 7 percent of the study area (3,000,000 hectares) and 3 percent of the areas identified as priorities for conservation (430,000 hectares) will fit this conservative definition of potentially overgrazed, an increase of 30 percent and 50 percent, respectively

Coupling Regional-scale Planning with On-the-Ground Adaptation Strategies

By overlaying the priority conservation areas (portfolio sites) with the grassland productivity and grazing intensity analysis (fig. 8.2), on-the-ground conservation and ecosystem-based adaptation strategies can be tailored to specific areas. We detail a few such strategies that could be used in the Mongolian grasslands here.

In areas with high grazing pressure and low grassland productivity, sustainable and climate-resilient grazing management is especially needed. Measurable benefits to both people and nature have resulted from grassland management programs in Mongolia's Gobi Desert (Leisher et al. 2012). Grazing intensity must fit the productivity of the grassland systems to prevent overgrazing. Productivity can vary widely from year to year based on climate, and so grazing management needs to be flexible and innovative to ensure the grasslands are not overgrazed when forage conditions are poor. A key management tool for achieving this is the use of grass banks.

Areas with high productivity and low grazing pressure could be candidates for grass banks. These areas would be set aside—that is, "banked"—for wildlife and not grazed during most years, but then would be made available for grazing in drought years when there is not enough forage for livestock (Gripne 2005). This can be thought of as a type of insurance policy to help herders and their livestock survive tough years. Herders would need access to water in these areas, or the forage would be cut and moved to areas with water that herders could access. Grassland reserves have been used in Mongolia for centuries, and this is the type of ecosystem-based adaptation strategy that can be implemented more broadly and intensively to help safeguard people and ecosystems against overgrazing when productivity is especially low. A recent study in the Gobi Desert by Hess et al. (2010) suggests that in drought years, protected areas can provide emergency forage or grass banks for herders if effectively managed for that purpose.

The task of measuring suitable forage in the Mongolian Steppe will require additional data and a refinement of the NPP classification. Using NPP data derived from satellite imagery, Mueller et al. (2008) found that areas of intermediate NPP, rather than high NPP, likely provide optimal forage for gazelles. Therefore, areas classified as highly productive (>2 SUs per hectare) may include standing biomass that is suboptimal forage for gazelles or livestock (Karl Didier, pers. comm.). Also, NPP values near the steppe/forest transition may be artificially high due to the presence of shrubs and trees. In the future, we hope to expand this analysis of potential grass bank reserves by (1) refining the classification of steppe NPP; (2) using more current and complete data for locations of herder camps; and (3) incorporating data for distance and access to wells and springs, which are essential determinants of suitability for livestock.

In all of these areas, monitoring of the grasslands systems is important, both on short and long time scales (Angerer et al. 2008). We suggest (1) an "early warning" system that can detect and communicate when summer drought is likely so herders can modify their grazing accordingly, and (2) long-term monitoring from a combination of remote sensing technologies and on-the-ground observations that could track the long-term health of the grassland system and provide feedback to modify grazing management plans, as needed.

Identifying the best set of on-the-ground strategies will take a combination of bottom-up and top-down approaches that utilizes strong partnerships between conservationists, development organizations, private industry, government agencies, and particularly the local communities that live in these ecosystems. Research has found that community-based spatial planning is an overlooked tool for balancing conservation and human development needs but may require certain critical factors including strong local leadership, an accountable financial management mechanism to distribute income, outside technical expertise for the zoning design, and community support (Leisher 2011). On the one hand, the appropriate implementation strategies in any given place will be context dependent and will require a bottom-up approach to planning with local people and partner organizations identifying what would be feasible and effective. At the same time, a top-down approach is needed to coordinate across actions implemented on the ground and to ensure that local conservation strategies fit within the broad-scale strategies being implemented at the large-landscape scale.

Here we have presented a regional analysis that identifies ways top-down, landscape-scale coordination can link to local on-the-ground conservation actions through spatially identifying where grass banks and grazing management strategies are most important. However, this top-down analysis needs to be integrated with bottom-up information about how the grass banks and grazing management could be implemented in specific places. This regional analysis is not an end point in planning, but rather provides a good starting point for the development of a coordinated ecosystem-based adaptation plan that links to local people and biodiversity on the ground across this system.

Opportunities

Overgrazed and degraded grasslands are not good for herders, ecosystems, or fish and wildlife. Herders have sustainably grazed these grasslands for centuries, and their direct reliance on natural systems vividly demonstrates why functioning ecosystems are important not only for wildlife, but also for human livelihoods. The survival of herders and wildlife depends on healthy, intact grasslands and is threatened by desertification due to combinations of poor management and climate change. Yet increased grazing pressure and climate change are pushing these systems beyond what they are capable of supporting. More specifically, dzuds and other climate-related impacts will continue to strike Mongolia and with greater intensity than in the past.

To help herders cope with these climate-change related challenges, they can use a number of ecosystem-based adaptation strategies. Conservation organizations have an opportunity in the Mongolian grasslands to demonstrate that it is in the best interest of multiple stakeholders both to have good grassland management and to protect nature in order to help make the region's herders more resilient to climate change. Since this approach addresses impacts to both people and nature, it expands the potential collaborative opportunities for conservation, including international development and aid organizations, government agencies, and philanthropic foundations, all of which are working to find ways to help the poorest and most vulnerable people adapt to climate change. In a broader sense, conservation organizations need to find ways to help promote the livelihoods of people, or they risk becoming marginalized as people's needs overshadow nature conservation in an increasingly resource-limited world with more people and a changing environment.

Ecosystem-based adaptation strategies can also operate in coordination with environmental mitigation of mining and petroleum extraction. Although mining and oil and gas extraction constitute a real threat to the grassland ecosystems and the pastoralist herders, they also represent an opportunity. With a huge portion of Mongolia leased for exploration, these extractive activities will occur regardless of what conservation activities are established on the ground. The Nature Conservancy is applying a "Development by Design" approach that blends landscape conservation planning with the mitigation hierarchy ("avoid, minimize, restore, or offset") to identify areas where mining should be avoided, and to design sites for compensatory mitigation, or offsets, for those mining and development projects that proceed (Kiesecker et al. 2010). Mitigation offsets from this development approach can be used to help implement regional conservation assessments and associated ecosystem-based adaptation strategies.

Conclusion and Recommendations

The Eastern Mongolian grassland steppe provides a rich example of the challenges and complexities for conservation in adapting to climate change, as well as the many opportunities. As in many systems around the world, people are an integral part of the Mongolian grasslands, and their livelihoods are closely tied to the health of the ecosystem. The challenge lies in balancing the needs of biodiversity with the needs of people in this ecosystem, particularly as climate change brings about further reductions of forage and water, and as people begin to respond to and cope with dwindling resources. However, an opportunity lies in ecosystem-based adaptation approaches of using landscape conservation to promote the resilience of both natural ecosystems and people to climate change impacts.

In conclusion, nature conservation and good stewardship of the land can help herders in Mongolia adapt to climate change, while ensuring the persistence of some of the most vast and intact grasslands in the world. Quantitative, landscape-level conservation assessments can be targeted to benefit both nature and people, by balancing resource development and grazing pressure with conservation goals. Then on-the-ground conservation strategies that use nature conservation—such as grass banks and sustainable grazing management—can be coupled with these regional plans to provide conservation for wildlife and climate adaptation for herders. Ultimately, for large landscape conservation to be successful, especially in the face of climate change, it must integrate a coordinated set of bottom-up and top-down approaches that accounts for the needs of both the biodiversity and people living in ecosystems around the world.

Chapter 9

Northern Great Plains, North America

ANNE M. SCHRAG AND STEVE FORREST

The Northern Great Plains in many ways represents North America's grassland future. It houses high-quality grassland and sagebrush habitat for a diverse assortment of species that are uniquely adapted to the region, and it supports farming and ranching livelihoods that have helped to shape the land. Because grasslands and shrublands are such complex systems that face myriad threats, the impact of climate change is often overlooked. Efforts to engage scientists and managers in climate change-scenario planning are under way. However, moving into the coming decades, it will be necessary for climate change to become a unifying theme in conservation in the region.

Introduction to the Region

The Northern Great Plains Ecoregion is North America's largest grassland ecoregion, spanning five states and two provinces and covering approximately 722,600 square kilometers, or about 25 percent of the Great Plains (fig. 9.1; Ricketts et al. 1999). The Northern Great Plains supports a relatively high level of species richness (Forrest et al. 2004) and of the thirty-nine endemic North American grassland vertebrates in the Northern Great Plains, 15 percent are listed as endangered or threatened by the US and/or Canada (Batt 1996). In addition to hosting native grazers, including the keystone species bison (*Bos bison*) and prairie dogs (*Cynomys ludovicianus*), the Northern Great Plains remain a critical breeding area for grassland birds and black-footed ferrets (*Mustela nigripes*), the most endangered mammal in North America.

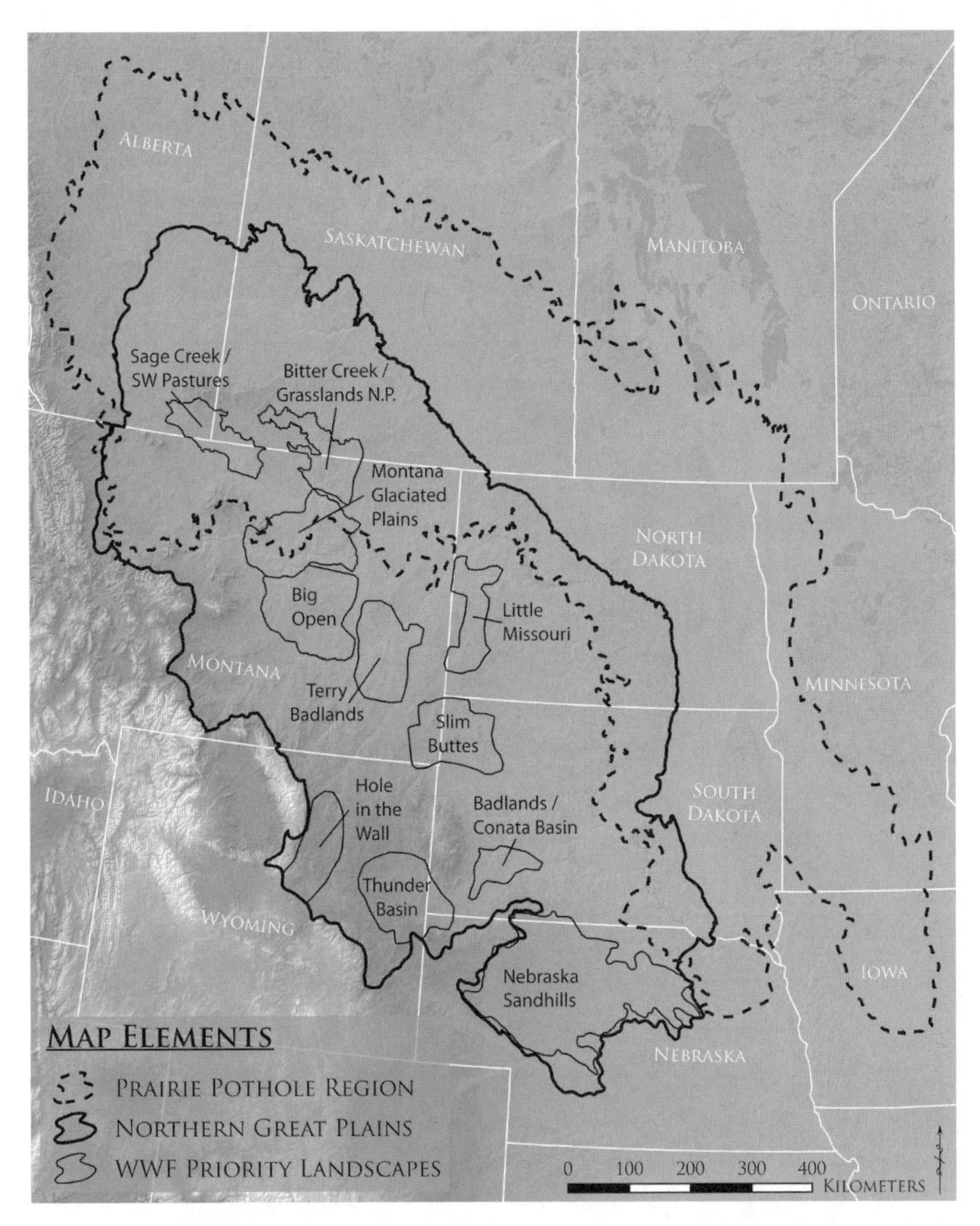

FIGURE 9.1 Map of the Northern Great Plains ecoregion and World Wildlife Fund priority landscapes.

Prairie species are adapted to highly variable precipitation, prolonged droughts, and periodic fire, all of which generally prevent the establishment of forests, except on some moist upland sites. Extensive areas of shrub-steppe occur, with Wyoming big sagebrush (*Artemesia tridentata wyomingensis*) being the most abundant species. The Northern Great Plains encompasses portions of two of the five major North American prairie wetland complexes: (1) the Missouri Coteau, the southeastern part of

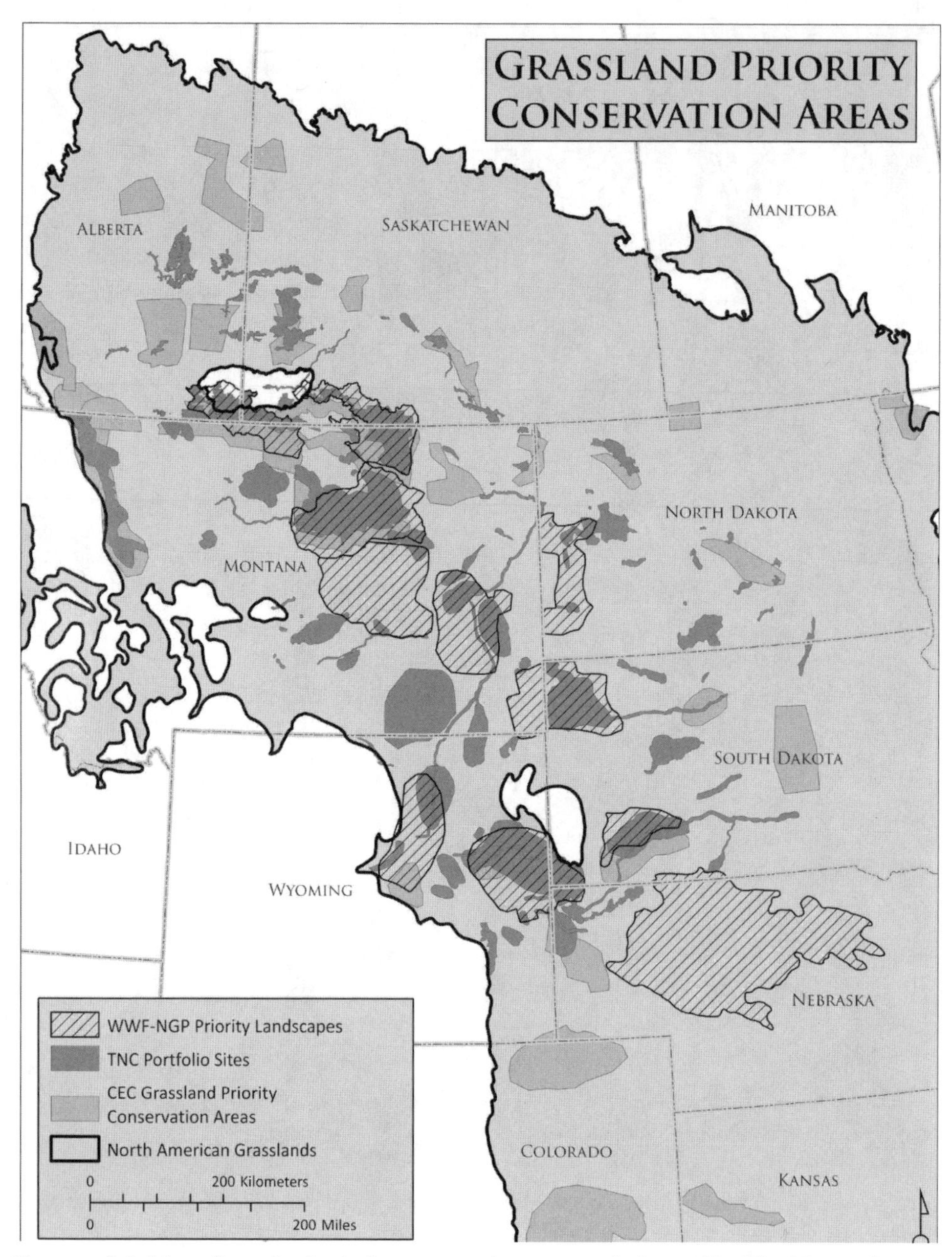

FIGURE 9.2 Map of grassland priority conservation areas as designated by The Nature Conservancy (TNC), World Wildlife Fund (WWF-NGP), and the North American Free Trade Agreement Commission for Environmental Cooperation (CEC).

what is more generally referred to as the Prairie Pothole Region (fig. 9.1); and (2) the Nebraska Sandhills Wetlands (Batt 1996).

Human presence in the ecoregion dates back 10,000 years. Plains peoples pursued abundant game or lived agrarian lifestyles in the fertile river bottoms. The mid-nineteenth century brought European settlers, who converted a substantial amount of the grasslands to annual crops. Today, there is a net out-migration of people from many parts of the region, and population loss is highest in counties that have low population density, are relatively remote, and lack natural amenities, such as fish and wildlife resources (McGranahan and Beale 2002).

Current Conservation in the Region

The largest landholders in the ecoregion are the US Bureau of Land Management, the US Forest Service, the US Fish and Wildlife Service, state and provincial land-management agencies (Crown lands in Alberta and Saskatchewan), the US National Park Service, and various tribal entities. The Great Plains have been described as one of the most altered ecosystems in North America (Bragg and Steuter 1996). Less than 16 percent of the Northern Great Plains ecoregion (11 million hectares) is managed primarily for natural resource conservation, with less than 1.5 percent managed solely to ensure conservation of biodiversity (IUCN classes 1–4; Forrest et al. 2004). Yet, about 57 percent of the Northern Great Plains grasslands are not currently tilled (40.1 million hectares), a significantly higher percentage than the Great Plains as a whole.

In contrast, wetlands in the Northern Great Plains, which are located along the eastern and northern borders of the region, may be more threatened due to their geographic overlap with prime farming lands. For instance, in the mixed-grass prairies of Saskatchewan, Alberta, and Manitoba, wetlands have been reduced by 10 to 40 percent from presettlement times (IISD 1990). In eastern South Dakota, 78 percent of wetlands are at risk of conversion to agriculture (Higgins et al. 2002).

Because of the high level of intactness, high biodiversity, and the large amount of public lands in the Northern Great Plains, several major efforts have identified conservation priorities in the region. These include *Ocean of Grass: A Conservation Assessment for the Northern Great Plains* (Forrest et al. 2004); *Northern Great Plains Steppe Ecoregional Assessment* (TNC 2000); and the North American Free Trade Agreement Commission for Environmental Cooperation's *North American Grassland Priority Conservation Areas* (Karl and Hoth 2005). All of these initiatives show congruence among their respective identified priority conservation areas, including the Montana Glaciated Plains, Conata Basin, Thunder Basin, Sage Creek, and Bitter Creek areas (figs. 9.1 and 9.2).

Two major initiatives are currently under way to conserve and restore some of the identified priorities in the ecoregion. Grasslands National Park in Canada has reintroduced bison and black-footed ferrets and is pursuing new land acquisitions that would potentially double the size of the area where bison can roam. The American Prairie Foundation has reintroduced bison on approximately 50,000 hectares of grassland habitat acquired through fee purchase and lease in the Montana Glaciated Plains (fig. 9.1).

In addition, several entities are working on incorporating connectivity into conservation planning. Montana Fish, Wildlife and Parks, along with a group of non-agency experts, is currently assessing connectivity for priority species by mapping and reviewing core habitat areas and corridor linkages as part of the Crucial Areas Planning System, which will be used to update Montana's State Wildlife Action Plan (MFWP 2010). In addition, various research efforts are under way to identify fine-scale species movements. A study funded by World Wildlife Fund and carried out by the University of Calgary established evidence of a pronghorn corridor between the Milk River uplands in Montana and the Great Sandhills of Saskatchewan. Both Alberta Game and Fish and Canadian Armed Services have recognized the existence and significance of this wildlife corridor, and managers are taking steps to alleviate identified pinch points by modifying fences to allow for uninterrupted movement (Jones et al. 2007). Other ongoing studies are identifying migratory pathways of greater sage-grouse in Saskatchewan (Tack 2006) and prairie rattlesnake in northern Montana and southern Alberta and Saskatchewan (Jørgensen and Nicholson 2007; Jørgensen et al. 2008; Jørgensen 2009).

Major Gaps and Impediments

Some major challenges to advancing conservation of lands and wildlife are (1) fragmentation due to development of all types, including energy development such as wind, oil, gas, and coal-bed methane; (2) invasive species; (3) wildlife disease; (4) federal programs and market forces that encourage tilling of grasslands; (5) grazing practices that are not conducive to biodiversity; (6) alteration of aquatic regimes; (7) direct take of wildlife; and (8) cultural resistance to change (Forrest et al. 2004).

Climate change may exacerbate threats in ways that are known; in other instances, we may not yet fully realize how great the impact climate change might have. No matter what, reexamining current conservation priorities and threats in light of climate change will be essential to strategic conservation in this region into the future.

Regional Effects of Climate Change

The climate of the Northern Great Plains is changing, with northern states and provinces experiencing greater and faster changes than southern areas. From 1951 to 2002, average annual temperatures increased by up to 2.6 degrees Celsius, with greater increases in the northern and eastern portions of the ecoregion (Mitchell and Jones

2005; Meehl 2007). Spring and winter temperatures are increasing more quickly than summer and fall temperatures. Average annual precipitation is increasing most in the southeastern portion of the ecoregion during spring and fall, by up to 130 millimeters in areas of South Dakota and Nebraska. Areas along the Montana-North Dakota border have seen decreases in precipitation by as much as 80 millimeters.

Over the next century, estimates suggest additional increases in temperature of 1.4 to 7.2 degrees Celsius depending upon the emissions scenario. Following past trends, the northern portion of the region is expected to experience the greatest amount of change. Predictions for precipitation changes are less certain, but suggest increases in the northern and eastern portions of the region and decreases in the southern and western portions of the region (Mitchell and Jones 2005; Schrag 2011). Frequency and intensity of extreme events are also projected to increase (Karl et al. 2009).

Vegetation in the Northern Great Plains has withstood droughts and floods for centuries and has coevolved with grazing mammals and fire to produce a highly diverse system that is naturally resilient to these stochastic events. Paleoecological data demonstrate that previous droughts led to decreases in productivity, increases in erosion, and shifts in species composition, whereas humid periods led to increases in productivity, abundant fuels for fire, and stabilization of soils (Clark et al. 2002). Future changes may come in a variety of forms, including changing species composition, directional shifts in movements (east-west or north-south), and range contractions. Studies suggest a possible east-to-west shift in the forest-prairie transition zone due to increasing suitability for woody species to inhabit current grassland and shrubland (Bachelet et al. 2003). Other models suggest a directional shift northward for many vegetation types, which may lead to novel vegetative communities (Thorpe 2010). Models for Wyoming big sagebrush and silver sagebrush (*Artemisia cana*) predict that these species will contract into the core of their range in Wyoming (Schrag et al. 2011). Cheatgrass (*Bromus tectorum*), an invasive species that decreases forage quality and increases fire frequency, has established itself on rangelands across the West and is predicted to continue to spread in the future (Bradley 2009).

Although there is some uncertainty associated with these predicted changes, many recommendations for maintaining resiliency will hold across scenarios. Management agencies must reevaluate their long-term goals for vegetative communities across the Northern Great Plains. Maintaining current composition may be impossible in the future. Invasive species such as cheatgrass will require heavy management due to their ability to drastically decrease ecological integrity. Protecting sagebrush communities must be prioritized due to the difficulty of restoring them. Working with private landowners to ensure a diversity of vegetation structures (e.g., bare ground and a mixture of grasses of various heights) through light-, medium-, and high-intensity grazing will be essential for maintaining biodiversity across the public-private continuum. If prairie vegetation communities move northward, habitat may become available in regions of southern Canada that currently have limited prairie habitat. Consequently, establishing conservation goals across international borders will be important.

Climate change will also directly and indirectly affect birds and other wildlife. Peterson (2003) used models to predict that bird species in the Great Plains are more likely than species in other ecoregions to experience shifts in, and reductions of, suitable habitat due to climate change (in some cases up to 35 percent). Specifically, greater sage-grouse (*Centrocercus urophasianus*) are expected to face contraction of sagebrush habitat throughout the Northern Great Plains and expansion of West Nile virus. Mosquitoes transmit West Nile virus to sage-grouse after temperatures have reached a certain level for multiple days in a row (Konrad et al. 2009), and these temperature thresholds will be reached earlier in the year and at higher elevations (e.g., Rocky Mountain front) in the next two decades (Schrag et al. 2011). Strategies that reduce mosquito populations, such as limiting the amount of standing water, must be implemented in areas that are at risk for increasing West Nile virus occurrence.

Another disease that may spread under changed climatic conditions is sylvatic plague, which decimates prairie dog colonies, thus removing the primary food source for black-footed ferrets. Studies have shown a positive association between plague outbreaks and the previous year's spring precipitation; there also appears to be some association between plague outbreaks and temperature, where warm days are positively associated with outbreaks, while hot days are negatively associated with outbreaks (Collinge et al. 2005; Snäll et al. 2008). In order to increase the resiliency of prairie dog populations to plague, conservationists dust prairie dog burrows with pesticides to rid the area of fleas that transmit plague, and plague vaccines may also be administered to prairie dogs and black-footed ferrets to protect them against the disease. These approaches have been successful in mitigating plague, but are resource-intensive due to the cost of the dust and the application to widespread prairie dog communities.

Climate change will also affect wetland and hydrological systems in the Northern Great Plains. The Prairie Pothole Region serves as the "duck factory" of North America—producing up to seven million ducks annually during high-precipitation periods. Climate extremes have led to increased numbers of distinct plant species in this region, with high-water events leading to a mix of open water and plants that emerge above the water surface (e.g., cattails). Meanwhile, droughts lead to increases in diversity and productivity by pulling new seeds from the seed bank and mobilizing nutrients (Johnson et al. 2005). Ducks have adapted to this variability by passing over areas that are experiencing drought on their annual migration routes, and studies suggest that protecting wetland complexes, which include both temporary and semipermanent wetlands, may be the most effective way to ensure future resilience. Currently, a federal-level initiative by researchers at the US Geological Survey, the US Fish and Wildlife Service, and regional universities is investigating the complex relationship between climate change, agricultural land use, and sedimentation rates in wetlands, and how these changes will affect wetland birds (Skagen 2009; USFWS 2010).

Prairie streams will also be affected by climate change. Small increases in temperature (1–2 degrees Celsius) and decreases in precipitation (5 to 10 percent) may lead to increased evapotranspiration, decreased surface discharge, and increased salinity. Native fish may suffer due to increased temperatures because they are already living at their thermal threshold. Warming groundwater and the diversion of surface water for others uses could lead to the complete loss of small prairie streams (Covich et al. 1997). If the total amount of available water decreases with climate change, increasing conflicts may arise among different land uses. Clear planning goals that take into account a variety of future scenarios will help alleviate potential conflicts among user groups and biodiversity values, and removal of artificial water diversions (e.g., dams, stock ponds) will restore natural flows and increase functionality of prairie streams.

The economic and social implications of climate change for this region could be significant. Changes in forage may have substantial impacts on grazers, including cattle, bison, elk, and other ungulates; some studies show that changes in the timing of precipitation may lead to decreased forage quality (Craine et al. 2009). These changes could have ripple effects for ranching and hunting. Pollinators are likely to be affected by mismatched timing between their migration and the flowering of plants (Memmott et al. 2007), which may impact agriculture in the region. Increasing nighttime temperatures may also lead to smaller fruits and grains, and some crops may experience an increased risk of episodic frost damage due to warmer temperatures, which lead to earlier establishment (Prasad et al. 2008). Water demand is predicted to outpace supply in many counties by 2050, leading to shortages for both human and agricultural uses (NRDC 2010).

All of these impacts on the human component of the ecosystem require increased communication and coordination among user groups. Forage productivity analyses have been performed for landscapes in Saskatchewan, Montana, and South Dakota, the results of which can be useful in explaining the potential impacts of climate change to private landowners and wildlife agencies. State and provincial agencies that manage fish and game populations are integrating climate change into their management plans (e.g., State Wildlife Action Plans in the United States), which will allow for more effective planning and management of fish and game populations into the future. And water-demand issues will need to be raised at a state/provincial to regional level in order to ensure that actions at source areas will maintain water availability for downstream users.

Approaches to Conservation under Climate Change

Climate change is a serious threat to the Northern Great Plains, one around which agencies and conservation organizations are beginning to organize. However, with respect to many species and processes in the region, climate change is not the only—and sometimes not the greatest—immediate threat to their persistence. Habitat fragmentation, land-use change, energy development, and conversion of grassland to crops are just some of the other threats facing this region (Forrest et al. 2004). Climate change may intensify the impacts of these threats, but the more imminent risk of habitat loss through direct conversion for other uses has, to date, overshadowed a focus on climate change.

However, some organizations and agencies in the Northern Great Plains are in various stages of undertaking vulnerability assessments for habitats and species within the region. For instance, World Wildlife Fund has been working with The Nature Conservancy to develop climate "exposure" maps for the region. These maps show historic trends in temperature and precipitation variables across the region, as well as predicted changes in the variables by the end of the century. These maps will help to guide the identification of areas that have (1) experienced the most and least amount of change over the past few decades, and that are (2) expected to experience the most and least amount of change over the next few decades. This information can help frame conversations about the need for climate adaptation actions and help set priorities for conservation at a regional scale (Schrag 2011).

In addition, two separate scenario-planning exercises have been undertaken in the region. The first was an effort by Wildlife Conservation Society, National Wildlife Federation, and World Wildlife Fund to assist Montana Fish, Wildlife and Parks with a scenario-planning exercise for two systems of concern: sagebrush ecosystems mainly in eastern Montana, and the Yellowstone River ecosystem. Agency managers were given a set of three possible future climate scenarios and asked to step through a process of identifying impacts, thresholds, management objectives, and actions (Miller et al. 2008). The results of this workshop will inform the upcoming planned revision of the Montana State Wildlife Action Plan.

A similar workshop was held with a group of wildlife and land managers from southwestern Saskatchewan as a part of the South-of-the-Divide Species at Risk Action Plan (where the name refers to an area in southwestern Saskatchewan that is south of the Northern Divide), a formal plan written by Parks Canada and collaborating agencies to ensure the long-term persistence of species at risk in the region. This workshop had a similar focus in that managers from a variety of agencies were asked to identify impacts, thresholds, management objectives, and strategic actions for each species of concern. Downscaled climate data for Saskatchewan was used as a basis for the scenarios, and vegetation models were developed to show potential changes in both vegetation type and productivity in the future (Thorpe 2010). Two overarching strategies emerged from the workshop: (1) maintaining a diversity of height structures among plants that can be used as habitat by grassland birds within systems will promote high biodiversity because it creates a mosaic across the landscape, and (2) finding incentives for private landowners to maintain wildlife habitat and tolerate species in lands adjacent to public lands will allow wildlife to move in response to climate change. The results of this workshop will be used as one part of a larger threat analysis for each species and will be integrated into the larger planning framework.

Aside from work on scenario planning, a better understanding of the impacts of climate change on species and processes is a priority in the Northern Great Plains. A more formal, region-wide assessment of the vulnerability of species, ecological processes, and economic impacts would be helpful in order to prioritize the use of resources. The US Department of the Interior's newly formed Landscape Conservation Cooperatives are well positioned to become a leader in this arena.

Roadblocks

Many of the roadblocks to integrating climate change into conservation planning and action in the Northern Great Plains are similar to those faced elsewhere. Cross-jurisdictional collaboration and agreement on management objectives and actions is difficult or sometimes impossible. Because climate change tends to be perceived as a comparatively less pressing threat to many species and processes, it is often difficult for agencies to direct time and financial resources toward the issue. Also, managers express frustration that the uncertainties associated with future climate models are either unknown or too high for them to feel comfortable making decisions based on these data.

Given societal and agency pressures, thinking outside the box and envisioning a new future for the plains is difficult. The mandate of many management agencies and their staff is either to maintain or restore the full suite of species that once roamed the plains, thus focusing attention backward instead of forward. Imagining a new ecosystem that may include species historically found outside of the Northern Great Plains—or in a different part of the plains—is difficult, and developing long-term plans that provide the flexibility to allow for these changes is complex.

Conclusion and Recommendations

Along with the roadblocks, however, come opportunities. In northern areas of the Northern Great Plains, climate change may actually provide more suitable habitat for some species in the near future. Climate change may prove to be a unifying theme among private and public lands, leading to collaboration across boundaries where other issues have been divisive in this landscape. Scenario-planning workshops have begun to be an accepted method for engaging natural resource managers and leading thinkers on climate change in a discussion of possible strategies for the future.

Current and predicted future impacts of climate change suggest a three-pronged approach for moving forward with conserving both the landscape and the way of life that presides in the Northern Great Plains:

1. Scientific research is the first need and the key for ensuring that on-the-ground actions are likely to prove beneficial. Region-wide analyses of vulnerabilities and establishing baseline data for monitoring changes are necessary for many aspects of prairie systems.
2. The next step is planning. A quantitative or qualitative analysis of sensitivity, exposure, and adaptive capacity can help shape priorities in a changed future and ensure that all aspects of the system have been considered. The key for ensuring that planning processes will lead to adoption of climate-ready practices on the ground is coordination and collaboration from the beginning and from all levels within agencies, organizations, and the private sector. Without that initial buy-in, it is all but impossible to guarantee that planning will actually lead to changes in management on the ground. Furthermore, to ensure that the most effective steps have been taken, monitoring must be implemented and continuously funded so that thresholds are recognized and actions are taken at appropriate times.
3. The final step is taking immediate actions on the ground that can help alleviate threats to wildlife and habitats. Many of these actions are delayed in the process of developing more and better science information.

Combining science, planning, policy and on-the-ground management actions, we can move into the coming decades with a clear sense of our conservation priorities under climate change. Coordination and collaboration will be the key to success as we tackle this great challenge in one of the most intact grasslands in the world.

Chapter 10

Washington State, USA

MEADE KROSBY, JENNIFER R. HOFFMAN,
JOSHUA J. LAWLER, AND BRAD H. MCRAE

Washington State's diverse ecosystems have been under intense pressure from extractive industries such as logging, fishing, and mining for at least a century. While these extractive economies have declined in recent years, only small, isolated patches of intact habitat remain in the densely populated Puget lowlands in western Washington and the heavily farmed and grazed Columbia Plateau in the east. The need to maintain and restore habitat connectivity has therefore become a primary driver of conservation science and action across Washington. As the climate changes, there will be an even greater need for connectivity across this fragmented landscape to enable species range shifts. Lessons from an effort to integrate climate change into a statewide habitat connectivity assessment suggest that a strong mandate for addressing climate change, engagement of diverse constituencies, and alignment of disparate efforts are critical for successful adaptation planning and action.

Introduction to the Region

The northwestern-most state in the contiguous USA, Washington is characterized by diverse geography, climates, and ecosystems (fig. 10.1). The Cascade Mountains divide the state, environmentally and demographically, into western and eastern Washington. To the west, the influence of the Pacific Ocean maintains a moist, temperate climate across the coniferous rainforests of the Olympic Mountains, the prairies and

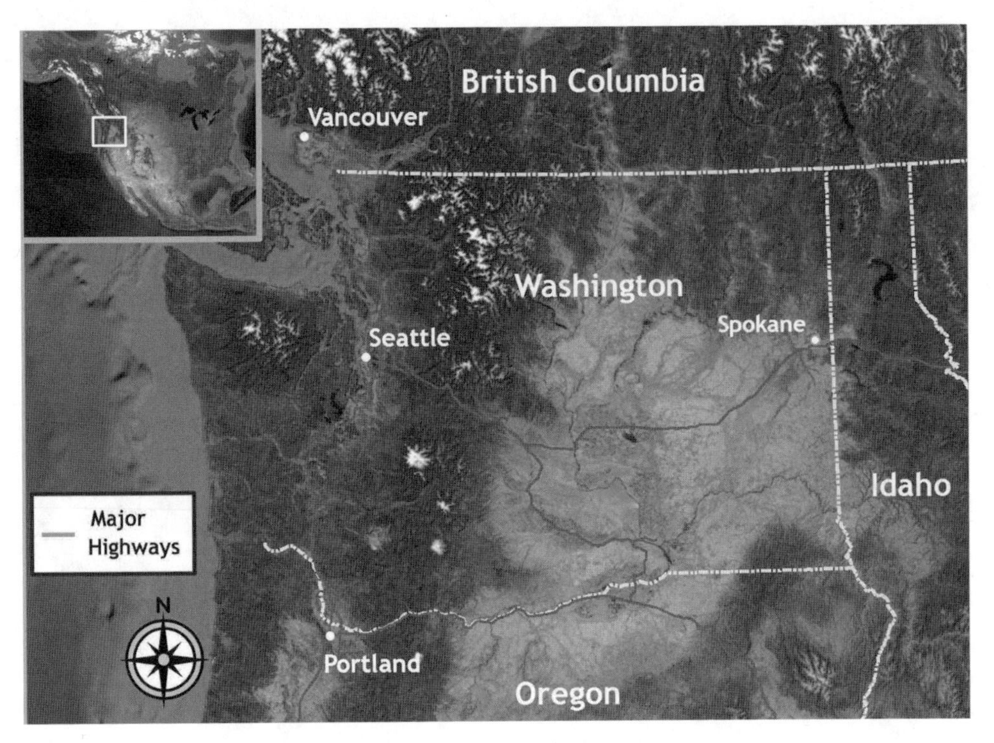

FIGURE 10.1 Washington State, USA.

forests of the Puget lowlands, and the western slopes of the Cascades. The majority of Washington's roughly 6.5 million inhabitants live along the Interstate 5 corridor in western Washington, with 60 percent of the state residing in Seattle alone (U.S. Census Bureau 2010). In contrast, eastern Washington is in the rain shadow of the Cascades and is predominantly rural and arid. Across the vast Columbia Plateau, most native shrub steppe has been converted to agriculture and rangeland, facilitated by irrigation from the Columbia River. Washington's economy has historically depended on its rich natural resources, particularly timber, fish, and minerals. Although still significant, these industries have undergone major declines. Today, agriculture dominates east of the Cascades, and technology and manufacturing have become increasingly important in western Washington, home to corporations such as Boeing, Amazon, and Microsoft.

Historical Overview of Conservation and Science Initiatives

In the broadest sense, the dramatic changes in Washington's environment since European settlement are the result of changes in land use and the loss of several key species and populations. Native Americans used fire to maintain desirable vegetation types and structures, to create clearings to attract game, and to maintain travel corridors. By the early 1800s, decimation of native populations by disease significantly changed the fire regime, which in turn led to major vegetation changes. The inception of commercial logging in the late 1880s, the increasing demand for timber beginning with World War II, and the subsequent shift from selective harvest to clear-cutting radically altered both forest type and forest cover. The commencement of large-scale grazing and agriculture further altered vegetation communities.

At the same time, heavy commercial fishing, intensive dam construction for hydroelectricity, and the impacts of clear-cutting on streams led to massive declines in salmon populations, including the complete loss of fifty Columbia Basin salmon runs. Beyond the cultural and economic losses, the loss of salmon runs further altered ecosystems around formerly salmon-bearing streams by reducing the import of nutrients. In addition to these impacts, the virtual extirpation of beavers for the fur trade led to widespread loss of wetlands and other hydrological changes, and the loss of wolves allowed populations of grazers to flourish, further degrading wetland and riparian habitat.

Within this context, the impetus for conservation has come from a variety of actors with a variety of interests. Native tribes have pushed for conservation of resources important for their culture and subsistence, particularly salmon and shellfish. The dustbowl and Depression years saw the creation of conservation programs focused primarily on agriculture and forestry, including the federal Civilian Conservation Corps, which ended with the onset of World War II, and Washington's conservation districts, which remain active today. Hunting and fishing communities have long been a voice for conserving wetlands and other game habitats. With the rise of the environmental movement in the 1960s and 1970s, the disappearance of northwest forests and resulting threats to its wildlife became an increasing source of conflict across Washington and the Pacific Northwest.

One of the most profound shifts in land management in Washington and the Northwest resulted from the 1990 listing of the northern spotted owl (*Strix occidentalis caurina*) as threatened under the Endangered Species Act. In 1991, a court injunction halted state and federal timber sales pending further studies of owl habitat and impacts, and 11.6 million acres outside of existing protected areas were set aside as critical habitat. Functionally, the spotted owl became an umbrella species for the protection of old growth forest throughout the Northwest. A Northwest Forest Summit held in April 1993 to resolve the intense conflict between the timber industry and conservationists led to the development of the 1994 Northwest Forest Plan. Covering 24.5 million acres in Oregon, Washington, and northern California, the Northwest Forest Plan provided the first mandate for coordinated ecosystem management among multiple agencies at large scales in the region. More recently, the listing of most remaining Washington salmon runs as threatened or endangered has led to further shifts in land and water management, affecting logging, agriculture, and development activities.

Current Conservation in the Region

Over 70 percent of Washington's 42 million acres consist of working forest, crop, and rangeland, much of it privately owned. The single largest land-use type is working forest, with over 42 percent of Washington's lands managed by federal, state, and private foresters. Eight million acres are currently managed or protected primarily for wildlife and recreation purposes. This includes the more than 30 percent of national forest lands that are now protected wilderness areas.

Only small, isolated patches of intact habitat remain in the densely populated Puget lowlands and the heavily farmed and grazed Columbia Plateau. Intensive clear-cutting has fragmented forests, a problem exacerbated by the checkerboard pattern of private and public land ownership resulting from mid-nineteenth-century government incentives to lure railroad companies to the West. Roads and power corridors further fragment Washington's landscape with 7,000 miles of highway and a growing network of transmission lines for traditional and alternative energy development.

Thus, the need to maintain and restore habitat connectivity has in recent years become a primary driver of conservation science and action across Washington. This has occurred in the context of more long-standing concerns around threatened species and habitats, development pressures, and the ecological impacts of industrial forestry, fisheries, and agricultural practices. More recently, climate change, and the interaction between climate change and other threats, has become recognized as a critical conservation priority for Washington and the Northwest.

Regional Effects of Climate Change

Although a handful of weather stations have reported cooling trends over the last century, warming has occurred across most of the region with average annual temperatures increasing by approximately 0.8 degree Celsius. Warming has been most pronounced in the winter months and weakest in the autumn months. In general, precipitation has also been increasing over the past century, with the largest relative increases recorded in the spring and in eastern Washington (Mote 2003).

Temperatures in the Pacific Northwest are projected to rise between 0.8 and 2.9 degrees Celsius by 2040 and between 1.6 and 5.4 degrees Celsius by 2080. Precipitation projections are more variable, ranging from −11 percent to +12 percent for 2040 and −10 percent to +20 percent for 2080 (Elsner et al. 2009). However, most models project increases in winter precipitation and decreases in summer precipitation. These changes in temperature and precipitation will have profound effects on both the physical and biotic systems of the Northwest, including two of the region's most important physical forces: hydrology and fire.

In Washington and the Pacific Northwest, both the timing and magnitude of stream flow are intimately linked to temperature and precipitation through winter snowpack (Hamlet et al. 2005, 2007; Mote 2006). Winter temperatures largely determine how much precipitation falls as snow versus rain. Thus, despite precipitation increases, warming temperatures have led to decreases in snowpack over much of the Pacific Northwest (Mote et al. 2005). Projected temperature increases will likely increase the proportion of winter precipitation falling as rain, increase the frequency of winter flooding, reduce snowpack, increase winter streamflow, result in earlier peak streamflow, and decrease late spring and summer streamflows (Hamlet and Lettenmaier 1999; Mote et al. 2003, 2005; Hamlet et al. 2007).

Given their iconic status in the region, there has been a strong focus on the potential impacts of climate change on salmon in the Pacific Northwest. Salmon are sensitive to climatic conditions at a number of different life stages. Salmon eggs are sensitive to the timing and magnitude of stream flow, and heavy winter and spring floods and flows may scour streams and dislodge eggs, washing them downstream. The timing of spring stream flows affects the ability of juvenile salmon to migrate from their natal streams to the ocean and the timing of that migration (Mote et al. 2003). Once they reach the ocean, the survival of the juveniles is highly dependent on the timing of spring upwelling. The capacity of adult salmon to return to their natal streams to spawn is also dependent on the timing and amount of stream flow. Furthermore, most adult salmon cannot survive for long periods in water temperatures greater than 21 degrees Celsius (McCullough 1999).

Overall, salmon in the Pacific Northwest face many threats, making it difficult to weigh the relative threat of climate change. However, at least for some runs, the most sensitive habitats appear to be those at mid- to high elevations where increases in the percentage of precipitation falling as rain and resulting reductions in snow pack are expected to be the greatest (Battin et al. 2007).

Diverse and highly productive forests are another iconic feature of Washington and the Pacific Northwest. The Pacific Northwest has two very distinct basic forest types. Most of the forests west of the Cascades are wet, dense, and dominated by conifers; eastern forests are dry, more open, conifer forests often transitioning into open, dry woodlands at lower elevations. Given that seedling establishment, tree growth rates, and disturbance regimes such as fire and insect outbreaks differ markedly between the two regions, the eastern and western forests are likely to be differentially affected by climate change.

At alpine treelines in both forest types, warmer temperatures and reduced snowpack may enhance growth and seedling establishment, potentially leading to expansions of forests upward in elevation into alpine zones. Decreased summer precipitation, decreased snowpack, and increased temperatures can also lead to upward shifts in the lowest elevations at which trees grow, in areas where shrub-steppe dominates at lower elevations (Neilson et al. 2005). In contrast, vegetation models have projected the expansion of forests and woodlands into the sagebrush steppe in eastern Washington and Oregon—driven in part by increased water-use efficiency associated with increased atmospheric carbon dioxide concentrations (Neilson et al. 2005), although there is some uncertainty about the magnitude of this effect.

In addition to such direct effects, climate change will likely have even larger indirect effects on these forests through changes in fire regimes. Climate change has already affected fire regimes in the western United States, with the frequency of large fires increasing fourfold from 1970–86 to 1987–2003, and the fire-season length increasing by an average of seventy-eight days in the same period (Westerling et al. 2006). East of the Cascades, as a result of projected temperature increases and decreases in precipitation and snowpack, models predict an increase in fire season length and in the likelihood of fires. These changes will likely result in changes in species composition, habitat availability, and the prevalence of insect outbreaks (McKenzie et al. 2004). Predicting changes in the fire regime west of the Cascades is more difficult (Mote et al. 2003), but recent projections anticipate increases in fire frequency there as well (Elsner et al. 2009). In addition, drier, warmer conditions and drought stress are also likely to directly lead to increased insect infestations and outbreaks.

There have been many research projects and tools developed to investigate climate impacts in Washington and the Pacific Northwest. Most—particularly those produced by the Climate Impacts Group at the University of Washington—have focused primarily on water resources, and more specifically on the vulnerability of human systems (Miles et al. 2000; CIG 2004; Elsner et al. 2009). The next section examines several projects that aim to assess climate change impacts, vulnerabilities, and adaptation strategies in the broader context of landscape conservation.

Approaches to Conservation under Climate Change

There are three principal drivers of climate change adaptation in Washington: (1) top-down mandates from the governor or legislature to provide climate change-impacts assessments and adaptation plans, (2) bottom-up efforts to adapt statewide or smaller-scale conservation efforts to climate change, and (3) the inclusion of Washington State in adaptation efforts beyond the state level.

There have been several recent directives in Washington to assess potential climate change impacts and response strategies for land and wildlife management. In 2007 the governor issued an executive order for state agencies to form Preparation and Adaptation Working Groups to develop climate change response strategies. The final report from these groups laid out a series of recommendations, including the need for a statewide climate impacts and vulnerability assessment (WCAT 2008). In response, the Washington State Legislature funded the University of Washington's Climate Impacts Group to undertake a statewide assessment, completed in 2009, of projected climate change impacts for multiple sectors, including forests, salmon, and coasts (Elsner et al. 2009).

In 2009, the Washington state government enacted a bill directing state agencies to build on the work done by both the Preparation and Adaptation Working Groups and the Climate Impacts Group to create an "integrated climate change response strategy." Four topic advisory groups were formed to assist in the state strategy development: Human Health and Security; Built Infrastructure and Communities; Ecosystems, Species, and Habitats; and Natural Resources (working lands and waters). Collectively, these groups were due to issue a final report by the end of 2011.

Statewide efforts are occurring in the context of a variety of regional efforts. The Western Governors' Association Initiative on Climate Change and Adaptation seeks to facilitate the sharing of climate adaptation information and experience throughout the western United States through its Climate Adaptation Working Group, Wildlife Corridor Initiative, and targeted projects on issues such as water and forests. In 2010, the Western Governors' Association released a scoping report on climate adaptation priorities for the western states (WGA 2010). For coastal systems, the West Coast Governors' Alliance on Ocean Health provides a framework for seascape-level conservation work that also requires states to consider climate change.

At the federal level, the Department of the Interior's new regional Landscape Conservation Cooperatives are tasked in part with coordinating and supporting climate adaptation efforts across large landscapes. Two Landscape Conservation Cooperatives cover portions of Washington: the North Pacific Landscape Conservation Cooperative extends over coastal lands west of the Cascades, and the Great Northern Landscape Conservation Cooperative includes the Cascades and eastern Washington. The cooperatives are in the early stages of development and it is not yet clear what role they will ultimately play. In addition, the US Forest Service and National Park Service have collaborated on vulnerability assessment and adaptation planning on the Olympic Peninsula and are beginning a similar process for four national park and national forest units in the North Cascades.

A particularly collaborative and wide-ranging effort to assess climate impacts and adaptation opportunities across the region is the Pacific Northwest Vulnerability Assessment. This project involves researchers at the University of Washington, The Nature Conservancy, the US Geological Survey, the University of Idaho, the National Park Service, the National Wildlife Federation, the Department of the Interior's Great Northern Landscape Conservation Cooperative, and three state wildlife agencies. The assessment covers an ecologically defined area that extends beyond the borders of Washington, Oregon, and Idaho and covers parts of Montana and much smaller portions of Wyoming, Nevada, and California. It will produce a database of species and system sensitivities to climate change, maps of downscaled climate projections, maps of projected changes in vegetation, and maps of potential shifts in focal species distributions. In addition, there will be a series of workshops for natural resource managers across the region to discuss the assessment products, their interpretation and applicability, and what other products might be useful to managers, planners, and policy makers. The project, slated to be completed in 2013, is expected to inform the revision of state wildlife action plans; aid in setting priorities for restoration, preservation, and management of lands; as well as the funding and management of species.

In addition to these stand-alone adaptation efforts, there are also many instances in which climate change adaptation has been incorporated into existing conservation planning efforts. Indeed, this is likely how most climate change planning and action across the region will occur. Experience globally shows that adaptation action often occurs not as a stand-alone endeavor, but rather through modifying existing conservation efforts, plans, and processes to include climate change adaptation among their goals (IPCC-WG2 2007). In the following section, we provide a brief case study of one such process: incorporating climate change into connectivity conservation planning in Washington.

Integrating Climate Change into Connectivity Planning

The Washington Wildlife Habitat Connectivity Working Group (WHCWG) was formed in 2008 with the mission of promoting the long-term viability of the state's wildlife populations through a science-based, collaborative approach to identifying opportunities and priorities to conserve and restore habitat connectivity (WHCWG 2010). The group is led by two state agencies—the Washington Department of Transportation and the Washington Department of Fish and Wildlife—but is open to all parties interested in supporting the group's mission. In 2010, the WHCWG completed an ambitious statewide habitat connectivity assessment based on intensive analyses of the connectivity needs of sixteen focal species and areas of high landscape integrity or low human footprint (WHCWG 2010).

As the statewide assessment was being completed, the WHCWG's Climate Change Subgroup began developing a rigorous framework for additional analyses that would incorporate climate change into the connectivity assessment. In particular, these new analyses aimed to identify areas likely to continue providing habitat and connectivity as climate changes, and to provide connectivity at the scales and along the climatic gradients necessary to accommodate climate-driven shifts in species' ranges. The primary initial roadblock to this work was the lack of established analytical approaches for integrating climate change into connectivity conservation planning. The handful of available approaches varied broadly in their use of models of projected climate change and species responses (e.g., Beier and Brost 2010; Vos et al. 2008; Rose and Burton 2009), and few (e.g., Phillips et al. 2008) attempted to identify linkages at the scale of the WHCWG analysis. None attempted to adapt an existing connectivity plan to climate change. The group therefore had to build an analysis framework from the ground up.

The most common approach to incorporating connectivity into climate adaptation strategies has been to use bioclimatic envelope models to predict climate-driven changes in species ranges; connectivity conservation is then directed toward areas that connect or overlap current and future ranges (e.g., Vos et al. 2008). Yet because species ranges are expected to shift individualistically in response to climate change, connectivity plans designed around range adjustments of individual species may benefit few others. Additionally, climate envelope models are fraught with uncertainty (Beier and Brost 2010). Models of future climate also were not available for Washington at the resolution required.

Given the uncertainties and caveats surrounding the use of future climate projections and bioclimatic envelope models, the WHCWG's Climate Change Subgroup decided to focus instead on methods for identifying climate-smart connectivity areas that rely on information about current climatic gradients, land-use patterns, and topography. The group's first analysis uses such data to identify areas providing connectivity along climatic gradients (Nuñez 2011; WHCWG 2011). The analysis models corridors between currently warmer and currently cooler areas of low human footprint, while minimizing the amount of temperature change encountered along these routes. Conserving such areas should protect the routes that species ranges are likely to follow as they track changing climate. Comparing results from this analysis with those from bioclimatic envelope models, such as those produced by the Pacific Northwest Vulnerability Assessment, should help identify conservation actions that are robust to model assumptions, while illuminating the relative merits of these two very different approaches.

Ultimately, the WHCWG's goal is to provide the information needed for policy makers, regulators, and organizations to incorporate climate-smart connectivity into their work. To this end, the WHCWG commissioned a study that identified more than 120 mechanisms through which their science-based products could be integrated into county, state, and federal land management and conservation policies and plans. One of these is the Arid Lands Initiative, a collaboration among agencies and nongovernmental organizations focused on developing and implementing a coordinated strategy for the conservation of Washington's arid lands. Addressing climate change impacts, which are expected to be relatively severe in this region, is among the group's top priorities. The Arid Lands Initiative is actively advising the WHCWG in its efforts to conduct a finer-scale connectivity analysis in the Columbia Plateau Ecoregion and will use the results of this analysis to identify shared priority areas for the implementation of strategic actions such as fire management, habitat restoration, and the identification of viable alternatives to development on working lands. By partnering with the Arid Lands Initiative, the WHCWG is ensuring that the science-based products they deliver will be relevant to the needs of implementers. They are also benefitting from the Arid Land Initiative's ability to organize input from local biologists and stakeholders in the region.

WHCWG outputs will also be critical to the implementation of the Washington statewide climate adaptation strategy, which is likely to emphasize increased maintenance and protection of habitat connectivity, and to the Western Governors' Association's aforementioned efforts to identify critical habitat and connectivity needs across the West. Washington, Oregon, and Idaho recently completed the first phase of a pilot project in collaboration with the Western Governors' Association to identify crucial wildlife habitat and habitat linkages across the Columbia Plateau. The effort to integrate climate change into this project has emulated the collaborative WHCWG model of engaging with diverse partners and will use many of the methods and tools developed by the WHCWG. Map products produced by the Western Governors' Association's crucial habitats mapping efforts will be made available through online, GIS-based decision support tools that may be used, in part, to help developers and state agencies identify areas where development can occur with minimal impacts to wildlife, now and into the future.

Conclusion and Recommendations

Our experience in Washington suggests that climate change is increasingly being incorporated into conservation at a variety of levels, and that this is proceeding via both top-down and bottom-up processes. In the case of the WHCWG, a postdoctoral researcher initiated formation of the group's Climate Change Subgroup and has been coordinating integration of climate change into the group's connectivity assessment. The Western Governors' Association mandate enlisted the necessary resources to ensure that climate-smart conservation planning occurred at a regional scale across the Northwest. Fortunately, there is no indication that the various adaptation efforts resulting from such divergent processes will conflict with each other, and experience so far with pilots of the WHCWG and the Western Governors' Association have shown that the two are building upon progress made by the other. Data provided by projects such as the WHCWG, Western Governors' Association pilot, and Pacific Northwest Vulnerability Assessment will also directly improve implementation prospects for high-level policy efforts such as the Washington statewide climate adaptation strategy. In many cases, overlap in individual participation among these different projects has directly facilitated communication among efforts, which in turn has raised awareness of potential funding sources and opportunities to share and build capacity. Collectively, these efforts also demonstrate that the threat of climate change is not so much changing the fundamental approach to how conservation is done as it is adding an additional dimension to existing conservation strategies.

In closing, we propose the following recommendations for large-scale climate adaptation processes. First, building individual projects around a highly collaborative approach offers multiple benefits that can offset the decreased efficiency of engaging a broad constituency. These benefits include increased relevance of the final product (in both form and content) to the intended users; increased buy-in by collaborating organizations and others for using the products in policy and management decisions; and increased scope of the project and products enabled by the broader technical and financial resources brought to the table. Second, coordination among disparate efforts is critical. It reduces duplication of effort, supports the holistic and integrated approach seen as central to successful climate adaptation, and stimulates both greater creativity and greater rigor in the development of climate-smart conservation and management. Coordination can happen informally, such as having a few key individuals participate in multiple efforts, and also formally, such as the mandated collaboration between the WHCWG and the Western Governors' Association pilot project. And finally, without either a mandate or motivated individuals willing to take it on, climate adaptation planning is unlikely to happen, even if its importance is widely recognized.

PART 4

Freshwater and Seascapes

These freshwater and seascape case studies emphasize the importance of considering surrounding or adjacent land management in advancing climate change-adaptation efforts. They also point to the importance of water and the impacts climate change will have on changing water regimes. Adaptive management is discussed in some of the case studies as a method for testing whether actions are actually leading to desired outcomes.

Chapter 11

Alps Freshwater, Europe

LEOPOLD FÜREDER, AURELIA ULLRICH-SCHNEIDER,
AND THOMAS SCHEURER

The "water tower" of Europe, the Alps provide immense quantities of freshwater both for humans and biodiversity. However, already impaired and not well conserved by protected areas, water courses are being additionally stressed by climate change. Conservation efforts to address climate change have important international vision and directives, but more attention must be paid to freshwater in this multijurisdictional context. Finding strategies that acheive sustainable water use for whole riverine systems, integrate biodiversity needs into adaptation discussions for human health and energy, restore aquatic ecosystem functionality, and strengthen cross-boundary coordination must be a priority.

Introduction to the Region

The European Alps form an arc of 1,200 kilometers in length from Nice, France, to Vienna, Austria (fig. 11.1). The ecoregion boundaries as defined by the Alpine Convention cover 191,000 square kilometers, belonging to (from west to east) France, Monaco, Italy, Switzerland, Liechtenstein, Germany, Austria, and Slovenia. Mont Blanc is the highest peak (4,807 meters) in the Alps. The Alps are an interzonal mountain system, situated between the temperate life zone of central Europe, with deciduous forests as potential natural vegetation, and the Mediterranean life zone, with evergreen forests of sclerophyllous trees. Both abiotic factors and historical influences,

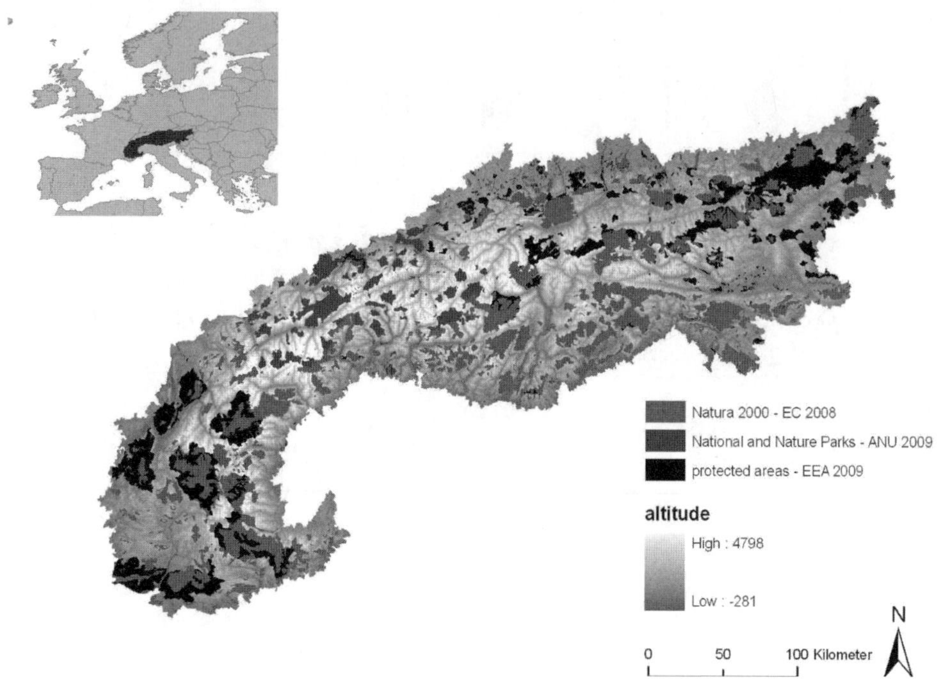

FIGURE 11.1 Protected areas in the Alps (Alpine area) according to the borders of the Alpine Convention, digital elevation model). Produced by A. Bou-Vinals, Innsbruck.

such as the Pleistocene glaciations and human influence since Neolithic times, have influenced ecological patterns in the region.

The landscape of the Alps is varied and characterized by the great diversity of its surface structures from meadows and forests to water bodies and open spaces. Throughout the course of a year or a life cycle, a vast number of different species of animals use various landscape features. As such, the interconnectedness—and therefore the accessibility—of the different elements and resources thus represent an essential basis for survival, distribution, and migration of organisms. This mosaic also contributes to the high biodiversity in the region. Because of the compression of climatic life zones with altitude and small-scale habitat diversity caused by different topoclimates, mountain regions are commonly more diverse than lowlands and are thus of prime conservation value. They not only support a high percentage of terrestrial biodiversity, but also a high level of ethnocultural diversity.

Alpine ecosystems are exceptionally fragile, as they are subject to both natural and anthropogenic drivers of change. These range from seismic events and flooding to global climate change and the loss of vegetation and soils because of inappropriate agricultural and forestry practices and extractive industries. Mountain biota are adapted to relatively narrow ranges of temperature (and hence altitude) and precipitation. Because of the sloping terrain and the relatively thin soils, the recovery of mountain ecosystems from disturbances is typically slow or does not occur.

One of the key environmental resources and services of the Alps is water. The Alps are recognized as the most important source of freshwater for continental Europe (Veit 2002). Although the area in Europe covered by the Alps is relatively small, the hydrological regime of the Alps has a crucial influence on the European water balance, and it supplies a disproportionate amount of water to the outer-alpine regions. The Alps significantly contribute to the discharge of major European rivers, such as the Danube (about 35 percent) or the Po (up to 80 percent). Especially in spring and summer, the lowlands of the Danube, Rhine, Rhone, and Po profit from alpine runoff (Permanent Secretariat of the Alpine Convention 2009). But the Alps also play a specific role in times of water scarcity, which may change with proposed climate change scenarios. About 4,000 lakes are located in the Alps, and about forty larger lakes in the vicinity of the Alps receive most of their water from alpine catchments (Veit 2002). Many of these waters developed during the end of the last Ice Age and are today very attractive for settlement and tourism.

Anthropogenic socioeconomic activities have a fundamental impact on water management. The Alps are home to about 15 million persons with an additional 60 to 80 million people visiting each year as tourists. Due to the landform configuration, human activities and anthropogenic impacts concentrate in the valleys and lower elevation landscapes that are also important to biodiversity. Besides settlements, roads act as obstacles and barriers. Transportation infrastructure in the Alps inckudes 4,239 kilometers of roads and 8,364 kilometers of railroads. (Kohler 2009).

Most valley bottoms have been heavily altered by human activities that impact freshwater systems (see Finlayson and D'Cruz 2005) (fig. 11.2). These activities include land drainage, dredging, flood protection, water abstraction and interbasin water transfer, building dams to create reservoirs, and digging new canals for navigation. The extent of human-made modifications is apparent, for example, in Switzerland, where the water network encompasses 65,300 kilometers of rivers and streams. Of these, 10,600 kilometers have been considerably altered by construction work and around 5,200 kilometers are culverted. Furthermore, about 88,000 artificial barriers with a height difference of more than 50 centimeters hinder the migration of fish. Since the late 1980s, sections have been restored, resulting in a slight increase in natural rivers and streams (Weissmann et al. 2009; FOEN 2009a, 2009b). A quite similar situation is evident for the Austrian rivers (Muhar et al. 1998).

As a result, freshwater biodiversity is at risk: the conservation status of freshwater species of European community interest is generally unfavorable, although there are some variations across biogeographical regions (European Topic Centre on Biological

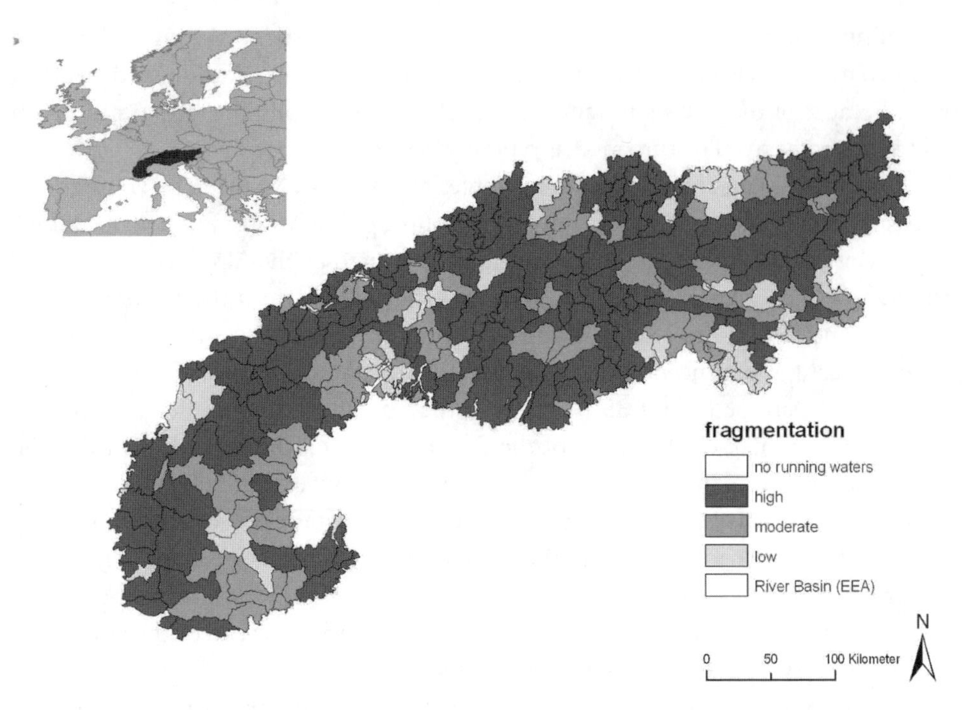

FIGURE 11.2 Extent of fragmentation of waterways within river basins in the Alps. Produced by A. Bou-Vinals, Innsbruck.

Diversity 2008). The assessments of the 51 amphibian species, 16 dragonflies, 54 freshwater fish, 12 freshwater mollusks, and 5 freshwater mammals listed in the Habitats Directive (see below) show that the situation is worst in the Alpine, Atlantic, and Continental biogeographic regions.

Historical Overview of Conservation and Science Initiatives

The legal framework for the protection of freshwater biodiversity in the Alps is mainly represented by the "Water Framework Directive" (WFD 2000/60/EC, http://ec.europa.eu) and other frameworks such as the "Wild Birds Directive" (Directive 2009/147/EC, http://ec.europa.eu) and the "Habitats Directive" (Council Directive 92/43/EEC of 21 May 1992 on the conservation of natural habitats and of wild fauna and flora, http://ec.europa.eu). The goals of the WFD are the protection and the improvement of the quality of all water resources such as rivers, lakes, and groundwater. Its incorporation into national laws is obligatory to all member states. The WFD represents a target-oriented and collaborative approach that changes the way of monitoring, assessing, and managing waters in European countries by introducing the concept of ecological status into legislation. Other goals include the introduction of river-basin management on a Europe-wide scale, pollution prevention and control, public participation in water management, and economic water-use analysis.

The principal actors affecting freshwater conservation range from international to local authorities. These range from the Council of Europe and the European Commission to national administrations and regions/provinces/counties (in most alpine countries nature protection falls within the competence of the regions) and to local communities and nongovernmental organizations. The Council of Europe has provided an international framework, supporting coordinating initiatives that facilitates the approach of developing a common safeguard for biodiversity and landscape. Main achievements of the Council of Europe are the Bern Convention on the Conservation of European Wildlife and Natural Habitats, the Pan-European Biological and Landscape Diversity Strategy, the Pan-European Ecological Network, and the European Diploma of Protected Areas, all of which set the framework for freshwater conservation in the Alps. Specific political framework for nature conservation is given to the alpine countries by the Alpine Convention with its framework convention (signed and ratified by all alpine countries and the European Union (EU)) and its protocol "Conservation of nature and the countryside" (one of nine implementation protocols), which has been ratified by six countries.

Within alpine countries, national ministries are ultimately responsible for enforcing the legislation on water resources and supply. Responsibility may be shared by more than one ministry, each covering different aspects. In practice, supervision of compliance is often fragmented, with much of the responsibility delegated to the regional level and, more frequently, to the local level. In most countries, too many actors are responsible for, and involved in, water management. The jurisdiction over water is often very fragmented and there is not always a single institution ensuring coordination between the different managing agencies. Also, wetlands management is often considered a nature conservation issue. This leads to uncoordinated actions in managing wetlands and missed opportunities for fully exploiting their positive role in water management. National wetland restoration policies are almost nonexistent, although the international framework should lead to national wetland protection policies.

Across most alpine countries public participation in water management is rather poor. The most critical obstacles to public participation are the lack of proactive information provisions to nongovernmental stakeholders and the impediments to enabling the active involvement of interested parties in decision-making processes. Stakeholders often lack both specialist knowledge and the capacity to get involved in decision making for water management measures. It is difficult for nongovernmental water stakeholders to contribute to and influence the decision-making process because the issuing of consultation documents and the involvement of interested parties often take place only toward the end of the process. There is often low transparency for specific projects.

Major Successes

The main aim of the EU Habitats Directive is to promote the maintenance of biodiversity by requiring EU member states to take measures to maintain or restore natural habitats and wild species with a favorable conservation status by introducing strict protection for those habitats and species of European importance. The EU Birds Directive provides an additional framework for the conservation and management of, and human interaction with, wild birds in Europe. These two directives form the basis of the European-wide ecological network Natura 2000 (fig. 11.1), which has been set up in order to protect the most seriously threatened habitats and species across Europe. Additional protected areas, such as Ramsar Sites and national parks exist within the alpine region.

Over the last twenty-five years the EU countries have built up a vast network of nearly 26,000 protected areas covering all the member states and a total area of more than 850,000 square kilometers, representing approximately 18 percent of total EU terrestrial area. The Natura 2000 network, the largest coherent network of protected areas in the world, is a testament to the importance that EU citizens attach to biodiversity. In addition, the Emerald Network was developed in 1998 under the Bern Convention. The Emerald Network is based on the same principles as Natura 2000, and represents its *de facto* extension to non-EU countries. Today, 1,000 larger protected areas (> 100 hectares) exist in the Alps, covering about 25 percent of the alpine area (http://de.alparc.org/die-schutzgebiete/zahlen-der-asg).

While Natura 2000 and the Emerald Network are now regarded by European member states (including the alpine states) as key instruments for conserving and managing species, ecosystems, habitats and landscapes, their focus has heavily emphasized terrestrial conservation. Traditional European protected areas rarely adequately protect freshwater. Rivers are regulated for flood and landslide protection and withdrawn for hydropower use. A new emphasis on landscape-level connectivity offers more opportunity for protection and buffering of freshwater systems.

The LIFE programme, the EU's funding instrument for the environment, began in 1992 to cofinance best practice or demonstration projects that contribute to the implementation of the Birds and Habitats directives and the Natura 2000 network. It does consider aquatic habitats, and several important projects in the alpine countries were carried out on freshwater ecosystems and habitats and associated endangered species, river restoration, and species conservation (http://ec.europa.eu/environment/life/). Some projects under the LIFE program support the implementation of EU water policy and, in particular, integrated river basin management.

The EU Water Framework Directive sets up ambitious objectives for water quality and protection. It relies on a river basin approach for water management, an approach from which alpine freshwaters benefit considerably. According to the WFD all water bodies should meet good ecological status by 2015, which is defined in terms of a set of biological elements (often common indicator species) that are to be expected in the different habitat types of a (geographical) region under clean and natural conditions. Consequently, the WFD is the key tool to protect and restore biodiversity in terms of creating appropriate chemical and hydromorphological conditions in the Alps. The WFD further specifies monitoring requirements for habitat and species protection areas.

In addition to the WFD, the directive on the assessment and management of flood risks (http://ec.europa.eu/environment/water/flood_risk) requires that EU member states assess risk of flooding for all water courses and coast lines, map the extent of floods as well as assets and humans at risk in these areas, and establish Flood Risk Management Plans by 2015. Further EU regulations have been introduced with a view to reducing water pollution by agriculture, industry, and urban wastewater. In non-EU European countries, like Switzerland, comparable policies and targets exist regarding water protection and management.

Substantial efforts have been made within the states of the Alpine Convention to characterize and assess the status of alpine waters (Permanent Secretariat of the Alpine Convention 2009). Biological quality elements, hydromorphological quality elements, general physical-chemical quality elements, and other pollutants such as priority list substances and surface water quantity are all monitored periodically and strategically aligned with possible pressures occurring within the catchments of alpine river systems.

The Alpine Convention, as the framework agreement for the protection and sustainable development of the alpine region, calls for the establishment of "a network of existing national and transboundary protected areas, of biotopes and other protected elements or those to be protected." (Alpine Convention 1994). A 2004 study concluded that a common approach for the entire alpine area was needed to guarantee the coherence of different national and regional approaches for creating an ecological network at the pan-alpine level (ALPARC 2004). Indeed, each of the alpine countries has already adopted different approaches at different levels: for example, the Swiss national ecological network at a national level; the German initiative, *BayernNetz Natur*, at the regional level; or the departmental ecological network of the French department of Isère at a more local level. As all these programs have as a general aim the support and protection of endangered species and degraded habitats, many concentrate on freshwater (especially river) systems, including their floodplains. However, all these activities are limited to administrative areas (countries, regions, departments, etc.), without taking into account a wider biogeographical context. Recently, three major efforts have been undertaken in the Alps to maintain biodiversity by establishing a pan-alpine ecological network (Heinrichs et al. 2010):

- The Ecological Continuum Initiative: This initiative aims to improve ecological connectivity in the Alps by offering an entirely new vision of protecting the natural environment of the alpine massif as a whole, from France to Slovenia. Among various activities for the establishment of ecological networks, aquatic ecosystems in the Alps are included (http://www.cipra.org/en/ecological-networks/ecological-continuum-initiative).
- ECONNECT—Restoring the web of life: The objective is to protect, maintain, and restore ecological connectivity in the Alps using pilot regions to demonstrate tools and methods to improve connectivity at a local level, ranging from analyses of physical and legal barriers to databases (http://www.econnectproject.eu). One activity considers the four-dimensional nature of alpine river systems (longitudinal, lateral, vertical, and temporal), and defines barriers for aquatic and water-associated animals and plants in these habitats. Habitat requirements, distribution, dispersal, and migration abilities of more than thirty mostly endangered species have been characterized and compared to the existing structural conditions in the riverine landscape network (L. Füreder, unpubl. data).
- Alpine Convention: This brings together and establishes communication among the relevant authorities and stakeholders in the countries of the Alps, as well as the alpine research community, to promote cross-border cooperation in the development of ecological networks (http://www.alpconv.org). It also serves to liaise with the European Union and the Convention on Biological Diversity.

Current Conservation in the Region

Today the European Alps are a largely protected area (~ 1,000 larger protected areas, 25 percent of the alpine region). The area of national parks and natural reserves specifically set aside for safeguarding biodiversity is considerable (7 percent of the area of the Alpine Convention). Most protected areas are at higher latitudes with about 72 percent of strongly protected nature reserves and core zones and national parks situated above 1,500 meters (Kohler 2009). Yet despite these protections, biodiversity is continuing to decline. The main reasons are the destruction of natural habitats and the deterioration of cultural landscapes associated with the fragmentation of vital areas for fauna and flora (Jäger and Holderegger 2005), phenomena that manifest themselves mainly outside the protected areas. For many species of animals and plants, protected areas in the Alps are too small.

The maintenance of biodiversity depends therefore not only on the preservation of natural habitats and traditional practices, but also on the interstitial areas that allow biological exchanges between these habitats. It is therefore important to implement ecological networks that allow the dispersal of plants and animals. The need to develop new approaches is even more relevant, given the recent phenomenon of climate change. This is an issue that requires a long-term management vision, something that is also necessary in the design of ecological networks, including for riverine systems.

As many parts of the Alps are unsuitable for human settlements, areas of permanent settlement with high population densities are found at the bottom of valleys. The effect was that alpine river systems, once naturally braided, were regulated to protect human settlements and infrastructure from flooding. Although this provided more suitable space for settlements, infrastructure, and land use, ecological diversity and river ecosystem integrity were lost.

The major river systems of the Alps, along with their tributaries, provide natural water resources for about 180 million people. The diverse topographic and biogeographic characteristics of the Alps especially support the unique riverine landscapes colonized by a specific fauna and flora, which are highly endangered through the hydromorphological alterations and ongoing exploitation of their water resources. In particular, in the Alpine Region, great efforts are under way to improve the ecological status of human-altered rivers and to protect habitats and associated biota with favorable conservation status, according to WFD and the Habitats and Birds Directives. At the same time, the ever increasing need of space for human activities and many projected hydropower facilities pose a major threat to the remaining rivers—which is particularly challenging against the background of the legal obligations for renewable energy production as a climate change mitigation strategy for the EU member states.

To counter these potentially disruptive forces, various actors and initiatives such as the WWF European Alpine Program have placed a strong emphasis on the conservation of alpine rivers and wetland ecosystems and their dynamics (Lassen and Savoia 2005). They intend to focus on the conservation of high-priority freshwater ecosystems in the Alps with the aim to stop artificial schemes in areas where large-scale geomorphic processes can still be observed, and where the rivers still have their specific natural retention capacity.

Regional Effects of Climate Change

In the Alps, a temperature increase of 2 degrees Celsius in the period between the late nineteenth and the early twenty-first centuries has been recorded, indicating the region's exposure to climate change (CEC 2009). According to the available climate models, regional alpine temperatures will steadily rise until the end of the twenty-first century with an increase between 2.6 degrees and 3.9 degrees Celsius. The projected pattern for precipitation shows no great change in total annual precipitation but notable changes in the seasonality of precipitation. The models project decreased precipitation in summer for most regions, and an increase in precipitation in spring and winter, the latter falling as rain rather than as snow thereby shortening the snow cover period. This will likely change the runoff pattern, increasing flows in winter and decreasing flows in summer (http://www.climate-and-freshwater.info).

Climate change is also predicted to lead to widespread melting of glaciers, a climbing snowline, and changes in the runoff regimes of rivers and water resource availability, based on analyses of climate changes in the Alps over the last 250 years (Auer et al. 2007). Future projections (Zisenis and Kristensen 2010) include the reduction in snow cover at lower altitudes; receding glaciers and melting permafrost at higher altitudes; changes in the hydrological cycle of mountain regions; altered water availability in elevated and surrounding regions; reduction in permafrost; increasing natural hazards and damage to high-mountain infrastructure; and reduction in the attractiveness of major tourist resorts.

Generally, mountain areas host many specifically adapted species and are very susceptible to changes. Hence, they will show the highest species loss; for example, 60 percent of the alpine plant species could be extinct by the end of the twenty-first century (Zisenis and Kristensen 2010). Only a few studies provide a coarse picture of the relationships between climate and the distribution patterns of freshwater species anywhere, let alone in the Alps (Heino et al. 2009). The lack of detailed climate change and freshwater biodiversity models of species' ranges in most freshwater groups across broad geographical regions severely hinders the prediction of species' responses to projected climate change. Given that a 4 degree Celsius rise in water temperature (i.e., well within the predicted limits for the near future) broadly corresponds to a 680 kilometer latitudinal shift in thermal regimes (Sweeney et al. 1992), changes in the ranges of freshwater species are very likely to occur (Poff et al. 2002). Although changes in the thermal regimes of lakes and streams are probably the most important factors modifying the distributions of freshwater organisms, other direct and indirect effects may also be active.

In river ecosystems, more intense flash floods and extended periods of low flow may also contribute to changes in biodiversity (IPCC-WG1 2001; Poff et al. 2002). Understanding the present-day determinants and future shifts in species' ranges and diversity gradients is extremely important, because geographical distributions ultimately determine the composition of regional species pools from which local communities are assembled (Tonn 1990). However, relevant information is not available for most alpine freshwater species.

There is evidence to believe that rivers will be among the most sensitive of all ecosystems to climate change. Ormerod (2009) argues that rivers will be sensitive to two indirect consequences of climate change. First, other pressures with which climate interacts have already impaired part of or whole river systems. Among them are eutrophication, organic pollution, sediment release, acidification, abstraction, impoundment, urbanization, hydropower development, flood-risk management, and invasion by exotic species (Daufresne and Boët 2007; Durance and Ormerod 2007). Second, climate will affect Alpine river conditions and processes by changing the human use of river catchments, riparian zones, and floodplains. Land management will increasingly be affected by climate change and its many socioeconomic consequences, including demand for agricultural products, water security, water scarcity, population displacement, and management for carbon sequestration. All of these will drive agricultural change and intensification, forestry practices, water resource development, and other land-use patterns over extensive areas.

Approaches to Conservation under Climate Change

There is a good framework available for the implementation of nature protection measures, which should be widely propagated and adequately used: from conventions such as the Convention on Biological Diversity, the Alpine Convention, and the WFD, to networks of protected areas such as the Pan European Ecological Network, Natura 2000, and the Emerald Network. While these nature protection measures were not originally conceived as a reaction to climate change, they reduce stress factors for flora and fauna and allow ecosystems to better react to climate change. However, classical nature protection measures seem to be no longer sufficient with respect to the projected effects of climate change, and the concept of protected-area "islands" has to be replaced by the concept of functional networks of multitudes of protected areas (Heinrichs et al. 2010).

Some progress has been made on developing policies and plans that highlight the importance of integrating climate change into biodiversity research and conservation actions across the European continent and the Alps. The European Commission adopted its first policy document on adapting to the impacts of climate change in 2007, which was based on the work and findings of the European Climate Change Programme (European Commission 2007). In its subsequent Biodiversity Action Plan assessment report, the European Commission (2010) stated that the EU will continue to highlight the connections between biodiversity topics and climate change and will further develop policies that reflect these connections in order to fulfill the objective of supporting the ability of biodiversity to adapt to climate change. In the White Paper on Adaptation to Climate Change (European Commission 2009) the Commission of the European Communities highlighted the importance of the development of "green infrastructure" in Europe to maintain and restore the integrity of ecosystems and the services they provide to human communities in the form of storm-water run-off management, clear water, clean soils, etc. The White Paper also addresses adaptation to climate change in water management, and directs water directors of all member states of the EU to address climate change in all river basin management plans. Furthermore, the Commission will present a document by 2012, currently entitled "Blueprint to Safeguard European Waters," which will assess climate change and human-made pressures on water and the environment and which will give deeper insight into their status and possible adaptation opportunities.

In 2009, the Alpine Convention adopted the "Action Plan on Climate Change in the Alps," which lays out a series of adaptation measures for the preservation of biodiversity and protection of water and water resources. With respect to biodiversity protection (both terrestrial and aquatic), the action plan recommends that (1) protected areas be demarcated and reinforced to create an interconnected ecological network; (2) management plans for protected areas be modified to take climate change into account; (3) species protection programs for endemic, endangered, and iconic species of the Alps are coordinated across the whole alpine range; and (4) ecological research and monitoring be established to detect species adaptations to climate change. With respect to water management, the action plan encourages the reinforcement of the WFD, the prevention of water shortages through water conservation and education, and the development of hydroelectric plants that are compatible with the natural ecology of streams.

While these policy actions represent important steps toward integrating climate change into conservation, they are just a beginning. There also needs to be a specific focus put on alpine river ecosystems in these efforts.

Roadblocks and Opportunities

Major impediments to protecting alpine systems from climate change stem from the lack of legal frameworks for building transboundary ecological networks (CIPRA 2010). Moreover, there is still great need for research and monitoring, particularly with respect to the aspects of nature protection measures that could have an impact on climate (CIPRA 2010). Some specific knowledge gaps include the following (Heinrichs et al. 2010):

- the effects of climate change on highly balanced alpine freshwater systems and the sensitive species and habitats they host;
- the current status and the adaptive capacity of endemic species and taxa with small home ranges;
- the potential for expansion of invasive alien species;
- the change or degradation of specific biotopes mainly or exclusively on account of climate change; and
- the dynamics of water body and residual water systems and their biotic and abiotic parameters.

Integrating Climate Change and a Freshwater Perspective into Protected Areas

Climate change adaptation is currently realized mostly through the existence of protected areas. Protected areas are delineated according to terrestrial and aquatic ecosystems of specific conservation value, including unique or endangered fauna and flora. Water courses and riparian biotopes inside these protected areas benefit in some way from the terrestrial protection status. There are also cases where water courses are directly protected, such as the Green and Blue Corridor project in France and the Danube River Protection Convention, the implementation of the WFD (which should ensure a good ecological status of freshwater rivers, lakes, and groundwater), and a few others. This type of protection of the water course cannot fully guarantee successful adaptation to future change. Hence, it would be recommended that climate change be accounted for by considering a catchment perspective when delineating a protected area, or redelineating existing protected areas by taking the catchment perspective into account. An integration of terrestrial and freshwater perspective into the delineation of protected areas should improve the adaptability of watercourses (Abell et al. 2007).

One way to take the catchment perspective into account in freshwater protection would be to enlarge existing heterogeneous protection areas. This would increase the likelihood that more parts of freshwater systems and catchments will be included (Heino et al. 2009). This would also be important when considering the connection between lowlands and mountain regions, as many aquatic species will try to adapt to climate change by migrating upstream.

Furthermore it is important to guarantee the undisturbed upstream and downstream movement of species by removing barriers or by making barriers permeable. The establishment and management of the connectivity of freshwater systems is of vital importance for climate change adaptation. Where connections between functional aquatic ecosystems are established, animal and plant species are able to move in and through otherwise hostile territory matrices. If there are no corridors to allow safe migrations, the consequences for the concerned animal can be severe, from local extirpations to complete extinction, and no chance of adapting to climate change at all. For species with limited dispersal abilities, it is possible to plan the reintroduction or reallocation of the concerned species into a restored environment or in environments that seem more able to host that species during change (Heino et al. 2009).

A very important topic is the interconnection of functional biotopes, because a functioning network is able to build a self-sustaining, resilient system. For this purpose it is important to connect biotopes by providing stepping stones and buffer zones around protected areas as corridors. The influence of barriers in the water course and between riparian areas and other wetlands has to be diminished. The longitudinal coherence of natural riverine biotopes (the riverine continuum) has to be greatly restored.

Conclusion and Recommendations

More European policies will be needed to help secure biodiversity adaptations in the face of changing temperature and water regimes. These are necessary to secure the coherence of the European Natura 2000 network. Particular focus should be to minimize any potential damage to biodiversity through unintentional side effects caused by climate change adaptation and mitigation measures. So far, freshwater ecosystems, in general, and rivers in particular, have formed only part of the debate about climate change and biodiversity (Ormerod 2009). To improve the extent to which climate change and freshwater biodiversity are considered in conservation in the Alps, we make the following recommendations:

- It is essential to incorporate biodiversity and climate change-adaptation policies specific to freshwater systems into adaptation discussions in the energy and health sectors, because healthy freshwater systems deliver crucial resources (e.g., energy, clean water) to human communities.
- Special emphasis should be placed on considering projected climate change effects in the implementation and updating process of the river basin management plans (as described in the EU Water Framework Directive). Furthermore, the management plans and goals of the WFD should reach beyond the EU and should also be communicated and propagated in non-EU countries.
- Special emphasis has to be placed on transboundary cooperation, as most larger water courses cross at least one national boundary. There has to be a collective effort, and positive developments upstream will have notable benefits downstream. Hence, integrated multinational and multi-institutional approaches can yield positive outcomes, representing a major tool for accomplishing freshwater conservation goals.
- The investment into "renaturation" (also referred to as restoration measures for reestablishing ecosystem functionality) is highly recommended as a cheap and cost-effective climate mitigation and biodiversity conservation measure, which allows even poorer countries to contribute their part, if expensive technical solutions are not possible.
- Protected areas have to be further maintained and should be extended by the establishment of buffer zones. New protected areas and existing ones have to take into account the catchments of the enclosing freshwaters.
- Natural hazards are predicted to increase with climate change in the Alps, which has to be taken into account through integrated risk management. The projected increase in precipitation events predicted for the Alps and the resulting severely increased runoff at peak times has to be considered in future

flood-risk management. New approaches for steering the use of hydropower in the Alps have to be found that do not excessively impair river ecology and river hydromorphology.
- The recommendations of the Alpine Convention's "Action Plan on Climate Change in the Alps" should be widely propagated and used.

The fate of riverine systems is a topic that concerns everyone, both upstream and downstream, because in every hydraulic system, actions on one site can have severe consequences on other sites (e.g., flood events due to floodplain modifications and accelerated water flows in rectified stream channels). Likewise, changes in the mountain areas, the upstream source locations of the water courses, will have effects at downstream locations in the plains and the delta areas. Stakeholders, citizens, and policy makers from the entire length of these water courses have to be involved since freshwater systems disregard national, legislative, and organizational boundaries.

As water stress is likely to increase along most stretches of the water courses, a strategy toward sustainable water use for whole riverine systems has to be set. The foundation for this has to also be tied to each person's impact on the environment, including austerity measures in daily water use, and it has to be scaled up to large multi-organization restoration projects implementing sustainable policies.

Chapter 12

Sundarbans Mangrove Forest, Bangladesh

BRIAN D. SMITH AND ELISABETH FAHRNI MANSUR

The Sundarbans are one of the most significant mangrove forests in Asia. Due to both lack of information and lack of capacity to conduct conservation planning in response to climate change, an essential component of future conservation work in the region will be to develop strong climate-sensitive indicators of conservation requirements. One of these can be freshwater dolphin populations, which integrate climate-related changes from the mountains to the sea and can thus inform adaptive resource management. Considering both the vital importance of the Sundarbans mangrove forest for sustaining livelihoods and biodiversity, as well as the region's extreme vulnerability to climate change, freshwater dolphin ecology offers a way forward to prioritize aquatic conservation activities in the region.

Introduction to the Region

Freshwater biodiversity is highly vulnerable to climate change—indeed, extinction rates for freshwater species match or exceed the rates projected for terrestrial taxa (Heino et al. 2009). Significant natural variability in freshwater discharge and sediment transport (both seasonal and year to year) determines the geomorphic and biological structure of running water environments, giving them a degree of ecological resiliency to accommodate environmental disturbances likely to be driven or exacerbated by climate change (Palmer et al. 2009). However, the ecological resiliency of most of the world's major running water systems has already been compromised by the construction of dams and major freshwater withdrawals (Nilsson et al. 2005), and this very much includes the Ganges–Brahmaputra–Meghna river system (Smith et al. 2000). Changes in flow regimes due to water development have also altered the salinity patterns, nutrient and sediment supplies, bottom topography, dissolved oxygen, and xenobiotic concentrations in estuarine systems with deleterious consequences for biological productivity and diversity (Sklar and Browder 1998). This means that any meaningful assessment of climate change in the Sundarbans must also incorporate consideration of upstream water development, the impacts of which may outweigh those related to climate change and/or be cumulative or synergistic with them (Vörösmarty et al. 2000; Xenopoulos et al. 2005). At the same time, the existence of upstream water engineering structures may provide potential mitigation opportunities through flow manipulations designed to maintain suitable salinity gradients and preserve essential habitat features, such as bars and midchannel islands that induce biologically productive counter currents (Smith and Reeves 2010).

Bangladesh is at the forefront of global concern about the implications of climate change (IPCC-WG2 2007) due to its low-lying, alluvial character and location at the downstream end of the Ganges–Brahmaputra–Meghna river system. Fed by glaciers, snowmelt, and rainfall in the world's largest mountain range of the Himalayas, this river system constitutes the world's second largest watershed, covering about 1.75 million square kilometers and supporting an estimated 10 percent of the global human population (Biswas 2008). These characteristics make Bangladesh particularly vulnerable to rising sea levels, extreme storms, and altered freshwater regimes due to both climate change and the growing upstream demand for hydropower and freshwater (Xu et al. 2009), with the Sundarbans mangrove forest bearing the brunt of all three.

The Sundarbans is the world's largest continuous mangrove forest, encompassing about one million hectares of which almost one third consists of a complex network of estuarine waterways ranging from a few meters to a few kilometers wide (fig. 12.1). Dynamic erosion and accretion processes, which determine its physiographic character, are buffered by the relative stability provided by the mangrove forest.

Shared by both Bangladesh and India, the Sundarbans play a vital role as a carbon sink and in tempering extreme climate events. The forest also provides substantial natural resources that support extensive biodiversity, including two threatened freshwater cetaceans: the Ganges river dolphin (*Platanista gangetica*) (fig. 12.2a) and Irrawaddy dolphin (*Orcaella brevirostris*) (fig. 12.2b), as well as densely populated human communities that are among the world's most economically impoverished. This chapter focuses on the eastern 60 percent of the Sundarbans lying within Bangladesh that receive freshwater mainly from the Gorai River, a distributary of the Ganges that has already undergone dramatic environmental changes with the construction of the Farakka Barrage (a low-gated dam) just upstream from the Bangladesh/India border.

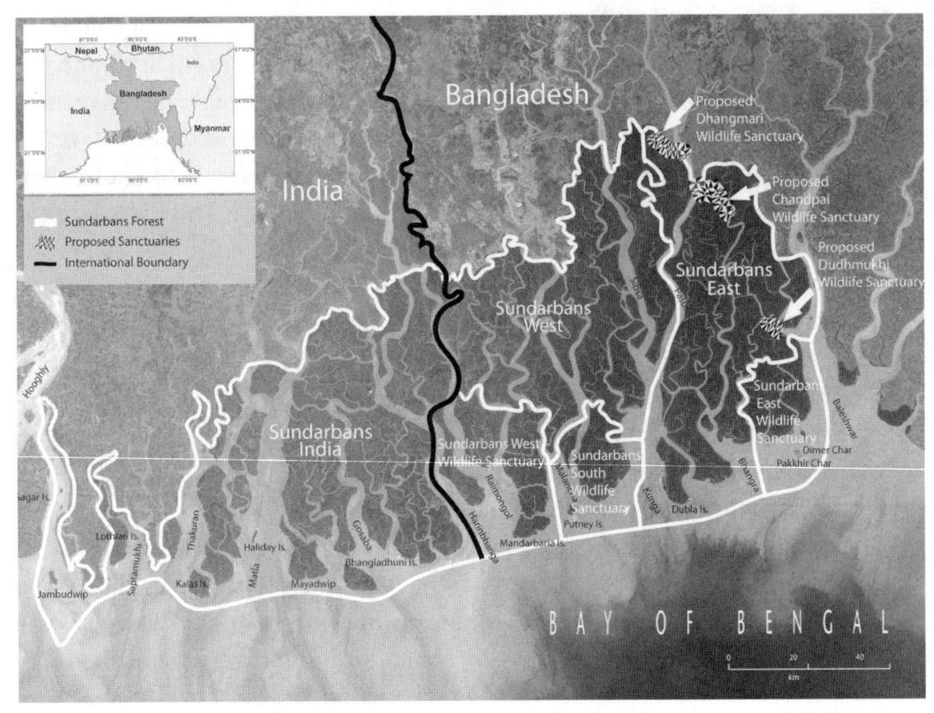

FIGURE 12.1 Map of the Sundarbans mangrove forest showing the existing and proposed wildlife sanctuaries.

The South Asian river dolphin is a relict, obligate freshwater species that is among the most anomalous in the Cetacean order, having adapted to a running water environment in the late Neogene when it remained in shrinking epicontinental seas of the Indo-Gangetic Basin (Hamilton et al. 2001). The Ganges subspecies (*P. g. gangetica*) ranges from the fast-flowing, cool-water reaches of major Himalayan and peninsular tributaries of the Ganges–Brahmaputra–Meghna river system to where these waters meet the sea in the mangrove channels of the Sundarbans. Both a declining range and the diversity and magnitude of ongoing and projected threats led to its classification as "endangered," according to the International Union for Conservation of Nature (IUCN) Red List ("the most comprehensive, objective global approach for evaluating the conservation status of plant and animal species"; see http://www.iucnredlist.org) criteria. In 2002, the Ganges river dolphin population in the Bangladesh portion of the Sundarbans was estimated to be approximately 225 individuals (Smith et al. 2006).

A member of the Delphinidae family, the Irrawaddy dolphin is regarded as a facultative (viz., nonobligate) freshwater species because its range includes coastal marine waters of the Indo-Pacific that receive freshwater input during all or part of the year, in addition to three large rivers (in Myanmar, Indonesia, and Cambodia and Lao

(a)

FIGURE 12.2
(a) A Ganges river dolphin (*Platanista gangetica)* and
(b) two Irrawaddy dolphins (*Orcaella brevirostris)* in waterways of the Sundarbans mangrove forest.

(b)

PDR, respectively) and two marine appended lakes or lagoons (in Thailand and India, respectively). The species is classified as "vulnerable" according to IUCN Red List criteria due to its declining distribution, the severe and persistent threats present throughout its range (particularly from bycatch and habitat degradation), and the low numbers of its assessed populations (10s to low 100s). The single exception to the latter is in Bangladesh, where in 2002 the population in the country's portion of the Sundarbans was estimated to number 451 individuals (Smith et al. 2006), and in 2004, the population in adjacent open estuarine waters was estimated to number 5,383 individuals (Smith et al. 2008).

The ecological requirements of freshwater dolphins are linked to the entire water cycle in all its complexity, from glacial melt and rainfall patterns to sea-level rise and its effects on salinity and sedimentation (Smith and Reeves 2010). These species require sufficient freshwater flow to allow movement between deep pools and the availability of hydraulic refuge from high velocity currents (Smith and Reeves 2000).

Historical Overview of Conservation and Science Initiatives

The Bangladesh government established the legislative framework for protecting the country's fauna, as well as for declaring protected areas, under the Bangladesh Wildlife (Preservation) (Amendment) Act, 1974. Protected areas in Bangladesh cover about 0.5 percent of the country, one of the lowest percentages in the world. The Sundarbans in Bangladesh is classified as a Reserved Forest where it is illegal for anyone to live, cultivate land, or graze livestock, although some forms of resource extraction are allowed, except within designated wildlife sanctuaries (Ahmad et al. 2009).

In 1977, three wildlife sanctuaries were established in the southern portion of the Sundarbans. These were Sundarbans West (715 square kilometers), Sundarbans South (370 square kilometers), and Sundarbans East (310 square kilometers). Together these sanctuaries cover about 23 percent of the reserved forest. In 1997, the United Nations Educational, Scientific and Cultural Organization (UNESCO) World Heritage Convention designated them as a single UNESCO World Heritage Site on the grounds that they served as outstanding examples of ecological and biological processes and contained important and significant natural habitats for threatened species of universal value (UNESCO 2011).

No specific provisions are in place in the Sundarbans Reserved Forest to protect aquatic biodiversity, except for prohibitions on the hunting and trade of threatened wildlife. Although fishing is not allowed in the three wildlife sanctuaries, the smaller channels that fall within their protection do not include priority habitat for freshwater dolphins. A proposal has been forwarded by the Bangladesh Forest Department to the Ministry of Environment and Forests to establish three new wildlife sanctuaries in the Sundarbans for the protection of Ganges river and Irrawaddy dolphins, as well as for other threatened aquatic-dependent wildlife, such as the river terrapin (*Batagur baska*), masked finfoot (*Heliopais personatus*), and small-clawed otter (*Aonyx cinerea*).

Bangladesh is a signatory to the Convention on Biological Diversity, Ramsar Convention on Wetlands of International Importance, and the Kyoto Protocol to the United Nations Framework Convention on Climate Change. A wide and disparate range of governmental, international, and private institutions are involved in supporting national biodiversity conservation commitments and climate adaptation priorities:

- The Bangladesh Forest Department, under the Ministry of Environment and Forests, regulates resource extraction, tourism, revenue collection, and law enforcement in the Bangladesh Sundarbans (Ahmad et al. 2009).
- In partnership with the forest department, local universities, tourism operators, and other nongovernmental organizations (NGOs), The Wildlife Conservation Society has pioneered the study and conservation of freshwater dolphins in waterways of the mangrove forest.
- The European Union is currently funding the Sundarbans Environmental and Livelihoods Security Project that supports sustainable use of natural resources, forests restoration and cyclone-proofing projects, and the development of an Information Management System to guide forest protection and management.
- The Integrated Protected Area Co-management Project, funded by the United States Agency for International Development (USAID), is assisting the Forest Department in developing a strategy for managing ecologically and economically significant protected areas, which includes the development of a Sundarbans Reserved Forest co-management plan.
- The Zoological Society of London and the Wildlife Trust of Bangladesh initiated the Sundarbans Tiger Project in 2005 to ensure protection for tigers (*Panthera tigris*) through research, capacity building, and increased conservation awareness.

A key challenge is to coordinate these different initiatives for achieving optimal conservation and human-welfare benefits.

The Government of Bangladesh has addressed the implications of climate change through the development of two plans mandated by the UN Framework Convention on Climate Change. The National Adaptation Programme of Action (MoEF 2009b) emphasizes food, energy, water, and livelihoods, as well as promoting respect for local communities in resource management and extraction. The Bangladesh Climate Change Strategy and Action Plan (MoEF 2009a) puts forth a vision of eradicating poverty and achieving economic and social well-being through a "pro-poor" climate change strategy. The government of Bangladesh has established a Climate Change

Resiliency Fund, with over US$100 million pledged by the UK, Sweden, Denmark, and the European Union in 2010 (Alam et al. 2011; CAN 2010).

Regional Effects of Climate Change

Compared to the oceanic environment, a running-water environment is more energetically demanding for dolphins since they must constantly swim to maintain their position along the channel's axis. A behavioral response to this environmental challenge is the dolphins' distinct preference for inhabiting countercurrent pools at channel confluences, divergences, and sharp meanders that provide refuge from hydraulic forcing and concentrated sites of biological productivity (Smith 1993; Smith et al. 2009). These "hydraulic nodes" within fluvial waterscapes are particularly sensitive to changes in flow regime caused by upstream withdrawals and altered precipitation, glacial melt, and evaporation patterns. In tidal channels, the environment is also significantly affected by physical forcing from sea-level rise, which increases salinity and sedimentation. Exceeding critical sea-level thresholds can initiate nonlinear and irreversible geomorphic processes in coastal systems with profound ecological consequences (Burkett et al. 2005).

To date, the only dedicated study on the implications of climate change for freshwater dolphins focused on the Bangladesh Sundarbans. The study found that, at a waterscape scale, both Ganges river and Irrawaddy dolphins prefer low salinity waters in the Sundarbans (Smith et al. 2009). However, both species also partition themselves such that Ganges river dolphins generally occupy the northeastern portion (which receives freshwater input from the Ganges with sightings characterized by a mean salinity of 1 and 6 parts per thousand in the high- and low-water seasons, respectively), while Irrawaddy dolphins generally occupy the southwestern portion (which receives only a small amount of freshwater from the Kobadek and Betna rivers that were cut off from the Ganges following construction of the Farakka Barrage, with sightings characterized by a mean salinity of 5 and 18 parts per thousand in the high- and low-water seasons, respectively). There is also a seasonally mobile band of distributional overlap for both species that shifts back and forth along a northeast-southwest axis following the Ganges flow regime.

This same study also suggested an indicator value for the dolphins to inform ecosystem management. At face value, this might appear unlikely since their life history characteristics are the opposite of what are generally considered necessary for scientifically defensible indicators of climate change—namely the occupation of a highly specific ecological niche. However, as large, mobile predators, the manner by which these animals satisfy their life history needs (e.g., movement patterns, habitat use, foraging behavior, etc.) may give them particular value for identifying ecologically significant attributes, such as local aggregations of biological productivity for site-based protection, and for monitoring long-term changes in these parameters. Using the informative properties of freshwater dolphins (i.e., observable adaptive responses), rather than simple measures of animal presence-absence or relative abundance, makes it possible to turn the concept of indicator species on its head. In other words, the very plasticity of freshwater dolphins, which is a feature that in some ways undermines their utility as conventional indicators, can be used for making certain ecological inferences at a waterscape scale (Smith and Reeves, forthcoming).

Climate Change and Water Development

The ecology of the Sundarbans is strongly influenced by the salinity and sedimentation dynamics of its waterways. These dynamics are in turn determined by seasonally and annually fluctuating freshwater flows and diurnal tides. The latter's upstream reach depends on the prevailing sea level, the rise of which is pushing the ecological baseline for freshwater dolphins and associated biodiversity farther upstream. Using high-resolution elevation data, Loucks et al. (2010) estimated that a 28 centimeter rise above 2000 sea levels, which is at the low end of global estimates of projected rise by 2090 (IPCC-WG1 2007; Rahmstorf 2007), would result in the inundation of 96 percent of the mangrove forest and tiger habitat in the Bangladesh Sundarbans. The flip side of this projected decline in forest habitat is the increase in aquatic area, although under very different environmental circumstances. The effects of inundation on terrestrial wildlife habitat are clear—no one expects tigers to live in water—and give reason for great concern. For aquatic species there may be opportunities, however, to adapt to more subtle, but not necessarily inconsequential, environmental changes in geomorphology, hydrology, and tidal forcing.

Existing hydrological models do not provide clear or consistent guidance on probable climate change-related outcomes for the Sundarbans, especially when combined with uncertainty about the magnitude and timing of upstream freshwater withdrawals. As already noted, this situation is made even more complicated due to the effects of sea-level rise on salinity levels and geomorphic structure in the Sundarbans. Integrating hydrological and sea-level models with the ecological, behavioral, and population characteristics of freshwater-dependent cetaceans will remain a major challenge (Smith and Reeves 2010).

Models of snow and ice storage in the Ganges watershed indicate declines of 0.22 ± 0.05 meter per year, resulting in a mean decrease in the upstream melt-water supply of about 18 percent between 2010 and 2050. Although these changes are substantial, they are less than the decline in melt-water supply would suggest due to partial compensation from a projected 8 percent increase in mean upstream rainfall (Immerzeel et al. 2010). Scenario-based models that provide a range of predictions under various, plausible freshwater-flow and sea-level rise conditions may be a more sound approach inasmuch as they can guide adaptive management according to environmental trends, as measured through long-term, rigorous monitoring designs.

Despite expressions of global concern, little attention has been given to understanding the waterscape-level impacts of climate change in the Sundarbans, especially in the context of both extant water development projects and plans for constructing additional water engineering structures in the Ganges system of India (Patkar 2004). The effects of such projects can be substantial. For example, after construction of the Farakka Barrage, salinity increased during the dry season in the north-eastern portion of the Sundarbans by almost 50 percent (Mirza and Qadar 1998).

Increased sediment deposition, resulting from reduced river discharges, has led to the closing off of distributaries (bifurcating branches that do not return to the mainstem, such as in deltas), which previously contributed to repelling salinity encroachment, and to a reduction in the flow capacity of larger channels, such as in the Gorai River. Sedimentation in the Sundarbans has also been accelerated by flocculation (IWM 2003). This process occurs when higher salinity levels increase the proportion of positively charged ions, which dilutes the repulsive forces of negatively charged ions and in turn promotes the formation of sediment "flocs" (aggregations of suspended particles) that settle on the channel bed at lower velocity flows, compared to unbound particles (Van Rijn 1993).

Approaches to Conservation under Climate Change

Lack of knowledge on the implications of climate change in the Sundarbans prevents the development of science-based, adaptive management strategies needed for conserving biodiversity and sustaining human use. Given limited financial resources, scientists are searching for "biological shortcuts" to make inferences about potential climate change impacts and appropriate management options. The Wildlife Conservation Society is developing a long-term program in the Bangladesh Sundarbans to monitor the fine-scale distribution and habitat use of freshwater dolphins in relation to environmental trends. This program uses data from a dolphin-sighting network established among local nature tourism operators and targeted research at sites identified as priority habitat for both freshwater dolphin species (B. D. Smith et al. 2010).

Currently, these sites are in the process of being declared wildlife sanctuaries by the Ministry of Environment and Forests, with specific provisions for using them as study areas both to monitor the ecological impacts of sea-level rise and changes in hydrologic regimes, and to test adaptive management strategies that incorporate sustainable human use. The declaration also calls for an institutionalized form of adaptive management for protected areas. Recognizing that long-term changes in geomorphology and hydrology can affect the habitat of freshwater dolphins, the declaration recommends that a review be undertaken every ten years to determine if the location, size, and configuration of protected areas may need to be altered to provide optimal protection for these threatened species and their associated biodiversity.

Roadblocks and Opportunities

The long-term implications of climate change in the Sundarbans have not yet been integrated into an adaptive management framework. This is due, in part, to the lack of knowledge or predictive models available that incorporate the full, or even partial, complexity of the region's physical-ecological dynamics. Science-based guidance is needed to develop adaptive approaches to cope with climate change. The development of these types of approaches is particularly important for waterways of the Sundarbans because the challenge of biodiversity conservation is growing with increasing human demands, and resource managers are already overwhelmed with addressing short-term threats (e.g., bycatch and overfishing) that will continue to intensify without concerted action, regardless of climate change impacts.

A partnership between the Bangladesh Forest Department and The Wildlife Conservation Society aims to address the implications of climate change in waterways of the Sundarbans by protecting key hydraulic nodes of biological productivity (i.e., countercurrents located at confluences, divergences, and meanders) and by developing a better understanding of potential environmental changes in order to facilitate the development of effective adaptive strategies. These hydraulic nodes correspond with priority habitat for Ganges river and Irrawaddy dolphins, and the animals will be used as an informative tool for adaptive management and an umbrella for biodiversity conservation.

As noted earlier, Bangladesh is at the forefront of global concern about climate change impacts. Without minimizing the potentially dire ramifications of these impacts, this concern presents the country with an opportunity to take a more holistic, ecosystem-based approach toward sustainable development within an adaptive management framework. This approach would regard biodiversity conservation as integral to socioeconomic development, rather than as an isolated component in a laundry list of considerations. It would be an iterative process whereby managers use the results of rigorous research and monitoring to guide the adaptation of local human communities in ways that enhance their resilience to expected changes in the ecosystem processes upon which they depend. It cannot be overemphasized, however, that Bangladesh can seize this opportunity only if it is able to obtain critical long-term technical and financial support.

Through their habitat preferences, movement patterns, and foraging behavior, freshwater dolphins connect changes in upstream hydrology with sea-level rise, and will respond to the ecological effects of both in the waterways where they live (Smith and Reeves 2010). These species can therefore provide managers with a biological shortcut for better understanding the waterscape-level impacts of climate change and guiding adaptive human responses. These responses could include (1) establishing protected areas whose location, size, and configuration could be altered according to long-term trends in habitat use by freshwater dolphins, (2) using these areas to test adaptive management strategies that aim to balance sustainable human use and the protection of threatened aquatic wildlife for broader application throughout waterways of the forest, and (3) manipulating upstream flows through existing upstream dams to preserve geomorphic and hydraulic features essential for maintaining high levels of biological productivity and native species diversity.

Conclusion and Recommendations

Human-induced climate change is expected to cause major changes to the ecology of the world's estuarine systems (Nicholls et al. 2007). Among the most profoundly affected will be the mangrove forest waterways where large Asian rivers meet the northern Indian Ocean. The elevated magnitude of climate change impacts in these environments result from their especially dynamic hydrologic and geomorphic character. Despite the vital importance of these waterways for supporting biodiversity and the subsistence needs of growing human communities, little attention has been given to unravelling their physical-biological complexities. This leaves a critical gap in the type of science-based knowledge required to balance biodiversity conservation with human needs in the context of climate change.

Additional research is needed to make better use of the informative value of freshwater dolphins in the Sundarbans. An adaptively managed, protected area network in core habitat for both Ganges river and Irrawaddy dolphins could function as a "living laboratory" for testing adaptive responses to climate change impacts. It could also provide a critical safety net for ensuring the long-term persistence of these threatened, iconic species and the ecological system upon which they, and a great deal of other biodiversity, depend.

Chapter 13

Vatu-i-Ra Seascape, Fiji

STACY JUPITER, CALEB McCLENNEN,
AND ELIZABETH MATTHEWS

The diverse natural resources of the Vatu-i-Ra seascape, and the human communities that rely on them for subsistence and income, face highly uncertain climate change threats. Climate models disagree on future projections of precipitation, El Niño-Southern Oscillation conditions, and tropical cyclone frequency and intensity. Adaptation options therefore focus on increasing the capacity of local communities to adaptively manage natural resources in ways that mutually benefit humans and biodiversity, and that are responsive to changing conditions. Drawing on existing ecosystem-based management principles and programs, governments and nongovernmental organizations (NGOs) in Fiji are incorporating climate change science into management plans and taking action to increase the resilience of coral reefs and social systems.

Introduction to the Region

Over 330 islands and atolls compose the Fiji island archipelago, home to approximately 35 percent of the coral reef area of all island nations in the southwest Pacific. Within this archipelago lies the Vatu-i-Ra Seascape, which covers over 20,000 square kilometers of relatively intact reefs, seagrass meadows, mangroves, rivers, and forests. Spanning the Bligh Waters passage between the country's two largest islands of Viti Levu and Vanua Levu, Vatu-i-Ra is considered one of Fiji's last great wild places (fig. 13.1).

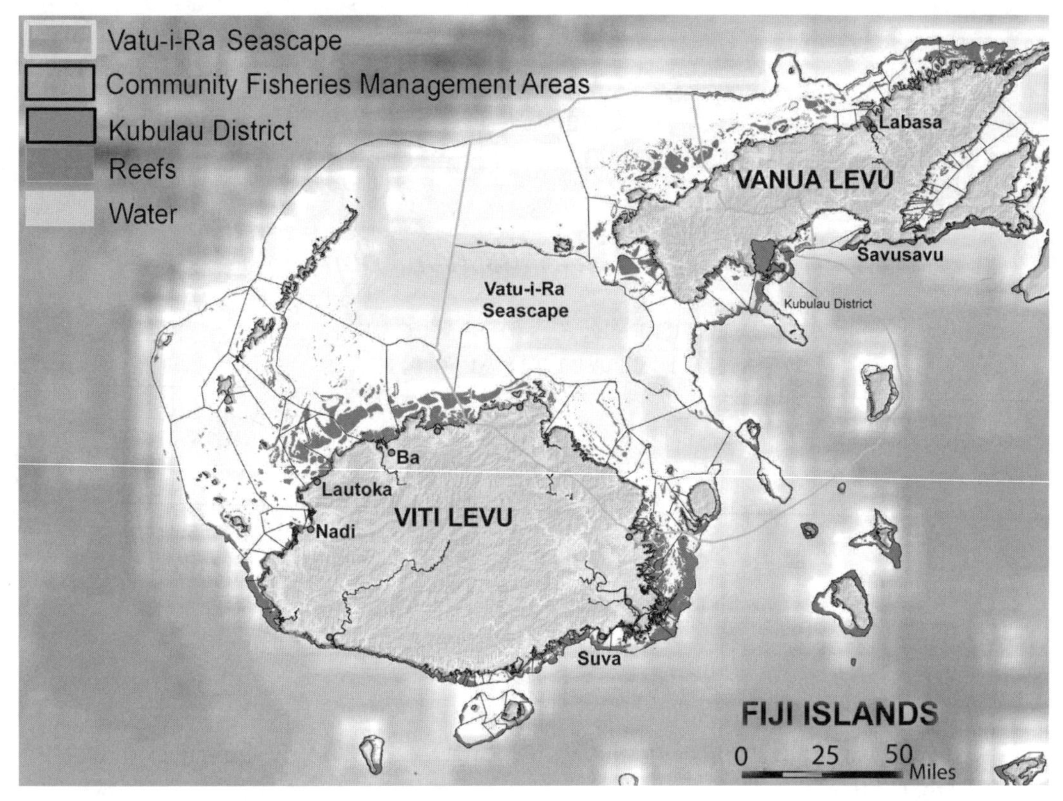

FIGURE 13.1 Location of the Vatu-i-Ra Seascape between the main islands of Viti Levu and Vanua Levu in Fiji.

The island watersheds of the Vatu-i-Ra include six out of the forty priority forests nationally nominated for conservation (Olson et al. 2009) and have well-preserved hydrological connectivity between the land and sea (Jenkins et al. 2010). The small islands scattered throughout the passage are home to important seabird colonies, nesting turtles, and the critically endangered and endemic crested iguana (*Brachylophus vitiensis*). Due to these unique features, participants at the Fiji Islands Marine Ecoregion assessment declared three areas of global significance, one area of national importance, and two areas of subregional importance within the Vatu-i-Ra (WWF 2004). This chapter discusses steps to incorporate climate change-adaptation measures across one of these sites of global significance, which includes the Kubulau District on Vanua Levu (fig. 13.1) and its traditional fisheries management area.

Since 2005, Kubulau has been the focus of a collaborative ecosystem-based management initiative, which supports the sustainable management of terrestrial, freshwater, estuarine, coastal, and marine ecosystems in the district. The land-based bound-

TABLE 13.1 Ecosystem-based management principles (after Clarke and Jupiter 2010b) used in the design of management strategies and implementation for Kubulau District and fisheries management area in Fiji.

Principle	*Use in Kubulau Marine Protected Area network design*
Adopt an integrated approach to ecosystem management.	Developed rules for outside protected areas and managed for threats from adjacent ecosystems.
Maintain healthy, productive, and resilient ecosystems.	Selected reserve locations in a range of habitats, ensuring inclusion of naturally productive and resilient features (e.g., reef adjacent to potential spawning aggregation sites in channels).
Maintain and restore connectivity between systems.	Ensured spacing of reserves is within known distance of larval dispersal (as defined by Jones et al. 2009).
Incorporate economic, social and cultural values.	Conducted resource mapping exercises to ensure costs were evenly distributed.
Involve stakeholders through participatory governance.	Established the Kubulau Resource Management Comittee (KRMC) and engaged a wide range of stakeholders in marine protected area network design consultations.
Recognize uncertainty and plan for adaptive management.	Included within ecosystem-based management plan provisions for amending rules as necessary and reviewing the entire document every five years.
Use all relevant forms of scientific, traditional, and local knowledge.	Initial marine protected area design based on biological and socioeconomic assessments, plus local knowledge of important reef features.

aries of Kubulau District (98.5 square kilometers) are defined by the traditional boundaries of local clans, while the boundaries of Kubulau's traditional fisheries management area (261.6 square kilometers) extend from the coast to the outer edge of the coral reefs. The approximately one thousand people of predominantly indigenous

Fijian origin living in the district are highly dependent on fishing and farming for subsistence and income (WCS 2009). With fewer employment opportunities in nearby towns, pressure on natural resources is expected to rise in the coming decades. Given the uncertainty of future climate change impacts, there is urgent need to strengthen existing management structures in order to regulate resource use and respond quickly to changing environmental conditions.

Historical Overview of Conservation and Science Initiatives

Although Fijian law holds national and provincial government agencies responsible for administering natural resource regulations, in practice customary governance systems exert significant influence over the use of natural resources. The *Bose Vanua* (high council of chiefs) has traditionally regulated decisions related to all aspects of resource management within Kubulau District (Clarke and Jupiter 2010a). In the early 2000s, some of the chiefs requested assistance from The Wildlife Conservation Society (WCS) to help prevent continued resource decline. WCS, in partnership with Wetlands International-Oceania, the WWF South Pacific Programme, and the University of the South Pacific, initiated an ecosystem-based management project in Kubulau in 2005. Using scientific and local knowledge, the conservation organizations assisted community leaders to develop a network of marine protected areas that combined a traditional practice of small, periodically harvested fishing closures with large, no-take reserves. WCS assisted the community decision makers to identify the protected area sites based on ecosystem-based management principles, including representation and replication to protect species, habitats, and critical processes, as well as cost sharing to maximize compliance (table 13.1; Jupiter and Egli 2011).

The protected area network covers 78.7 square kilometers (30 percent of the traditional fishing grounds) and includes seventeen smaller, traditional closures and three large, no-take district reserves (fig. 13.2). The protected areas cover large portions of Kubulau's reef, seagrass, estuarine, and mangrove habitats, including 13 percent of the total mangrove area (750 hectares) in the district. Although none of the marine protected areas have been formally recognized under Fijian legislation, the fisheries department and police recognize conditions placed on fishing licenses banning fishing for "trade or business" from inside the reserves (Clarke and Jupiter 2010a). One legally protected terrestrial area (40 hectares) is located on Namenalala Island, and a 500 hectare Kilaka Forest Reserve has additionally been proposed on lands presently managed as a community conserved area.

Coupled with the establishment of the protected area network, the Kubulau communities formed the Kubulau Resource Management Committee to manage the protected area network and areas within the broader seascape matrix. The committee is primarily responsible for coordinating implementation of management activities and

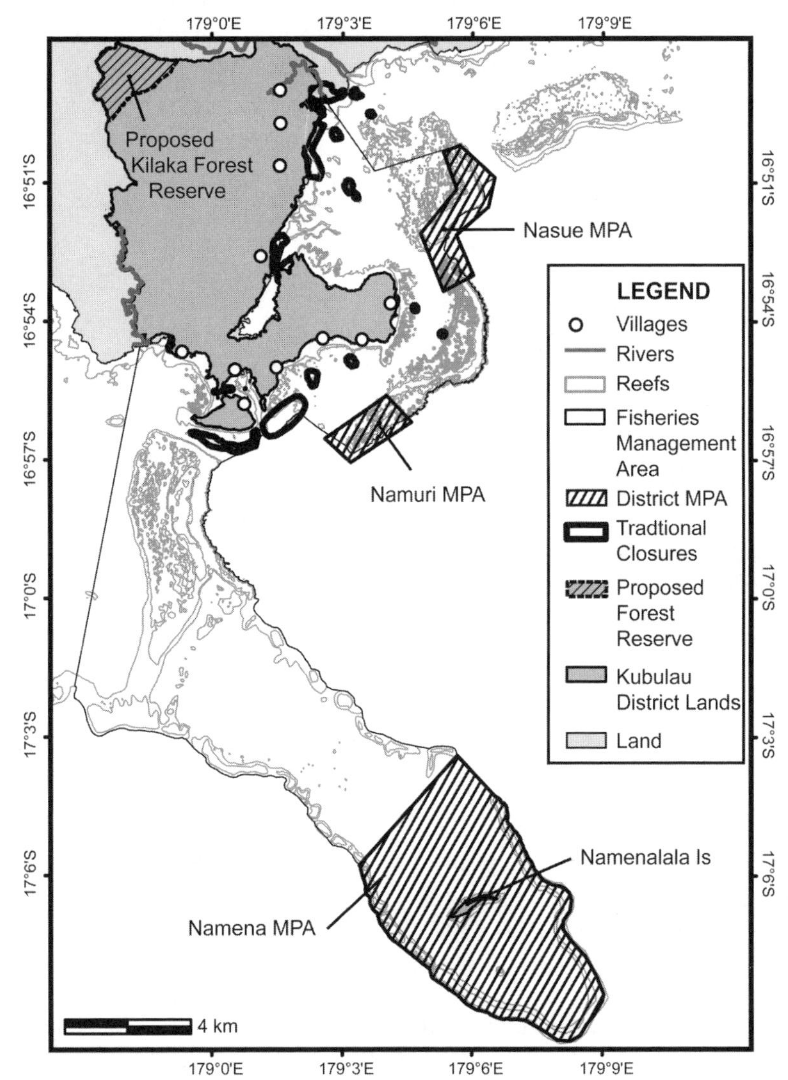

FIGURE 13.2 Kubulau District and adjacent fisheries management area showing the locations of the district Marine Protected Areas, traditional marine closures, the Namenalala conservation area, and the proposed Kilaka Forest Reserve.

enforcement, ensuring broad awareness of management rules, assessing proposed activities, and managing funds (WCS 2009). In February 2009, the committee began to consolidate separate draft management plans for the Namena Marine Reserve, the Kilaka Forest Reserve, and the Kubulau fisheries management area into a single plan. In July 2009, the *Bose Vanua* endorsed the result of this process: a comprehensive ridge-to-reef ecosystem-based management plan for Kubulau—the first of its kind for Fiji.

The Kubulau plan contains key management rules regulating actions in terrestrial, freshwater, coastal, and marine habitats, with provisions that depend upon

TABLE 13.2 Examples of rules from the Kubulau, Fiji, ecosystem-based management plan (WCS 2009), with indications of the rule source, whether it is a national or community rule, and local management actions.

Rule	*Exceptions*	*National*	*Community*	*Management Actions*
Hunting birds, or collecting their eggs, is prohibited.	Licensed hunting of fruit pigeons (15 May to 13 June)	X[1]		Report breaches to Department of Primary Industry.
Dumping waste in rivers and streams or on the banks of rivers and streams is prohibited.		X[2]	X[3]	Report breaches by commercial or industrial facilities to Department of Environment. Report other breaches to Kubulau Resource Management Committee (KRMC).
Laying nets overnight in mangrove areas and around mangrove edges is prohibited.	Fishing approved by village chiefs on special occasions.		X[4]	Monitoring by fish wardens. Report breaches to KRMC.

Notes:

[1] *Birds and Game Protection Act*, part III, 56. Note: the act also allows hunting of barking pigeons, but this species is now protected under the Endangered and Protected Species Act 2002.

[2] *Litter Decree* 1991, s8.

[3] Kubulau Management Plan Review Workshop, January 2007; Kubulau Management Planning Workshop, February 2009.

[4] Kubulau Management Planning Workshop, February 2009.

whether each rule stems from national legislation or community resource managers (table 13.2). Activities and gear that are subject to these management rules both inside and outside protected areas include clearing, burning, or farming adjacent to riverways; harvesting of rare and endangered species; placement of livestock enclosures; and the mesh size of fishing nets. The plan provides procedures for adapting rules at the level of district, village, and clan, as well as protocols for enforcement of national laws and community rules (WCS 2009). Although the plan does not yet carry any legal weight, strong respect for the *Bose Vanua*'s traditional authority ensures a certain measure of compliance (Clarke and Jupiter 2010a). The committee has assigned responsibility for carrying out individual management actions to various groups within the communities, including community health and hygiene groups, village government spokespersons, village development committees, local fish wardens, village chiefs, and the committee members themselves.

The 2009 annual lessons-learned meeting of the Fiji Locally Managed Marine Area network, which hosted community managers from across the nation, highlighted the process used to create the Kubulau Plan. Presented as a model for other sites in Fiji, lessons from Kubulau were incorporated into a guidebook for ecosystem-based management planning and implementation tailored for communities and resource managers in the tropical Western Pacific (Clarke and Jupiter 2010b). Widely distributed and utilized in the country, the success of this model is attributed to strong community support and scientific data showing positive biological responses to management—although not all of the data indicated positive outcomes (Egli et al. 2010; Jupiter and Egli 2011).

Conservation organizations are currently developing recommendations to reduce conflicts within the marine protected area network through maximizing the equitable spread of opportunity costs among fishers (Adams et al. 2011). At the same time, WCS is developing techniques to include reef resilience considerations into these recommendations in order to improve future recovery from climate change impacts. On the terrestrial side, conservation organizations have recently completed ecological surveys to assess how both catchment land cover and riparian forest width influence in-stream fish communities and biophysical characteristics of streams. These data will be used to identify and designate terrestrial and freshwater corridor zones in Kubulau and adjacent districts, with the aims of (1) protecting highly endemic freshwater communities (Jenkins et al. 2010); (2) preserving the ecosystem services, such as clean water, on which communities rely for livelihoods; and (3) reducing future risk of downstream disturbance from flooding, which may be exacerbated through Fiji's changing climate.

Regional Effects of Climate Change

Developing the most effective management plans and structures for Vatu-i-Ra communities will require understanding of projected climate impacts to Fiji and the South Pacific. Current data indicate that while some aspects of climate-related environmental change in the region can be predicted with confidence, other areas remain highly uncertain.

Data for air and sea surface temperature are reasonably robust. In Fiji, air temperatures have increased approximately 0.6 degree Celsius over the past fifty years (uncited information in this paragraph comes from unpublished data of the Fiji Meteorological Service). Surface air temperatures are predicted to increase by at least 2.5 degrees Celsius by 2100 over 1990 levels (Lal 2004). Meanwhile, rainfall and tropical cyclone trends are less certain. Precipitation in the southwest Pacific is strongly linked to El Niño-Southern Oscillation cycles, with cooler, drier periods during El Niño conditions, and warmer, wetter conditions during La Niña phases in the region (Salinger et al. 1995). Rainfall in Fiji has increased by approximately 10 percent during the wet season over the past 100 years, but there has been no corresponding increase during the dry season. Local climate simulations are evenly divided between a more El Niño-like future and a more La Niña-like future, and Intergovernmental Panel on Climate Change (IPCC) projections of future precipitation patterns are still highly uncertain, with models indicating either increases or decreases in rainfall up to 10 percent over most small Pacific islands. Scenarios for future tropical cyclone frequency and intensity are equally uncertain, with some models predicting no change, and others predicting higher frequency or intensity (IPCC-WG2 2007). Overall, given the uncertainty of the data, development of accurate predictive models for the region has been challenging; despite some important advances toward developing locally relevant future simulations, model outputs are not yet available to inform local-level planning and adaptation processes (Barnett and Campbell 2010; Fiji Government 2005; Koshy 2007; Lal et al. 2008).

Given the high ratio of shoreline to land area across many South Pacific islands, coastal ecosystems and infrastructure are highly susceptible to damage from sea-level rise. The South Pacific Sea Level and Climate Monitoring Project was initiated in 1991 with funding from AusAID with the aim of generating long-term accurate records of variance in sea level (Barnett and Campbell 2010). Under the project, twelve Sea Level Fine Resolution Acoustic Measuring Equipment installations across the South Pacific measure water level, air and sea temperature, atmospheric pressure, and wind speed and direction at six-minute intervals. While the sea-level instruments are producing highly accurate and precise data, it will take at least another decade to establish reliable trends of sea-level change (Barnett and Campbell 2010). In the meantime, separate long-term, tide-gauge data between 1950 and 2001 from six island records show a mean rate of sea-level rise of 2.0 millimeters per year from monitoring sites across the tropical Pacific (Church et al. 2006). Sea-level rise impacts will not affect islands uniformly: for example, Fiji islands exhibited a diversity of tectonic processes throughout the Holocene and Quaternary, with active uplifting, subsidence, or maintenance of stable conditions occurring in different areas (Nunn and Peltier 2001). Moreover, it is often difficult to tease out effects of sea-level rise from naturally occurring geomorphological processes that can and have been disrupted by anthropogenic alterations to coastlines (Webb and Kench 2010).

As new data on sea-level rise are used in conjunction with climate simulations from finer-scale models, sector-specific vulnerability assessments for South Pacific islands may become more robust. However, judging from past climate policy development in the Pacific, even improved models may be of limited utility for decision making. Barnett and Campbell (2010, 79) therefore suggest that attention should move away from dependence on quantitative climate scenarios with uncertain outcomes and focus instead on collating the large amount of existing data on social and environmental conditions that "could be meaningfully used to assess vulnerability and adaptation to climate change in the region."

Changing Socioeconomic and Environmental Conditions

As Pacific island populations have historically lived on extremely vulnerable low-lying atolls and narrow coastal island margins, many traditional societies developed high levels of social capacity to cope with climate extremes (Barnett 2001). Practices to ensure food security included crop diversification (including selective breeding for resilient strains) and surplus production and preservation (Barnett and Campbell 2010). Historical resource management systems often operated across watershed units from ridge to reef managed by traditional leaders with strategies that included temporary prohibitions on resource use and negotiated access rights (Berkes et al. 1998). In times of environmental and short-term climate crises, intercommunity cooperation through trade links, traditional social networks, and clan alliances was essential for survival (Campbell 2006).

Following colonization by European powers, these strong traditional ties eroded in many areas, with displacement of populations to urban areas in highly vulnerable coastal areas and subsequent degradation of natural infrastructure (Barnett 2001). Natural infrastructure includes habitats such as mangroves and floodplains that provide important services such as storm surge protection and flood water regulation (Dudley et al. 2010). These shifts have resulted in increased vulnerability to climate disturbance. For example, high density settlements on Fiji's floodplains of the Ba, Labasa, and Nadi rivers, plus heavy alteration of watersheds for agriculture and logging, contributed to flood damage to crops and infrastructure estimated at approximately US$155 million following heavy rains in January 2009 (Fiji Prime Minister's office, unpublished data). The costs associated with impacts to downstream ecosystems have not been assessed, but evidence from nearby countries suggests that repeated inundation of high sediment and nutrient-rich floodwaters may cause shifts in coral reef ecosystems to unfavorable states—viz., where coral is replaced by macroalgae and other noncalcifying organisms—which would have strong, indirect effects on inshore fisheries (Jupiter et al. 2008).

Fisheries are the third largest natural resource sector in Fiji in terms of gross domestic product (GDP) (Teh et al. 2009). Many reef fish species that form the backbone of an artisanal fishing economy are strongly dependent on complex coral reef structure for habitat. Climate-related disturbances such as increased storm patterns, coral bleaching, and ocean acidification will lead to increased coral mortality that decreases reef complexity and may result in strong negative effects on reef fish assemblages (Graham et al. 2006). Models of global, sea-surface temperature rise indicate that the thermal tolerances of many reef-building corals to bleaching will likely be exceeded globally every year within the next few decades (Hoegh-Guldberg 1999), although responses will vary based on regional oceanographic conditions. Recent sea surface temperature simulations indicate that twenty-first century anomalies will be lowest for the Polynesia region (including Fiji) under nearly all of the IPCC climate scenarios (Donner 2009). To date, Fiji's reefs have recovered strongly from mass bleaching incidents in 2000 and 2002 (Lovell and Sykes 2008), although more research is required to adequately understand the mechanisms promoting recovery.

Approaches to Conservation under Climate Change

Climate change adaptation can be considered either as a new dimension added to previously existing threats in the matrix of site-based challenges for conservation (McClanahan et al. 2008) or as a new paradigmatic organizing structure under which conservation priorities are but one component (World Bank 2010). The distinction is critical since climate change adaptation has started to define the funding and technical structure of programs across the Pacific that focus on human communities, on coastal and marine conservation and management, or on both. It should be emphasized that in the developing country context, significant new programs, offices, and even organizations will emerge with a focus on improving humanity's chances of adapting to change in the critically vulnerable littoral zones. Consequently, to avoid policies and plans that either compete or contradict each other across overlapping jurisdictions—or that do not support optimal conservation strategies—it is essential that new climate change policies are well coordinated in their integration into existing administrative structures. For example, in Fiji both the National Disaster Management Office and the Department of Environment are currently responsible for implementing climate change adaptation actions; without better coordination between the two, there is a high risk that climate adaptation funding will be directed to activities that are detrimental to biodiversity conservation, such as seawall construction and dredging.

One way to avoid such conflicts is to link climate change adaptation to biodiversity conservation within existing management frameworks, specifically approaches that incorporate ecosystem-based management. For instance, across the Vatu-i-Ra Seascape, climate change is being viewed as a set of uncertain and cumulative threats that can be addressed by building social and ecological resilience within the existing ecosystem-based management framework. Although mainstreaming biodiversity conservation into existing management frameworks has historically proven very difficult

(Swiderska 2002), ecosystem-based adaptation to climate change provides an exceptional opportunity for conservation because the benefits lead to direct improvements in key provisioning (e.g., food and water security) and regulating services (e.g., storm surge protection, flood control) on which Pacific Islanders rely (Dudley et al. 2010). However, ecosystem-based approaches to adaptation are not a panacea. They must be coupled with policies and practices to increase and/or restore social resilience, which includes diversification of food and income sources, mobilization of current or traditional social networks, enhancement of knowledge transfer and redistribution systems, and poverty alleviation (Barnett 2001).

Largely due to both a lack of understanding of climate change impacts and little awareness of the costs of inaction, many Fijian communities neither perceive climate change as a significant threat nor prioritize the need to manage resources for the potential and unknown future impacts of climate change. Even with enhanced quantification of future losses to agriculture, fisheries, and water services due to climate change, it may still be difficult for local communities to plan for these future climate threats rather than to deal with more immediate and pressing concerns, such as health and economic development. To overcome these challenges, thereby reducing the impacts of climate change and more effectively mainstreaming the need for biodiversity conservation, the following two general approaches can be implemented: (1) maintaining or restoring ecosystems as natural infrastructure, and (2) building capacity for adaptive management.

First, the concept of natural infrastructure is based on the strong link between biodiversity conservation and the capacity of humans to adapt to climate change, a connection increasingly understood by Fijians as irrefutable. In order to mainstream this approach, integrative land-sea conservation initiatives should capitalize on a high demand for technical and scientific expertise to fix real management problems facing a population or a region. Often this need is linked directly to a failure of ecosystems to provide services that once were robust, such as the case of fishery production or shoreline protection. Improving both understanding and availability of information about how various ecosystem services change in relation to climate impacts will be critical (McClanahan et al. 2009), as will the ability to utilize traditional and local knowledge to better inform these models (Mimura 1999). "No regrets" actions, such as options that favor protecting ecosystems and using living shorelines rather than built structures to protect coastal areas, may be easier for communities to accept given the uncertainty of climate change impacts. Successful implementation will additionally rely on the development of innovative payment mechanisms to reward local resource owners and managers for preserving important infrastructure-oriented services, such as wave attenuation, erosion control, and flood protection. To better incentivize resource management actions that may have prohibitive initial costs, but in the long term will provide substantial return in the continued delivery of ecosystem services, there is a need to innovate financing mechanisms such as those that have proven successful in terrestrial habitat protection for water and carbon.

Second, greater attention should be placed on capacity building for adaptive management by local level institutions and organizations in order to promote flexible and responsive policies and management strategies (Berkes and Folke 1998). For example, increased understanding of the ability of communities to adapt to extreme climatic disturbances, such as the 1998 El Niño and the 2010 mass bleaching and coral die-off in the Andaman Sea west of Thailand, can provide feedback to design improved management measures that will offer increased resilience to future climate change-related perturbations. Because more flexible societies are able to respond more quickly, they can both react faster to negative events and take advantage of positive opportunities (McClanahan et al. 2009). Several ways to improve this type of social resilience include (1) strengthening existing social networks for information distribution and improving physical infrastructure for communication, particularly in remote areas; (2) diversifying sources of income, food, water, fuel, and fiber to maintain supplies during unpredictable climate changes; and (3) increasing participation in developing management plans and policies to increase likelihood of implementation (Barnett 2001).

Increasing Coral Reef and Social Resilience

Although the importance of a national scale vulnerability assessment for Fiji was recognized in the early 1990s, no specific vulnerability or needs assessment has been explicitly conducted for the Vatu-i-Ra. Partial assessments of the predicted impacts of sea-level rise have been conducted on the Viti Levu side of the Vatu-i-Ra (Gravelle and Mimura 2008), but not across the entire seascape. Conservation partners are instead focusing on managing for the predicted but uncertain impacts of flooding and coral bleaching through implementation of community-based natural resource management strategies under an ecosystem-based management framework. At the same time, the partners are increasing social capacity to adapt to climate fluctuations by strengthening community-based management structures, improving communications networks, and improving collaboration among a range of sectors and partners.

With respect to improving the resilience of Kubulau's coral reefs to climate-related disturbance, WCS adapted the reef resilience assessment methodology of Obura and Grimsditch (2009) to the local Vatu-i-Ra context. A survey team collected data on reef fish assemblages, coral population structure and recruitment, benthic cover, shading, flushing, and topographic complexity from a range of sites across representative reef habitats. These data have been combined with habitat maps from the Millenium Coral Reef Mapping Project (Andréfouët et al. 2006) within a decision support tool (e.g., Marxan software) to explore options for reconfiguring the marine protected area network that increase resilience to coral bleaching and spread risk by increasing representation of various coral reef habitat types. These outputs were coupled with recommendations for heightened protection across the entire fisheries management area of species such as grazers and top predators that confer higher resilience to reef communities when the results were presented back to Kubulau communities during an adaptive management workshop in July 2011. Meanwhile, in order to reduce downstream impacts from climate-related flood disturbance, WCS used the results from the riparian and in-stream surveys to develop recommendations for improved catchment management, including minimum sizes for riparian forest width, to preserve ecosystem integrity and essential services, such as safe water provisioning.

To strengthen social resilience in Kubulau, conservation organizations are piloting a new communications tool, the Community Educators Network, to help the Kubulau Resource Management Committee deliver conservation and management messages to their constituents. Through tailored workshops, the committee members learn how to draw upon both traditional ecological knowledge as well as scientific information to communicate effectively in the village setting, particularly to target groups such as women and youth who have been previously underrepresented in past management planning workshops. To date, the training has resulted in increased enthusiasm for coral reef conservation, increased community organization, and improved awareness of how to mitigate threats to reefs. In the near future, this model will be adopted across additional sites within the Vatu-i-Ra Seascape and the Fiji Locally Managed Marine Area network.

Improved communication through social networks also needs to be coupled with improved communications infrastructure. Although mobile telephones are currently affordable and widely available, the signal strength is extremely weak and there is very little electric power generation in Kubulau. Consequently, the committee chairman often walks tens of kilometers between villages to deliver messages and organize meetings and enforcement patrols. Further, safety warnings, such as those of potential tsunamis, are not reaching community members in vulnerable and remote coastal locations. In response, enhanced communications networks have become a priority.

Finally, in order to assist communities to diversify sources of fuel, food, and fiber, WCS is conducting surveys to document historical resource-use patterns in response to past climate fluctuations in order to help communities modify behavior to adapt to future climate shifts. For example, within the past two generations, the people of Kilaka village in Kubulau used to sustainably harvest an introduced plant (*Xanthosoma sagittifolium*), which was naturalized along marshy riverbanks. After the tuber of the plant was collected for food, the leaves were put back into the soil, thus providing needed enrichment. More recently, however, farmers have cleared the river banks to plant cash crops such as cassava and kava, leading to perceived declines in soil fertility and adjacent in-stream water quality, which is likely to worsen with increased erosion from climate disturbances. Understanding such historical practices can help identify adaptive management practices that include shifting commercial-scale agriculture away from riverbanks and restoring traditional cover types for improved ecosystem services.

Scaling-up Adaptation across the Vatu-i-Ra Seascape

As management structures and plans are strengthened in Kubulau, conservation organizations are beginning to work in adjacent districts, thus building management capacity to respond to environmental and climate change across the broader Vatu-i-Ra Seascape. Actively developing partnerships with national and international groups are coming together in order to harmonize efforts for ridge-to-reef management implementation at a larger scale as a means toward climate change adaptation.

On a national scale, Pacific island decision makers are often hesitant to act on IPCC model outcomes based on evaluation criteria used in developed countries that are subsequently generalized with uncertain outcomes for the Pacific (Barnett 2001). As an alternative to developing and implementing fixed policy that may or may not fit future climate scenarios, managers are looking for adaptively managed rules-based schemes with flexible measures that can react to a range of climatic futures (Berkes and Folke 1998). Protected area networks managed by communities and government agencies within an ecosystem framework provide such a flexible alternative to strengthen ecological resilience to climate disturbance.

The Fiji Protected Area Committee has been working to compile a list of priorities both for geographical expansion as well as additional protection and management rules for vulnerable sites across the country. This Protected Area Committee is a technical advisory arm to the National Environment Council and is composed of members from government, NGOs, and private business across multiple sectors. While its primary goal is to meet Fiji's national biodiversity targets under the Convention on Biological Diversity, in selecting sites the committee is taking into account factors such as ecological integrity, connectivity, and important regulating services, such as water provisioning and shoreline protection, all of which will be essential components of effective climate change adaptation. Management actions promoting climate adaptation will be incorporated into management plans for new sites and will be adapted into revised plans for existing sites.

Conclusion and Recommendations

Effective conservation of the Vatu-i-Ra Seascape will require full consideration of the existing and predicted impacts of climate change. Acknowledging that significant knowledge gaps are inevitable but improving understanding of existing and future predicted climatic conditions remains imperative. Existing conservation efforts have focused on developing an ecosystem-based network of protected areas and management across a linked land-sea system. The ecosystem-oriented strategy is highly congruent with the paradigm of ecosystem-based adaptation, creating significant opportunity for the mainstreaming of biodiversity conservation in existing national and local priorities and planning frameworks. Specific technical approaches, including improving design of resilient marine reserve systems, scaling up watershed protection, and implementing a host of relevant technical threat reduction actions across the seascape matrix as agreed to by resource managers, will provide a foundation for both conservation and adaptation. A singular focus on adaptation to predicted impacts is much less relevant to communities in the Vatu-i-Ra and across the Pacific than management of corollary present-day issues, such as loss of food and freshwater provisioning services, coastal protection, and inundation. To better enable implementation of specific adaptation strategies over time, improving overall socioecological resilience and adaptive management capacity is critical. Fortunately, because Fijians and other Pacific island societies have a long history of adaptively managing resources through extreme climate events, their traditions provide a solid foundation on which they can incorporate new, flexible management strategies to support biodiversity conservation in a rapidly changing world.

Chapter 14

Wider Caribbean Region

MARIANNE FISH

The Wider Caribbean Region's unique marine and coastal areas support some of the world's most productive and diverse ecosystems and provide numerous products and services for human communities. In combination with the host of stressors already affecting many areas, climate change is likely to lead to further declines in the health and sustainability of coastal ecosystems throughout the Caribbean. A collaborative approach to accelerating climate change adaptation in the region is using the migratory hawksbill sea turtle (*Eretmochelys imbricata*) as an umbrella species to highlight the importance of conservation at a seascape level. Through rigorous adaptation planning, encouragement of innovation, and shared learning, the Adaptation to Climate Change for Marine Turtles initiative is building regional capacity to address the challenges of climate change.

Introduction to the Region

More than 7,000 islands and cays lie strewn across the wide coastal shelves of the Caribbean Sea. Interspersed by deep oceanic basins, these shallow coastal seas are a mosaic of interconnected estuaries, beaches, wetlands, mangroves, seagrass beds, and coral reefs, which together support high levels of biodiversity and endemism. Over 90 percent of fish, corals, and crustaceans are unique to the region (Burke et al. 2011). The Wider Caribbean Region comprises twenty-eight mainland and island countries (fig. 14.1). Despite cultural variation across national and subnational boundaries, the

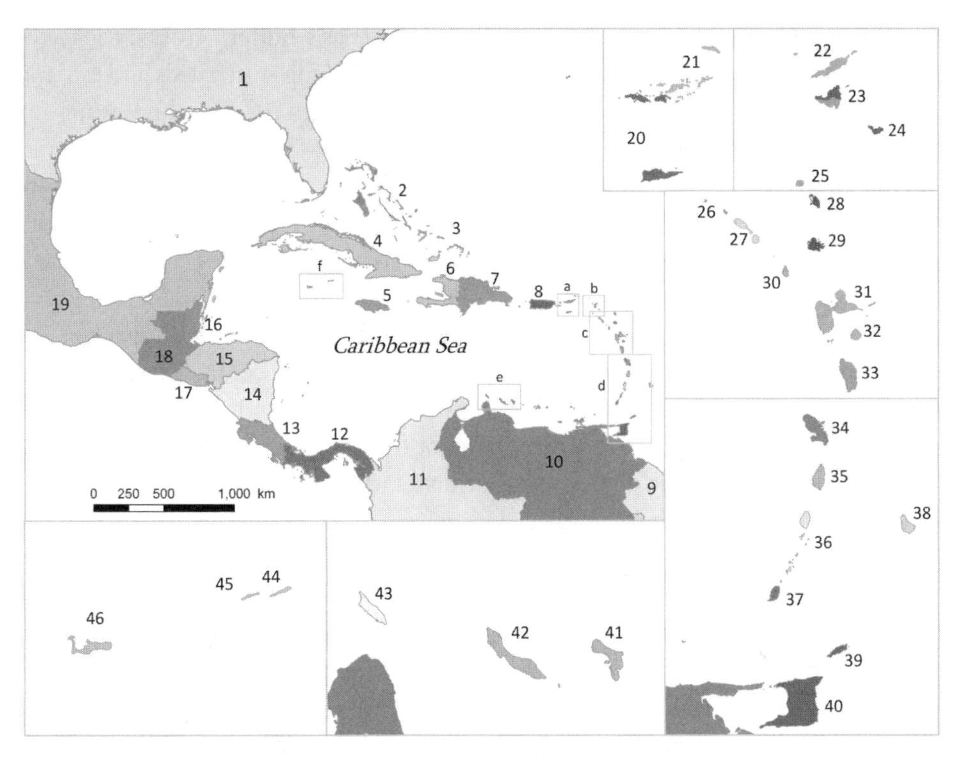

FIGURE 14.1 Islands and countries of the Wider Caribbean: (1) USA, (2) the Bahamas, (3) Turks and Caicos Islands, (4) Cuba, (5) Jamaica, (6) Haiti, (7) Dominican Republic, (8) Puerto Rico, (9) Guyana, (10) Venezuela, (11) Colombia, (12) Panama, (13) Costa Rica, (14) Nicaragua, (15) Honduras, (16) Belize, (17) El Salvador, (18) Guatemala, (19) Mexico, (20) US Virgin Islands (USVI), (21) British Virgin Islands (BVI), (22) Anguilla, (23) St. Maarten/St Martin, (24) St. Barthélemy, (25) Saba, (26) St. Eustatius, (27) St. Kitts and Nevis, (28) Barbuda, (29) Antigua, (30) Montserrat, (31) Guadeloupe, (32) Marie-Galante, (33) Dominica, (34) Martinique, (35) St. Lucia, (36) St. Vincent and the Grenadines, (37) Grenada, (38) Barbados, (39) Tobago, (40) Trinidad, (41) Bonaire, (42) Curaçao, (43) Aruba, (44) Cayman Brac, (45) Little Cayman, (46) Grand Cayman.

region has a strong, shared marine culture. Home to the second largest barrier reef system in the world, more than 43 million people live within 30 kilometers of a coral reef (Burke et al. 2011), and coastal and marine resources are vital for community livelihoods. Coral reefs and their associated biodiversity provide income and nutrition to millions alongside other services such as shoreline protection and generation of sand. Both reefs and seagrass beds are important fishing areas and seagrasses have many traditional domestic uses as roofing material, house insulation, fertilizer, and medicine. A wealth of species forage in these areas, including invertebrates, fish, manatees (*Trichechus manatus*), and green sea turtles (*Chelonia mydas*) (Heck et al. 2008).

Throughout the region, beaches and wetlands act as a buffer between the terrestrial and marine realms. Mangroves play a vital role in linking watersheds and estuaries with nearshore habitats (the coastal area extending seaward from the high-water line to the start of the offshore zone at approximately 20 meters depth). The complex structure of mangrove root systems stabilizes coastal areas, protecting coastal land and infrastructure from damage by strong waves and winds during storms and hurricanes. These roots also offer protective nursery and breeding grounds for numerous commercially important fish and invertebrate species (Mumby et al. 2004). Mangroves also trap pollutants, land-based sediment, and debris, preventing their transport into the nearshore where they could smother fragile seagrass and corals (Harborne et al. 2006) and act as an atmospheric carbon sink, thereby playing a role in mitigating climate change (Donato et al. 2011). Although not typically known for their high biodiversity, beaches are home to a wide range of taxa and support a number of vertebrate visitors, including shorebirds, reptiles, and mammals (Schlacher et al. 2008). Their most important economic role is as the foundation for the region's beach-based tourism industry (Dixon et al. 2001).

Caribbean economies are highly reliant on natural resource-dependent activities such as tourism and fisheries. The region is prone to natural disasters, and small domestic markets are particularly vulnerable to natural hazards that may affect the quality or availability of these resources. For example, the economic impact of the 2004–05 hurricane seasons was estimated at US$4 billion, causing damages greater than the annual Gross Domestic Product (GDP) for some countries (Boruff and Cutter 2007).

Historical Overview of Conservation and Science Initiatives

Reliance on marine resources has led to extensive coastal settlement and a subsequent decline in marine ecosystem health. Overexploitation; pollution; land clearing and modification; and destruction of mangroves, seagrass meadows, and inshore reefs for tourism development and other commercial activities are widespread (Crain et al. 2009). Despite their ecological and economic importance, mangroves are being cleared at an unprecedented rate for agriculture, development projects, conversion to aquaculture ponds, and wood harvesting. Mangrove area in the Caribbean has declined by 1 percent per year since 1980 (FAO 2003), and the health of remaining mangroves is threatened by upstream pollution and overexploitation of mangrove resources. An 80 percent decline in hard coral cover has been observed in some areas (Gardner et al. 2003) and more than 75 percent of the region's reefs are threatened (Burke et al. 2011). Approximately 35 percent of fish stocks are overexploited (FAO 2007), and concern over invasive species, such as lionfish, is increasing. The sandy beaches that form the basis of Caribbean tourism are also showing signs of degradation as a result of sand mining for construction, beachfront development, and the widespread use of coastal armoring to protect beachfront infrastructure. The Coast and Beach Stability in the Caribbean and Sandwatch programs have been monitoring beach change since the 1980s, and a recent summary of data from eight islands showed average erosion of a half meter per year (Cambers 2009). The degraded state of many habitats has reduced resilience to natural disturbance, and postdisturbance recovery is limited in many areas.

Twenty percent of the region's 26,000 kilometers of coral reefs lie within 285 marine protected areas (Burke and Maidens 2006). However, only 6 percent lie in marine protected areas with effective management. The main constraints to effective protection arise from information deficiencies, inadequate institutional and legal frameworks, scant policy enforcement, low levels of support from local communities, and a lack of long-term financial resources. Marine parks are rarely self-financing, with a few exceptions in Bonaire, Saba, and the British Virgin Islands (Burke and Maidens 2006).

Managing the complexity of issues and interests in the coastal zone presents an ongoing challenge. Although many countries have made progress in establishing policy, institutional, and legal frameworks for environmental management, they often lack the necessary resources, capacity, and political support for implementation. Coastal management is further complicated due to the fact that coastal zones typically comprise a highly diverse range of ecosystems and processes that fall under the jurisdiction of separate government agencies (be they for land use and planning, agriculture, fisheries, environmental protection, tourism, etc.). While some countries have established specific agencies to manage coastal areas, conflicts of interest between government bodies often hinder progress. Beyond government, numerous environmental nongovernmental organizations (NGOs) and community organizations at local to regional levels in the Wider Caribbean represent an increasingly powerful force in inciting changes in policies and actions. Nations in the Wider Caribbean are increasingly working together through coalitions based on issues of shared interest, a regionalism that serves to increase the capacity of isolated small island countries. Regional initiatives include the Caribbean Community, the Association of Caribbean States, and the United Nations Caribbean Environment Program.

Thanks to regional collaborations, steps are being taken to increase protection of marine ecosystems in the region. For example, in May 2008 the Caribbean Challenge was launched with the aim of protecting the health of Caribbean islands and waters. Under this challenge, nine countries (Antigua and Barbuda, the Bahamas, Cayman Islands, Jamaica, Grenada, the Dominican Republic, St. Kitts and Nevis, St. Lucia and St. Vincent, and the Grenadines) have committed to setting aside almost 21 million acres (20 percent) of their coral reefs, mangroves, seagrass meadows, and other important habitat for sea turtles, whales, sharks, and other wildlife by 2020. This effort would nearly triple the amount of marine and coastal habitat currently under protection in the Wider Caribbean. The objectives of the challenge are to create networks of marine protected areas; to develop national-level, climate change adaptation demonstration projects; and to establish sustainable funding sources for management, geographical expansion, and monitoring.

Regional Effects of Climate Change

Climate change, in combination with a host of other current stressors, is likely to lead to further declines in the health and sustainability of coastal ecosystems. As low-lying coastal nations highly dependent on natural resources, Caribbean countries are only too aware of their vulnerability to the impacts of climate change. Many nations in the region are urgently seeking to better understand and to prepare for those impacts.

Regional temperature and precipitation projections have been developed for the Caribbean by the Providing Regional Climates for Impacts Studies Caribe project. Relative to the average for the baseline period of 1960–90, temperature increases of between 1 and 5 degrees Celsius by the 2080s are projected for the region. Warming is predicted to be relatively higher in the summer months (May–October) and is highly dependent on location and emissions scenarios. Increases in temperature will be greater over land than over the sea, so larger land masses, such as the Greater Antilles and Central America, are likely to exhibit greater warming (Taylor et al. 2007).

Changes in temperature are already affecting marine ecosystems in many ways. As temperatures rise, shifts in the range and distribution of individual species, and the timing of life history events such as migration and breeding, are leading to changes in community structure and dynamics. Latitudinal and depth-range shifts of marine species (e.g., fish, invertebrates, marine turtles) in response to climate change have already been observed in some locations (Harley et al. 2006). Species reacting at different rates to changing conditions is leading to decoupling of trophic interactions with consequences for food web structure (Edwards and Richardson 2004). For sedentary species such as corals that cannot respond to changing conditions by moving to more suitable areas, stress responses such as bleaching are occurring with increasing frequency (Hoegh-Guldberg et al. 2007). Changes in temperature also affect ocean currents as well as patterns and duration of upwelling events, changes that could have significant consequences for nutrient cycling, larval dispersal, and oxygen availability in surface waters.

Other climatic variables are harder to predict on a regional level due to the influence of local factors on their rates of change. Precipitation projections are much more spatially variable than those for temperature. Projections of changes in precipitation show a general decrease in rainfall over the region as a whole, ranging from –25 to –50 percent by the 2080s. Aruba, Bonaire, and Curaçao (the ABC Islands) and the Lesser Antilles south from St. Kitts (fig. 14.1) show the largest relative decrease. The islands of the northern Caribbean (western Cuba and southern Bahamas) and Costa Rica and Panama are exceptions to the overall drying trend, with up to 25 percent more precipitation (Taylor et al. 2007). Changing freshwater flows will alter water availability for coastal communities as well as sediment/pollutant/nutrient inputs to nearshore areas, with potential impacts on coral reefs, seagrass beds, and mangroves.

In coastal areas, sea-level rise and storm surges may lead to erosion and inundation of coastal land and increased salinity of freshwater sources. Projections for global average increase in sea level are highly variable, ranging from 0.5 to 2.0 meters over the twenty-first century, and regional projections are further complicated by local land movement, tectonics, and oceanographic processes. There is also uncertainty surrounding how the frequency and intensity of hurricanes will be affected by increasing temperature, although most models predict an increase in frequency of the most intense storms (Knutson et al. 2010).

An additional threat to marine systems is ocean acidification. As a major planetary carbon sink, the ocean's absorption of carbon dioxide has already led to a 0.1 unit decrease in ocean pH, and a further reduction of 0.3 to 0.5 pH units is expected over the next century (Orr et al. 2005). Local changes in the acidity of coastal waters depend heavily on water temperature and the upwelling of low pH water (Feely et al. 2008). Lower pH levels due to ocean acidification decrease the availability of carbonate in the ocean and have serious implications for corals, mollusks, and other organisms that use carbonate to build skeletons and shells. Lower pH also affects reproduction, larval growth and development.

Approaches to Conservation under Climate Change

Climate change presents a unique set of challenges to marine ecosystems, and is consequently altering the way we approach conservation. The effectiveness of traditional approaches to marine conservation may become limited as the effects of climate change become more apparent. New approaches to marine conservation must take into account the broad, regional scale at which marine processes and species movements occur. In addition, management needs to take a more flexible and adaptive approach that is responsive to changes in unforeseeable physical conditions and species responses and that allows for frequent reevaluation of goals based on observed changes. For example, design of marine protected areas that addresses climate impacts might include flexible boundaries and buffer zones, networks of marine protected areas, and connections between resilient and more vulnerable areas.

Over the past two decades, a number of initiatives have addressed the impacts of climate change (see box 14.1). Since 2004, the Caribbean Community Climate Change Centre has acted as a regional hub, supporting the region's efforts to address the impacts of climate change by improving knowledge and fostering adaptation

BOX 14.1 ADDRESSING CLIMATE ADAPTATION: THE REGIONAL CARIBBEAN RESPONSE

- **1994: Barbados Programme of Action**

The United Nations Programme of Action on the Sustainable Development of Small Island Developing States, more commonly referred to as the Barbados Programme of Action, is a policy document produced at the Global Conference on the Sustainable Development of Small Island Developing States, held in Barbados in 1994. It is the only internationally approved program specific to Small Island Developing States that has been collectively and unanimously endorsed by the organization. The vulnerability of Small Island Developing States to climate change, climate variability, and sea-level rise was prominent in the Barbados Programme of Action, and countries responded by developing a proposal for the development of regional activities.

- **1998–2001: Caribbean Planning for Adaptation to Climate Change**

Goal: to build capacity in the Caribbean region for adaptation to climate change impacts. A Global Environment Facility-funded project coordinated by the Organization of American States responded to mandates in the Small Island Developing States Barbados Programme of Action at the national, regional, and international levels. The project supported the development of national programs to address the challenges of climate change and the formation of National Climate Committees in all participating countries.

- **2001–2004: Adaptation to Climate Change in the Caribbean**

Goal: to sustain activities initiated under the Caribbean Planning for Adaptation to Climate Change project and to address issues of adaptation and capacity building not undertaken by that initiative, thus further building capacity for climate change adaptation in the Caribbean region. The Canadian International Development Agency-funded initiative addressed gaps in the previous program and facilitated establishment of the Caribbean Community Climate Change Centre in Belmopan, Belize.

- **2003–2007: Mainstreaming Adaptation to Climate Change**

Goal: to mainstream climate change-adaptation strategies into the sustainable development agendas of the small island and low-lying states of CARICOM (Caribbean Community). This program aimed to create an enabling environment to enhance adaptation programs across the region, with a focus on the tourism, agriculture, fisheries, and infrastructure sectors.

- **2006–2011: Implementation of Adaptation Measures in Coastal Zones**

Goal: to implement specific pilot adaptation measures that address the impacts of climate change on the natural resource base (Dominica, St. Lucia, St. Vincent, and the Grenadines). Project activities include the design and implementation of adaptation measures, piloting renewable energy options, supporting land-use planning and management, and reducing anthropogenic stress on national parks and key natural habitats.

projects. The centre coordinates information exchange; provides policy advice and guidelines to Caribbean Community countries; facilitates community projects; and provides training, consultancy services, and funding (CCCCC 2009). The centre is also working with the University of Oxford on CARIBSAVE, a program addressing the impacts and challenges surrounding climate change, tourism, the environment, economic development, and community livelihoods across the Caribbean. Another organization in the region that works on complementary programs is the Caribbean Natural Resources Institute, which directs two programs of direct relevance to marine conservation and climate change entitled "Coastal and marine governance and livelihoods" and "Climate change and disaster risk reduction."

Despite over a decade of collaborative regional action to address climate change, there are still many knowledge gaps, and few adaptation measures have been applied widely. Most efforts in the region have focused on assessing vulnerability and building capacity to collect baseline data through monitoring programs. While these activities are essential, there is a pressing need to begin identifying and testing adaptation actions that can serve as models for similar action elsewhere and to strengthen adaptation capacity in the Wider Caribbean as a whole.

Adaptation to Climate Change for Marine Turtles Project (ACT)

In 2007, the World Wildlife Fund (WWF) embarked on a regional effort toward climate change adaptation for coastal and marine ecosystems in the Wider Caribbean. The first phase of the program was a two-year project (2007–09) to develop a regional approach to climate adaptation using hawksbill sea turtles (*Eretmochelys imbricata*) as an umbrella species for coastal/marine ecosystems. Marine turtles are useful umbrella species for examining climate vulnerability and adaptation at a seascape level since they depend on a range of interlinked coastal and marine habitats, including beaches for reproduction and healthy coral reefs and seagrass beds for feeding. Due to their highly migratory nature, many marine turtles will cross the jurisdictional waters of multiple nations during their lifetimes. As such, these species not only symbolize marine connectivity at local, regional, and global scales, but also highlight the importance of management at a seascape level. Marine turtles are also highly charismatic species and symbols of ocean health, frequently serving as flagships for conservation issues in the marine realm. As such, they can serve as a conduit for familiarizing people with climate change as an issue.

The Adaptation to Climate Change for Marine Turtles (ACT) initiative was launched in December 2007 with the overall aim of building capacity in the Wider Caribbean to reduce threats from climate change to coastal ecosystems used both by marine turtles and by local communities. One of the first steps was to decide what was feasible within a two-year time frame that would result in useful outcomes for conservation projects around the region and moved from identifying impacts toward taking adaptation action. The project involved using key adaptation planning steps (see fig. 2.1 in chap. 2) in regard to three principal questions:

- How will climate change affect sea turtles and their habitats?
- Which locations in the Wider Caribbean Region are particularly vulnerable?
- What can we do now, given our current level of knowledge, to reduce the vulnerability of marine turtles and the habitats that they rely on to the negative impacts of climate change?

The latter question incorporated the subsidiary issues of what adaptation options are possible, and which of these measures can—and should—be implemented, given our current level of knowledge.

Another objective for the program was to present the answers to these three initial questions in a user-friendly format. A decision was made to produce an Adaptation Toolkit for coasts that would present the findings and lessons as a series of reports, manuals, and an online interactive map of regional climate projections (WWF 2009). The toolkit was designed to assist coastal managers and adaptation practitioners with taking action, while also providing a nontechnical overview for educators, policy makers, and the media. In addition to the toolkit, WWF and its partners established a collaborative network for individuals and organizations working in fields relevant to coastal adaptation.

Adaptation Options for Marine Turtles

Changing conditions in the ocean and along coastal interfaces have the potential to exert strong effects on sea turtles (fig. 14.2; Hawkes et al. 2009). Marine turtles have faced large fluctuations in climate during their evolutionary history, but populations already under pressure from overexploitation, fisheries bycatch, and habitat modification are likely ill-equipped to adapt to current changes (Huntley 2007). There are also many unknowns about how sea turtles will respond to climate change, and their ability to adapt to both expected and unexpected changes. Attempting to address all these knowledge gaps could easily lead to analysis paralysis, but putting off action is potentially more serious, and costly, than trial by error. In order to move forward, we asked, given what we do know now, what are the adaptation options available to us?

The literature on adaptation options relevant to coastal and marine systems and wildlife is abundant and growing (see Cambers 2009; CDERA 2003; Heinz Center 2008; El Raey et al. 1999; EPA 2009; Jallow et al. 1996), and many of the options identified in this literature have potential benefits for marine turtles. Examples include protecting refugia (e.g., effectively managed nesting beaches and networked protected

Projected change		Habitat changes	*Potential* impacts on sea turtles
Higher temperatures	Land	Higher beach temperatures	•Female-biased hatchling sex-ratios •Egg mortality •Shorter incubation time: - smaller hatchling size - increased locomotor performance
	Ocean	Higher ocean temperatures	•Faster growth rates •Changes in distribution due to increase in thermally suitable habitat •Intra and inter-annual nesting periodicity reduced •Earlier onset of nesting/extended nesting season •Higher incidence of disease •Changes in predator distribution/abundance •Altered prey availability
Changes in precipitation patterns		Increased precipitation -cooler sand temperatures	•Nest flooding from higher groundwater •Smothering of reefs and seagrasses from sediment run-off: -damage to foraging habitat •Male-biased hatchling sex ratios •Longer incubation time: -reduced swimming ability -larger hatchling size
		Decreased precipitation -higher sand temperatures	(see above)
Increased hurricane intensity		High precipitation	(see above)
		Storm surge	•Nest flooding •Nests eroded/washed away •Nests smothered by redistribution of sand •Removal of beach vegetation •Debris limits access to beach •Seagrass removal/reef damage: -loss of foraging areas
Changes in ocean currents		Altered strength and location of sea-surface currents	•Shifts in migration patterns •Altered distribution of juveniles
Sea-level rise		Beach erosion Coastal squeeze	•Reduced nesting site options •Density-dependent effects: -intra/inter-specific nest destruction - higher density of predators - increased nest infection
Ocean acidification		Reduced coral calcification	•Loss/modification of foraging habitat (hawksbill)
		Higher seagrass productivity	•Increase in foraging habitat (green)

FIGURE 14.2 Potential impacts of climate change on sea turtles in the Wider Caribbean Region. Source: Hawkes et al. (2009)

areas), direct species management (e.g., using hatcheries to control nest temperature), and land management (e.g., using construction setback regulations to facilitate movement of coastal land in response to sea-level rise).

There are many theoretically possible adaptation measures, but any particular location will face logistical, financial, technical, and cultural constraints that restrict which adaptation options are actually feasible to implement. In order to determine realistic options that are currently available for reducing the vulnerability of marine turtles and their habitats to climate change, the ACT program engaged experts from across the region in a review of possible options. Screening, using expert judgment, is a commonly used approach for a preliminary evaluation of adaptation options or identifying those for further research. Although subjective, it is a quick means of determining which options are practical.

The theoretical list of measures was screened by means of a structured online survey. The self-administered survey was sent to a group of 105 experts, including conservation practitioners, academics, and coastal managers, who either hail from countries throughout Latin America and the Caribbean or have experience working in this region. Respondents were selected based on their knowledge of sea turtle biology, coastal management, climate change, and local expertise in order to evaluate whether particular measures would be feasible. Four physical changes were addressed: increasing beach temperatures, sea-level rise/storm surge, increasing ocean temperatures, and ocean acidification. A list of fifty-three theoretically possible adaptation measures was compiled and the experts were asked to determine which they would recommend, based on a number of criteria:

- Effectiveness: How effective would this measure be in achieving the overall aim of reducing hawksbill vulnerability to climate change?
- Technical feasibility: Does the technology and/or expertise exist to carry out this measure? Could this measure be implemented at a local or, in some cases, national scale?
- Financial/logistic feasibility: Are there sufficient resources available to carry out this measure? How much would it cost to implement this measure and who would pay?
- Risks: Are there any risks involved in carrying out this measure? Could there be any detrimental impacts on sea turtles, the ecosystem, or local communities? Might the results of implementing this measure be unacceptable?
- Considering all the previous criteria, is this a practical option to put into place now, and would you recommend it?

From the screening process a list of twenty recommended measures emerged that the majority of experts considered both feasible and low risk (Fish and Drews 2009;

Fuentes et al. 2012). Overall, there was greater support for measures that involve land protection/management, monitoring, and legislative changes than direct species manipulation. It could be inferred from the respondents' comments that reluctance to adopt more experimental approaches is a reaction to lack of knowledge about natural adaptive responses of sea turtles and a fear of inadvertently reducing the population viability of already critically endangered species.

Many of the recommended options offer clear additional benefits beyond conservation of marine turtles, particularly those involving land protection and management. One such example is the enforcement of setback regulations that determine how close to the high-water mark new infrastructure can be built. Setting buildings back from beaches leaves space for sand to shift landward in response to sea-level rise, thereby maintaining beach area over time. Since beach-based tourism is fundamental to many small island economies, measures that enable beaches to persist—whether for marine turtle nesting or tourists—can entail long-term economic benefits.

The project also supported the testing of adaptation measures at Playa Junquillal in Costa Rica, a nesting site for the critically endangered eastern Pacific leatherback turtle (*Dermochelys coriacea*). This small, community-run conservation project has taken a multipronged approach to climate adaptation, implementing adaptation actions at many levels to address both immediate and future impacts. Some of the measures include reforestation of an area behind the beach to stabilize it and provide shade to nesting areas, use of a sea-level rise model to design building setback regulations for inclusion in land-use planning, and moving nests to a hatchery where nest temperatures can be controlled by shading.

Local Action, Regional Collaboration

Because the success of any conservation action depends on local applicability and buy-in, local conditions will limit which adaptation measures are practical. While many conservation practitioners and coastal managers are anxious to take action to address the impacts of climate change, they frequently lack the requisite support and technical expertise. For organizations looking to incorporate climate change vulnerability assessment and adaptation into their work, the limiting factors have been identified as (1) lack of access to information on climate change and impacts, (2) lack of knowledge about methods, (3) lack of adequate information in an appropriate format to inform other stakeholders, (4) lack of resources or political will to bring about change, and (5) conflicting interests. In response to these needs, the focus of the second phase of the ACT project is on overcoming these barriers to implementation:

- capacity-building workshops to provide the background information, resources, and training needed to start taking action;

- technical support for pilot projects carrying out vulnerability assessments and identification, prioritization, and implementation of adaptation measures;
- development of new tools and resources to meet the requirements of adaptation practitioners, including manuals, technical summaries, and outreach and awareness materials;
- support for the incorporation of adaptation into national legislation, strategies, and/or local marine turtle conservation and management plans;
- coordination of pilot projects through a regional network of coastal and marine adaptation practitioners; and
- a forum for those taking action at the local level to share lessons learned to inform activities at other sites.

To move forward with coastal adaptation, action is needed at local, national, and regional levels alongside sharing of lessons learned. In this respect, the ACT program benefits immeasurably from collaborative relationships with several established regional networks, such as the Wider Caribbean Sea Turtle Conservation Network and the Western Hemisphere Migratory Species Initiative.

Conclusion and Recommendations

To effectively address the impacts of climate change in interconnected, large-scale seascapes, a regional approach is needed for both ecological and management reasons. Ecologically, mobile marine species such as sea turtles, which have the potential to shift more readily than their terrestrial counterparts in response to changing environmental conditions, highlight the importance of tackling climate adaptation at broad scales. Regional adaptation efforts are more likely to encompass the full range of habitats used by marine species and therefore offer greater protection to wide-ranging or migratory species.

From a management perspective on sea turtles in the Wider Caribbean Region, a regional approach has been advantageous inasmuch as it has enabled pooling of experience. Since climate adaptation is new to everyone involved, successful management approaches are likely to arise through innovation, trial and error, and learning by doing. Taking a regional approach is beneficial to those starting out because they have a pool of knowledge to learn from, both from those whose efforts have not been successful (since they can look for solutions to the challenges they have encountered) and from those who have succeeded (because of the ecological links between areas).

In order to move forward with coordinated regional adaptation activities, conservationists must recognize the advantages to investing the time in creating a culture of sharing lessons learned. In this respect, the strong collaborative, regional relationships already in place in the Caribbean are the first step toward ensuring that the Wider Caribbean is ready to address and adapt to the challenges of the coming decades.

Acknowledgments

This work was made possible by funding from the John D. and Catherine T. MacArthur Foundation. The contributions and input of Lucy Hawkes and Carlos Drews to this project are gratefully acknowledged. The author would like to thank the WIDECAST network, ACT initiative network, and all other survey respondents for taking the time to share their opinions and expertise.

PART 5

Montane Landscapes

A shared theme across all of these mountain landscapes is connectivity. Mountain regions, because of their extreme temperatures, high elevations, and relative inaccessibility, still house large intact areas, and mountain chains lend themselves well to the application of connectivity principles. During this time of climate change, mountains may offer more options for species to move and find suitable climate conditions and resources, because species can move to different elevations, aspects, or latitudes by moving along mountain ranges, assuming connections are not severed.

Chapter 15

Altai-Sayan, Eurasia

YURI BADENKOV, TATYANA YASHINA, AND GRAEME WORBOYS

The high mountains of the Altai-Sayan Ecoregion are centered on the borders of Russia, Mongolia, Kazakhstan, and China, in the heart of Eurasia. One of the challenges of incorporating climate change adaptation into conservation action is that temperature and precipitation forecasts vary somewhat across the Altai-Sayan. Understanding how this will affect ecosystem services and species such as the snow leopard (*Uncia uncia*) and argali sheep (*Ovis ammon*), the largest of the native sheep, has led to the recent establishment of enhanced monitoring and global change-research programs. At the same time, establishment of a transboundary Mega Connectivity Conservation Corridor has risen as a priority that will likely help many species in the region during this time of climate change, and work is ongoing to make this transboundary corridor a conservation reality.

Introduction to the Region

The Altai-Sayan Ecoregion occupies 1,065,000 square kilometers in the center of Eurasia, straddling the great continental divide separating Russia (62 percent of the ecoregion's extent), Mongolia (29 percent), Kazakhstan (4 percent), and China (4 percent). The width of the region is more than 1,600 kilometers in the west-east direction and 1,300 kilometers north-south. It is characterized by a mix of ecosystems, including alpine tundra, forest, steppe, and desert biomes, with the latter two dominating in Mongolia and China.

The Altai-Sayan includes a number of major mountain ranges (fig. 15.2) including the Altai, Salair, Kuznetsky Alatau, Western Sayan, Eastern Sayan, and Tannu-Ola mountains. The ranges are separated by large depressions, including the Kuznetsk, Minusinsk, Tuva, and the Great Lakes Basin Depression on the Mongolian border with Russia. Serving as a great watershed divide between the Arctic Ocean, Pacific Ocean, and internal-drainage areas of Mongolia, the Altai-Sayan Mountains serve as the water tower for the Katun River, and as headwaters of the mighty Ob River (the fifth longest river in the world) and Russia's Yenisei River (one of the ten longest rivers in the world). Glaciers are an important source of this fresh water; the largest mountain glaciers (over 12 square kilometers) are found in the Altai Mountains, and over one hundred cirque and hanging glaciers are found in the Eastern Sayan Mountains. One of these large glaciers is the Brothers of Tronov glacier found at Mount Belukha.

In 2010, the population of the Altai-Sayan (excluding the Chinese portion) was about five million. Over 60 percent of the population lives in thirty-one towns, and the average rural population density is about 2.4 persons per square kilometer (WWF 2009). The Altai-Sayan is inhabited by many peoples, including Russian, Mongolian, Kazakh, Uigur, Altai, Tuvinian, Shor, and Khakassian, as well as having a variety of religions. The region is important in terms of its historical cultural values, particularly as the home of the Skythian, Turkic, and Mongolian civilizations. A significant proportion of the Altai-Sayan still remains economically undeveloped, with social challenges that include poverty, unemployment, and a lack of alternative sources of income. In Mongolia, the majority of people are shepherds who practice traditional grazing. In the Russian and Kazakhstan portions, the population is mainly engaged in farming, cattle breeding, mineral extraction, and the timber industry, all of which reflects a heavy dependence on natural resources.

Historical Overview of Science and Conservation Initiatives

As one of the least disturbed and transformed large forest and steppe natural areas in the world, the Altai-Sayan presents an outstanding opportunity to conserve globally significant biodiversity (fig. 15.1). The region is one of the world centers of plant diversity, with some 3,726 registered species of vascular plants, 700 of which are threatened or rare, and 317 of which are endemic. The fauna consist of 680 species, 6 percent of which are endemic. The ecoregion is the northern range of the natural habitat of the snow leopard and the argali (or Altai mountain sheep), both of which occupy habitat in all four countries. The global significance of the Altai-Sayan has been recognized through the designation of two United Nations Educational, Scientific and Cultural Organisation (UNESCO) World Heritage Sites. The first, the "Golden Mountains of Altai," encompasses five protected areas in Russia's Altai Republic. In 2003, the transboundary area situated on the border between Mongolia and Russia was in-

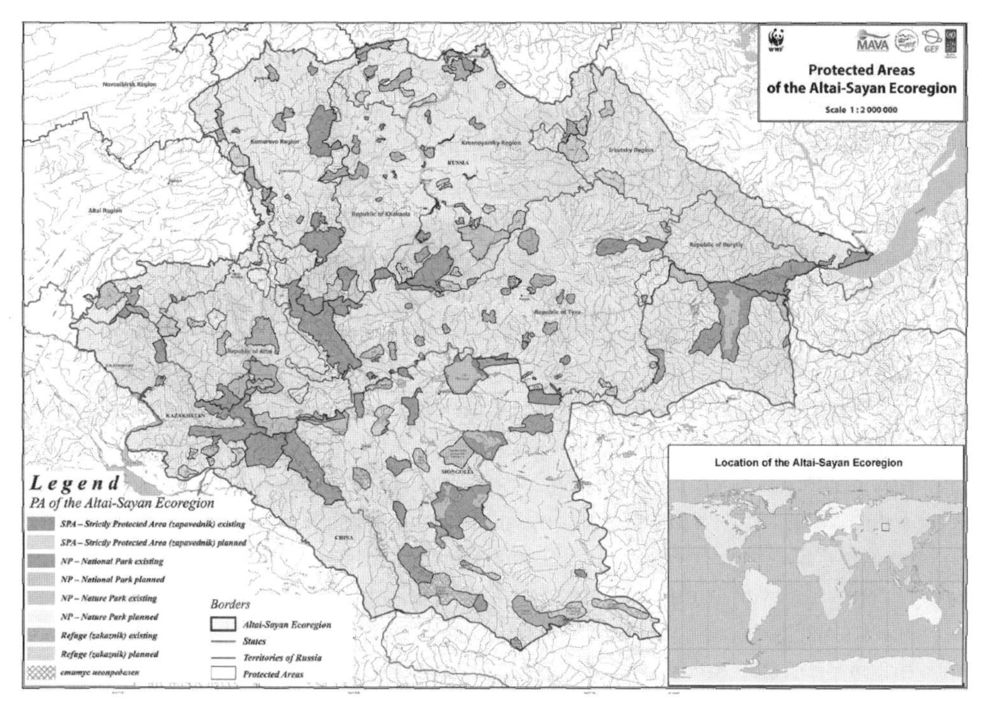

FIGURE 15.1 The Altai-Sayan Ecoregion is located in the transboundary area of China, Kazakhstan, Mongolia, and Russia (and parts of the Altai Republic).

scribed as the second: the Uvs-Nuur Natural World Heritage Site, made up of twelve protected areas representing the major biomes of eastern Eurasia. The Altai-Sayan is also one of the world's 200 priority ecoregions included in the World Wildlife Fund (WWF) Living Planet Campaign.

The Altai-Sayan Ecoregion's biodiversity suffers a number of threats. Poaching is a serious problem, affecting such species as musk deer (*Moschus moschiferus*) for its male scent glands, and the brown bear (*Ursos arctus*), lynx (*Felis lynx*), and sable (*Martes zibellina*) for their furs. WWF (2009) estimates that approximately several dozen argali sheep and snow leopards and 100 falcons are lost to poaching annually. Illegal trading is stimulated by high demand for animal derivatives used in traditional medicines in China and Southeast Asia.

Mining impacts some protected areas in the Mongolian part of the Altai-Sayan, as well as particular populations of endangered species (World Bank 2006). A number of proposed infrastructure projects for the Russian part of the Altai-Sayan pose a threat, including railway construction, a transboundary gas pipeline from Russia to China, and the construction of a large hydroelectric power station in the Altai mountains.

Deforestation in the Altai-Sayan is caused by industrial-scale exploitation of timber, as well as forest fires. Fire particularly affects protected areas in the Russian and

Kazakhstan parts of the ecoregion, a result of which has been the replacement of primary coniferous forests with secondary soft woods in the northern part of the region. Between 1997 and 2001, 1,630 forest fires in the Altai-Sayan Ecoregion burned about 50,000 hectares of forests (about 10,000 hectares per year). Of this, 35,000 hectares were in Russia's Tyva Republic. Forest fires are mainly caused by people, such as tourists and forest workers, and by local people burning hay fields, and sometimes by natural causes (lightning). Severe storms and lightning-caused fires are increasing in frequency.

Overgrazing is an environmental concern for the Mongolian and some of the Russian portions of the Altai-Sayan. Mongolia's transition to a market economy has caused some adverse effects on the state of the environment and grazing ecosystems. One of these effects relates to changes in grazing lands due to the cessation of seasonal grazing, which has been practiced by Mongolian nomads for millennia. In the past, seasonal herding led to a distributed impact on all grazing lands of the region. Today, an increase of livestock concentrated close to settlements and water sources has caused overgrazing and even complete degradation of grazing lands in some places.

In response to these threats, WWF has proposed an expanded and interconnected network of protected areas (described as an *econet*) for the Altai-Sayan, based on a detailed gap analysis (WWF 2009). Focusing on the Russian, Kazakh, and Mongolian parts of ecoregion, this system of reserves aims to build on and interconnect existing protected areas. The network would help establish a number of interconnected protected areas with a mega connectivity conservation corridor along the catchment divide (fig. 15.2). The Russian portion of the proposed network was officially approved for implementation by regional authorities of the Altai-Sayan (WWF 2009), and WWF-led negotiations with appropriate management bodies in Kazakhstan and Mongolia were in process in 2010 (table 15.1).

Current Conservation in the Region

The potential for transboundary protected area management cooperation in the Altai-Sayan is extremely high. This is principally because important conservation areas in Russia, Kazakhstan, Mongolia, and China are contiguous transboundary protected areas, resulting in a permanently protected connectivity corridor that helps to conserve wide-ranging species, including the argali and snow leopard (fig. 15.2).

In the past, protected area managers along the proposed corridor rarely if ever collaborated or even communicated with each other. Since 2004, significant progress in improving those relationships has been made, and transboundary cooperation between protected area managers of Russia, Kazakhstan, and Mongolia has been formalized for two parts of the Altai-Sayan, including two designated biosphere reserves and a UNESCO Transboundary World Heritage Site.

TABLE 15.1 Major protected areas of the Altai-Sayan Ecoregion. The proposed Altai-Sayan transboundary mega connectivity conservation corridor interconnects these areas.

IUCN Category	Number of protected areas				Area of protected areas (,000 ha)				Management objectives
	RU	*KZ*	*MN*	*CN*	*RU*	*KZ*	*MN*	*CN*	
Category I Strict nature reserves	9[1]	2	4	–	2805.4	159	1207.1	-	Strict protection of the environment; monitoring and research; ecological education of the local population
Category II National and nature parks	3	1	4	1	453.0	643.5	1504.4	235	Conservation of the environment; provision of visitor facilities and traditional land use

Key: RU–Russia; KZ–Kazakhstan; MN–Mongolia; CN–China

[1] Including core zones of four UNESCO Biosphere Reserves

In 2010, the legislative program for the establishment of transboundary protected areas included government implementation plans. United Nations Development Program-Global Environmental Fund (UNDP-GEF) funding enabled the design and implementation of the bilateral intergovernmental agreements establishing the transboundary protected areas between Russia, Kazakhstan, and Mongolia. Cooperative management targets for the proposed Russia-Kazakhstan transboundary protected area have been identified by protected area managers and other senior officials:

- coordination of ranger service actions, including conducting joint patrols to increase capacity for law enforcement in the protected area;
- joint monitoring and research efforts to understand the effects of climate change on ecosystem services of the protected areas;
- joint public awareness campaigns; and

FIGURE 15.2 Mount Belukha, 4,506 meters, and the Brothers of Tronor glacier snout, part of the Altai-Sayan Mega Connectivity Corridor (photo: Graeme L. Worboys).

- support for sustainable livelihoods in the transboundary area (Butorin and Yashina 2009; WWF 2009).

This conservation work also includes investing in local communities. The aim is to reduce threats caused by poaching and illegal trade of medicinal plants and animal derivatives and to provide effective conservation measures for the globally endangered argali and snow leopard. The approach is to provide alternative livelihoods for local communities in the Russian, Mongolian, and Kazakhstan portions of the Altai-Sayan under the framework of the UNDP-GEF projects on biodiversity conservation.

In addition to the protected areas, which are governed by the respective ministries and agencies of all four countries, there are a number of stakeholders involved in the conservation of Altai-Sayan's biodiversity (table 15.2).

TABLE 15.2 Key stakeholders contributing to biodiversity conservation in the Altai-Sayan Ecoregion

Key stakeholder(s)	*Biodiversity conservation contribution*
Ministries, government agencies	Protected area establishment and governance, including transboundary cooperation of protected areas.
Institute of Geography, Russian Academy of Science, and Katunskiy Biosphere Reserve	Connectivity conservation approach implementation in Altai-Sayan context (2009–10). Research/monitoring strategy development for climate change studies in Russian Mountain Biosphere Reserves (2003–9).
UNESCO (World Heritage Center)	Reactive monitoring of the state of two World Heritage Properties in the Ecoregion; designation of new World Heritage Sites; provision of managerial guidelines for the World Heritage Properties.
UNESCO-MAB Program and its national committees	Coordination of Biosphere Reserve activities; designation of new Biosphere Reserves; piloting climate change-related field projects in Biosphere Reserves, including GLOCHAMORE and GLOCHAMOST.
Research institutions	Designing protected areas network; conducting monitoring and research activities in protected areas; implementation of the connectivity conservation approach.
UNDP-GEF projects (Russia, Mongolia, and Kazakhstan)	Development of the climate change-adaptation strategy in Russia and Kazakhstan; field projects targeted to increasing capacities of protected areas to prevent and fight forest fires, as well as field projects on adaptation of local communities to climate change.
WWF	GAP-analysis and design of protected areas network; field projects supporting protected-areas management and alternative livelihoods for local communities.
IUCN-WCPA Mountains Biome and Connectivity Conservation	Promotion and facilitation of the connectivity conservation approach.

Key: IUCN–International Union for the Conservation of Nature; GEF–Global Environment Fund; GLOCHAMORE–Global Change in Mountain Regions; GLOCHAMOST–Global and Climate Change in Mountain Sites; MAB–Man and the Biosphere Program; UNESCO–United Nations Educational, Scientific and Cultural Organisation; UNDP–United Nations Development Program; GAP–Gap Analysis Program; WCPA–World Commission on Protected Areas

Regional Effects of Climate Change

Average temperatures in the Altai-Sayan have increased over the past century. Weather observations commenced at the Barnaul weather station in 1835, and over the last five decades of this period to 2010, the average annual temperatures have increased by 2.85 degrees Celsius. Temperature changes over the last fifty years across the ecoregion are heterogeneous, with the least warming detected being +1.45 degrees Celsius at elevations above 2,000 meters and the most warming being +3.5 degrees Celsius at lower elevations within intermountain depressions (Kharlamova 2010). More specifically, the Russian Hydro-Meteorological Service has measured an increase of 1.5 degrees Celsius in the annual average temperatures of the Altai-Sayan area since the 1970s. Total annual precipitation has not changed significantly in the same period, although seasonal anomalies have been observed. The main effects of climate change on the biophysical environment of the Altai-Sayan are the mass retreat of glaciers as well as intense droughts and desertification in the eastern parts of the region.

Climate change will likely alter the ecosystems of the Altai-Sayan in predictable ways. Recent studies have shown that the glaciers of the Altai Mountains have retreated by 19.7 percent from 1952–2004. At the same time, large glaciers of the Belukha massif have lost about 15 percent of their area (Galakhov and Mukhammetov 1999; Narozhny et al. 2006) (fig. 17.2). This accelerated glacier melting is changing the hydrology of the region's high-altitudinal catchments, with melting water forming more than 50 percent of the total discharge of rivers in the Altai-Sayan's upper and middle-elevation zones.

An increase in air temperature may cause upward shifts of altitudinal belts (alpine, subalpine, and montane) as well as significant changes in vegetation and habitat patterns, especially at high altitudes. As with global trends in forest growth, which entail an upward shift in forest vegetation by 30 to 60 meters in the last sixty to eighty years across different regions of the world, the upper tree line in the Katunskiy Biosphere Reserve has shifted upward by 60 to 100 meters during the last 120 years (Patrusheva 2010).

Future Changes

January temperatures within the next fifty years in most parts of the Altai-Sayan are forecasted to increase by an additional 2 to 3 degrees Celsius, although warming may even be more pronounced in the western areas. At West Sayan the temperature increase is predicted to be as much as 3 to 4 degrees Celsius, and around Mount Belukha and on the western spurs of the Altai Range the temperature is forecasted to increase by 4 to 5 degrees Celsius. According to the climate models, the Uvs Nuur Lake region may be a separate area of similar high levels of winter warming (by 3 to 4 degrees Celsius), although this requires more specific research (WWF 2001). In the Mongolian part of the Altai-Sayan, the average annual temperature may increase by as much as 1.8 to 2.8 degrees Celsius in just the first quarter of the twenty-first century. It is possible, however, that winter temperatures may increase by 3 degrees Celsius, while summer temperatures may change very little. The warming forecasted for this ecoregion is much larger than that forecasted for neighboring regions to the north, east, and south. The second quarter of the twenty-first century may see a rate of warming that is twice as fast as the warming in the first quarter. At the same time, an increase in precipitation by 20 to 40 percent is also forecasted, especially during winters in the western part of the region. However, according to some global climate-forecasting models, this increase may also be replaced by a corresponding decrease in precipitation in other parts of the Altai-Sayan (WWF 2001).

All climate change models forecast an intense glacial melt and subsequent retreat of glaciers into the mountains (WWF 2001). Within the next fifty years, it is predicted that the total area covered by glaciers in the temperate latitudes of the Asian continent will decrease by 25 percent. It is not yet clear how intensive the melting in the Altai-Sayan Ecoregion will be, but all data indicate that the process will be rapid and that the high mountain nature zone (natural alpine area) may shrink by half (WWF 2001). Considerable melting of the permafrost layer is also expected, and in some remote areas it may even disappear completely (IPCC-WG2 2007). However, there are still no reliable numeric assessments for these predictions.

With these changes, some species will need to move to survive. The movement of boreal vegetation species from the vast lower elevation taiga massifs in the northern part of the Altai-Sayan into the higher elevation areas farther south could result in habitat fragmentation, thereby forcing animals into high and rocky mountain landscapes. Research indicates a potential 3 degree Celsius warming (IPCC-WG2 2007), and this could lead to a loss of 10 to 60 percent of the mammal species in boreal mountain ecosystems like the Altai-Sayan (WWF 2001).

In the short term, the most unfavorable forecasts over the next fifteen to twenty years relate to significant warming in the spring and winter months that could result in a greater frequency and intensity of heavy snow and floods (WWF 2001). The possibility of midwinter thaws is also likely to increase, which are not yet typical of the region and thus may become a new important factor affecting the survival of plant and animal species. The northern part of the ecoregion will likely see more unfavorable winter conditions for ungulates and predatory mammals, forcing them to undertake large seasonal migrations. Moreover, a few years of particularly heavy snowfalls could result in the high mortality of animal species.

Winter conditions in the south, especially in the southeastern part of the Altai-Sayan, have become less harsh (warmer) for the mammals of the area (WWF 2001). However, due to the forecasted intensification of the summer dry period, ungulates may be forced to spend more time on high mountain pastures, thus increasing competition with domestic stock. Increased habitat fragmentation and the reduction of particular alpine nature zones into "islands" are forecasted in the longer term. The alpine and permanent snow-covered areas are forecasted to decrease considerably. This is inevitable since forests are expanding upslope. Potentially, within thirty to fifty years the upper elevation edge of the forest could rise significantly by 15 to 150 meters (WWF 2001). The length of the vegetation growth period for most plants will also increase by one to two weeks, particularly in spring (WWF 2001). However, the process of soil formation on the rocky surfaces of glacial melt areas will not keep pace with the capacity of forests to shift in altitude, and edaphic factors will ultimately limit upward forest expansion.

The most serious deleterious effects on natural areas are expected in the Mongolian part of the Altai-Sayan, where the total area of tundra and forest ecosystems is likely to decrease by 4 to 14 percent (WWF 2001). The total exhaustion of migration possibilities (to suitable habitats) in the mountain ecosystems is very possible within the next couple of decades. Assessments according to the Holdridge Life Zone Classification scheme also indicate a considerable shift of deserts toward the north (Leemans 1990; IPCC-WG2 2007). Thus, by the middle of the twenty-first century the territory of steppes could potentially decrease by 7 percent in the area of the Great Lakes Basin and its surrounding territories, while total desert area could increase by 13 percent (WWF 2001). In all scenarios, the conservation of the flagship species, the argali and the snow leopard, will demand landscape-scale and costly management intervention efforts to maintain connectivity and thus access to enough habitat to maintain viable populations.

Approaches to Conservation under Climate Change

In the most recent decade, organizations and individuals have focused on research, monitoring, and implementation of a landscape-scale connectivity corridor and other emerging conservation priorities under conditions of climate change. A number of assessments are informing how to move forward, given climate change. In 2001, WWF was the first group to comprehensively research and publish a report on the impact of climate change for the Altai-Sayan (WWF 2001). In 2004 the UNESCO Man and the Biosphere Group 6 (MAB-6) at the Institute of Geography of the Russian Academy of Sciences, in cooperation with the Federal Ministry of Natural Resources in Moscow, initiated a long-term National Global Change Research Program in Mountain Biosphere Reserves. This initiative included MAB reserves within the ecoregion. It has been influenced by the International Human Dimension Program, the International Geosphere Biosphere Program, the Mountain Research Initiative, and the Russian National Committee of MAB, all stressing the need for, and importance of, coordinated global change research. They also recommended the restoration of traditional cooperation between biosphere reserves and the research institutions.

The Russian Federal Ministry of Natural Resources supported the initiative financially and politically. In 2010, WWF-Russia initiated a gap-analysis for biodiversity conservation needs in the Altai Republic of Russia with special focus on climate change issues.

From 2005 to 2007, UNESCO-MAB and the Mountain Research Initiative commenced the first stage of an "Global Change in Mountain Regions" (GLOCHAMORE) program for four mountain biosphere reserves, including the Katunskiy Biosphere Reserve of the Altai-Sayan. The study assessed the ecosystem services of the biosphere reserves and the impact and potential impact of climate change on these services. Based on this, the key objectives of environmental monitoring within the biosphere reserves were identified, and monitoring was implemented.

The second stage of the GLOCHAMORE approach was established for 2009–12 and titled "Global and Climate Change in Mountain Sites" (GLOCHAMOST). Part of this approach—"Coping Strategies for Mountain Biosphere Reserves"—was implemented in two Russian mountain biosphere reserves, again one of them being the Katunskiy Biosphere Reserve in the Altai-Sayan, and the research was based on the theoretical framework provided by the GLOCHAMORE Research Strategy (GLOCHAMORE 2005). A pioneering GLOCAMORE-GLOCHAMOST effort to outline a climate change-adaptation strategy at the level of individual protected areas was performed for Katunskiy Biosphere Reserve by Dr. Tatyana Yashina. The key objectives of the adaptation strategy include the following:

- refining the monitoring program to detect the signals of the effects of climate change and land use on the natural ecosystems;
- strengthening the partnerships with research institutions for modeling changes in hydrology, biodiversity, and ecosystems;
- providing migratory routes for large mammals by implementing the concept of connectivity conservation and strengthening cross-border cooperation with Katon-Karagaiskiy National Park (Kazakhstan);
- developing educational and interpretative programs on climate change-related issues for different target groups;
- implementing model projects on (1) alternative energy supply (equipping ranger stations and high-altitudinal apiaries with solar batteries, minihydropower schemes, and solar cookers), (2) sustainable tourism (focused on the most stressed areas in the transition zone), and (3) diversification of vulnerable economies; and
- developing a more detailed climate change-adaptation strategy for Katunskiy Biosphere Reserve, as this is the first experience of developing protected area-adaptation strategies, at least in the Russian portion of the Altai-Sayan.

Core zones of biosphere reserves provide excellent natural laboratories in as much as they are relatively undisturbed by human activities; as such they constitute field observatories of global change impacts on the environment. Researchers in the Katunskiy Biosphere Reserve have implemented this model, adapting the GLOCHAMORE Research Strategy and developing special monitoring systems targeted to trace climate change impacts in the vulnerable high-altitudinal environment in the Altai-Sayan. This monitoring focuses on observations of meteorological parameters and the condition of vulnerable ecosystems, their components, and species. This includes glacier status, snow cover, biodiversity of alpine ecosystems, the condition of high-altitude catchments, the location of the upper and lower tree lines, and the number of key species in different types of ecosystems.

The key targets of UNDP's biodiversity adaptation strategy for the Russian portion of the Altai-Sayan include the following:

- conservation of populations of vulnerable species and their migration corridors;
- conservation of vulnerable ecosystems by the extension of the protected areas network and reducing nonclimate stressors;
- implementation of an adequate monitoring system for the status of biodiversity and threats in the protected area network;
- wide dissemination of climate change-related information in the region and its interpretation for different target groups; and
- incorporation of forecasted climate change impacts into the practice of land-use planning.

In 2010, the government of Germany's International Climate Initiative funded a two-year project in the Russian and Kazakhstan parts of the Altai-Sayan focused on extending the protected areas network, increasing the management capacities of existing protected area managers to prevent and fight forest fires, and developing climate change-adaptation strategies and appropriate monitoring systems. In Russia, national climate change policy initiatives were introduced with the adoption in 2009 of the Climate Doctrine of the Russian Federation (http://archive.kremlin.ru/eng/text/docs/2009/12/223509.shtml). This document identified the development and implementation of long-term measures for mitigation and adaptation to climate change as priority actions for state politics and governance. However, by 2010, climate change was not considered to be a key organizing factor in biodiversity conservation in Russia at the governmental level, nor in the other countries of the Altai-Sayan.

Transboundary Connectivity Conservation

In 2008, discussions commenced concerning the application of large-scale, transboundary, connectivity conservation management for the Altai-Sayan. This was a di-

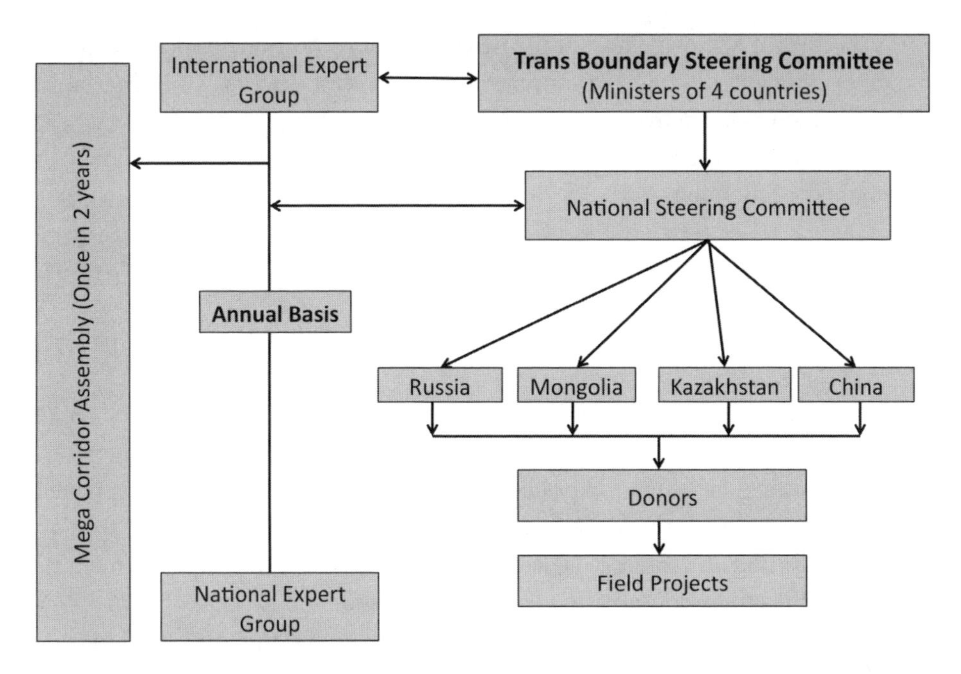

FIGURE 15.3 A governance model proposed for the Altai-Sayan Transboundary Mega Connectivity Conservation Corridor.

rect response to the threat of climate change to biodiversity. In 2010, Russian, Mongolian, and Kazakh participants of the international workshop, "Climate Change and Connectivity Conservation in the Altai-Sayan," agreed on a common vision and adopted an Action Plan for Connectivity Conservation for the Altai-Sayan (Worboys 2010).

In the context of climate change threats, the vision is to conserve these lands as a single mega connectivity conservation corridor. The Vision for the Altai-Sayan Mega Connectivity Conservation Corridor recognizes the special character of the Altai-Sayan (and Baikal) ecoregion: that it is of immeasurable significance to the culture and traditions of local people who express a strong desire for the region's conservation and protection. The many protected areas and natural areas interconnecting these lands offer an opportunity to conserve lands of immense ecological significance that extend over many degrees of latitude and longitude in the high mountain catchments of the Altai-Sayan.

In the action plan, a governance model for the Altai-Sayan megacorridor was also developed (fig. 15.3). Representatives of Kazakhstan, Mongolia, and Russia approved this proposed governance model at the 2010 Ust Koksa Workshop. By 2011, it had yet to be formally agreed to by the four governments represented in the transboundary area. The governance model envisages a cooperative four-nation steering committee responsible for achieving and implementing an agreed upon cooperative plan for the connectivity corridor. The committee would be supported at a ministerial level and would receive advice from experts and biannual meetings of stakeholders.

The initial steps of a process to achieve the Altai-Sayan Mega Connectivity Corridor include the following:

- conducting a comprehensive feasibility study for the connectivity conservation approach in the Altai-Sayan Ecoregion;
- identifying the boundaries of the megacorridor based on the principles of regional landscape zoning;
- developing a megacorridor web-atlas as an interactive tool box for connectivity conservation management and regional sustainable development;
- developing an adaptation policy model for connectivity conservation management and sustainable development;
- establishing an institutional and legislative basis for practical implementation of connectivity conservation management; and
- analyzing connectivity conservation management best practices and the establishment of working partnerships with other connectivity conservation areas.

The Altai-Sayan Mega Connectivity Conservation Corridor initiative has been inspired by the IUCN World Commission on Protected Areas (WCPA) (Mountains Biome) and supported by the management of a number of protected areas. The Institute of Geography of the Russian Academy of Science with official support of the IUCN-WCPA (Mountain Biome) has taken the leadership in consolidation of research and managerial efforts for establishment of the Altai-Sayan Mega Connectivity Corridor with its potential future extension further eastwards to Lake Baikal.

Conclusion and Recommendations

Large-scale transboundary connectivity conservation and its effective management are the main priorities for the conservation of the Altai-Sayan during a time of climate change, although other conservation mechanisms, such as new protected areas, are also critically important. The Altai-Sayan Mega Corridor is an important initiative that will permit the free movement of the argali, snow leopard, and other species between habitats that are effectively and cooperatively managed in the highest border areas and will help facilitate the survival of such species. It is a challenging initiative, for its effective implementation requires tremendous efforts in research, cooperative on-ground management, cross-border governance, and high-level political cooperation. Political support, at least at the ministerial level, is needed in four countries and is crucial for its success. The benefits of the Mega-Corridor include the conservation of endangered species, the conservation of the headwater catchments of some of the world's mightiest rivers, the enhancement of research, and positive cross-border conservation management that is adaptively responding to climate change. The 2010 Ust Koksa Workshop was a vital step toward the cooperative management of a new Altai-Sayan Mega Corridor that will help conserve two exceptional and endangered species.

It is recommended that the Altai-Sayan Mega Connectivity Corridor be formalized at the highest levels of governments and that cooperative on-ground planning for transboundary management be undertaken as early as possible by China, Kazakhstan, Mongolia, and Russia. With the agreement of the parties, there is also the opportunity to invite the IUCN World Commission on Protected Areas to help facilitate the early establishment of the corridor.

Chapter 16

Great Eastern Ranges, Australia

IAN PULSFORD, GRAEME WORBOYS, GARY HOWLING, AND THOMAS BARRETT

Australia has made a commitment to proactively conserve biodiversity in the face of climate change at a large-landscape scale. The first of a series of such efforts was the New South Wales government's decision to announce the Great Eastern Ranges Connectivity Conservation Corridor (hereafter referred to as the corridor) as a major land-use adaptation response to forecasted climate change effects in the region. To protect the corridor, the Great Eastern Ranges Initiative is an evolving partnership of organizations from Australian governments and nongovernmental organizations to businesses and private landowners.

Introduction to the Region

The Great Eastern Ranges (GER) forms a north-south corridor for more than 2,800 kilometers of essentially natural lands that extend along the eastern side of the Australian continent from the Atherton tablelands in north Queensland to the Australian Alps in southern Australia and beyond (fig. 16.1). Incorporating the two major geographic features of the Great Dividing Range and the Great Escarpment of Eastern Australia, the corridor interconnects and embeds an archipelago of protected areas that includes three World Heritage sites (Pulsford, Worboys, Evans, et al. 2010; Pulsford, Worboys, and Howling 2010; Mackey et al. 2010). This mostly unfragmented landscape helps conserve forest, woodland, and shrubland habitat across more than

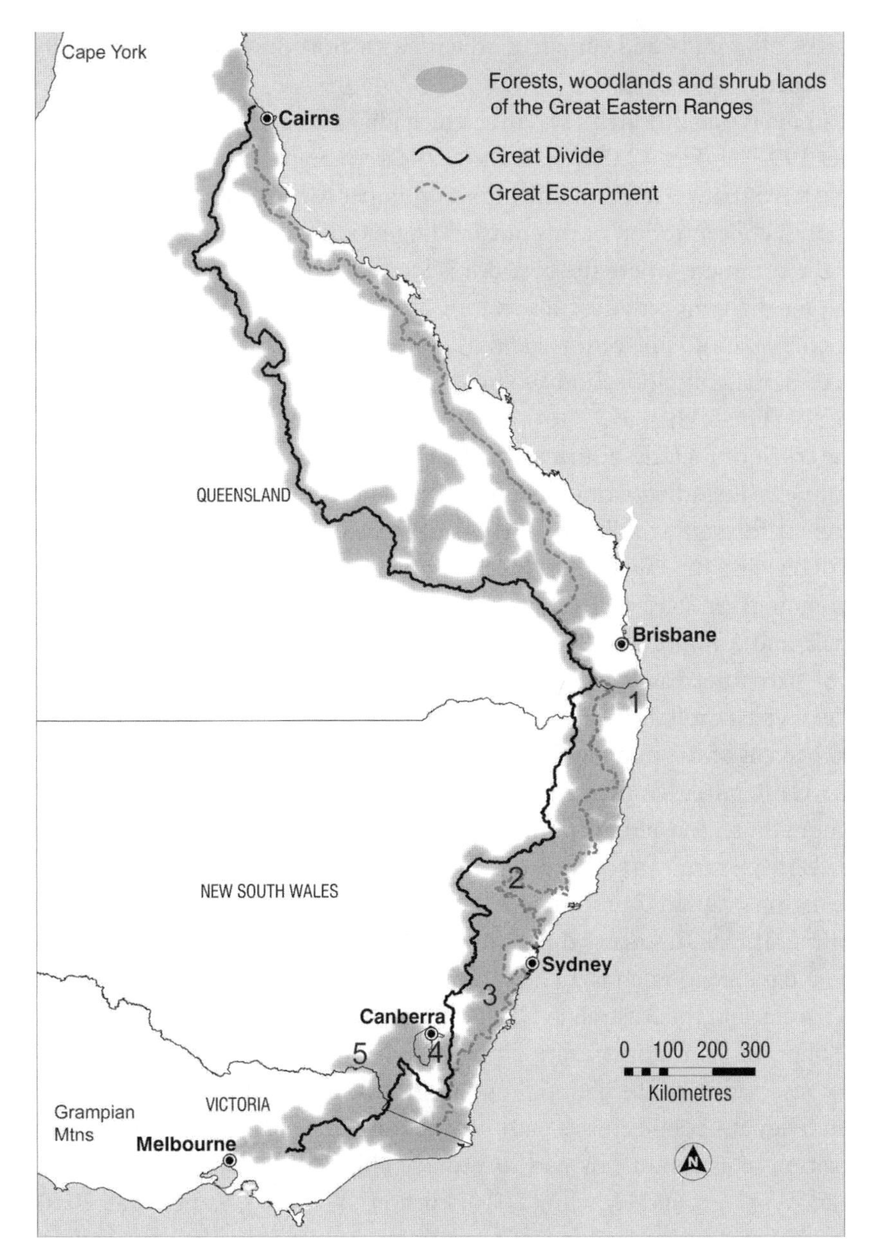

FIGURE 16.1 Potential connectivity of forests, woodlands, and shrubland of the proposed Great Eastern Ranges Corridor. There is a possibility of extending the corridor in the north to Cape York and to the Grampian Mountains in southwestern Victoria. NSW regional partnership areas: 1) Border Ranges, 2) Hunter Valley, 3) Southern Highlands, 4) Kosciuszko to Coast, and 5) Slopes to Summit. Source: Department of Environment, Climate Change and Water.

28 degrees of latitude and climate gradients with altitudes ranging from 100 meters to 2,228 meters.

Australia is one of Earth's seventeen megadiverse nations (Mittermeier and Mittermeier 1997) and its 22,000 flora and 6,794 vertebrate fauna species include 1,350 endemic terrestrial vertebrates, the highest number for any nation (Chapman 2005). The greatest concentration of this outstanding biodiversity is found along the eastern side of the continent where the corridor is located. The interconnection of protected areas along the connectivity conservation corridor assists species survival through habitat conservation, protection against habitat fragmentation, and the facilitation of species dispersal, migration, and evolutionary processes. This is particularly important at a time of climate change.

The corridor includes fourteen of the eighty-five terrestrial bioregions found on the continent, including tropical, subtropical, and cool temperate rainforests, grand tall eucalypt forests, woodlands, shrublands, grasslands, and alpine habitats (IBRA 2010). Receiving the highest levels of precipitation on the continent, the corridor's lands contain Australia's greatest concentrations of primitive flowering plants, birds, mammals, and amphibian species. Incorporating the headwaters or parts of the catchments of sixty-three east-coast rivers and the catchment areas of dams for all of Australia's east coast capital cities and major regional towns, they provide clean water to over 90 percent of the population of eastern Australia.

The GER initiative provides a national vision for interconnecting landscapes throughout the region and introduces a new land-use dynamic for Australia as a whole. It lies in close proximity to the greatest concentration of Australia's population, including the nation's capital Canberra and the state capital cities of Brisbane, Sydney, and Melbourne. It includes many different types of land tenures including private, community, and indigenous lands. In New South Wales (NSW) and potentially in Queensland, Victoria, and the Australian Capital Territory (ACT), the Initiative has created a unique opportunity for organizations in all public and private sectors to work together for the conservation of the nation's unique and precious natural heritage. This has enabled individuals, communities, and private organizations to contribute to effective conservation of natural lands and to provide effective responses to climate change (Pulsford, Worboys, and Howling 2010; Pulsford, Worboys, Evans, et al. 2010; Mackey et al. 2010).

Historical Overview of Conservation and Science Initiatives

The origins of this corridor concept date back to 1991 (Worboys 1996; Hamilton 2010), but its connectivity conservation design has been underpinned by eighty years of protected area establishment during the twentieth century, pioneered by conservationists, visionaries, and governments in three states and the ACT (Pulsford, Worboys, and Howling 2010; Pulsford, Worboys, Evans, et al. 2010).

Following decades of conflict over timber harvesting, Australian governments in 1992 adopted the Australian National Forest Policy, and established a comprehensive regional assessment process for harvesting areas including the Eden Region in the southeast of New South Wales in 1995. This resulted in the establishment of a "comprehensive, adequate and representative" reserve system (Commonwealth of Australia 1997) including a new South East Forests National Park in 1997 (Worboys 2005; Pulsford, Worboys, and Howling 2010; Pulsford, Worboys, Evans, et al. 2010). Combined with Victoria's Coopracambra National Park, this new park achieved 150 north-south kilometers of interconnected protected areas along the Great Escarpment where previously there had been an archipelago of smaller protected areas embedded in a landscape of state-owned production forests. It was the basis for a much longer connectivity corridor achieved under the Southern Regional Forest Agreement undertaken by the New South Wales Carr Labor Government and Australian Government in 2001, and further work extended the protected area corridor to 350 kilometers along the escarpment and then to 600 kilometers, from the Victorian Border to the Hunter Valley in central NSW (Pulsford, Worboys, and Howling 2010).

The initial corridor proposal of the NSW and Victorian section was first published in 2004 (Pulsford et al. 2004). At the 2006 meeting of the Environment Protection and Heritage Council of Australia and New Zealand, the Environment Minister for New South Wales, Bob Debus, proposed a 2,800 kilometer Australian Alps to Atherton (A2A) connectivity conservation corridor. The other environment ministers agreed to consider the matter, establishing an interstate government agency working group to provide further advice. On February 24, 2007, Minister Debus announced in the *Sydney Morning Herald* the "Alps to Atherton" connectivity conservation corridor as a major adaptation response to climate change and that the NSW Environment Trust had allocated A$7 million over three years to help implement the NSW section (Pulsford, Worboys, and Howling 2010; Woodford 2007).

From 2007 to 2010, a team under NSW's Department of Environment, Climate Change and Water (DECCW) achieved wide community awareness of the GER vision for the 1,200 kilometer NSW section through the development of partnership agreements, allocation of conservation grants, a website, films, presentations, workshops, educational information, and critical research reports. Although still formative in 2010, the Initiative continues to draw in new partners, including government agencies, nongovernment organizations (NGOs), community groups, and private citizens. Nationally, a working group of senior officers from state, territory, and the Commonwealth governments has assisted the Initiative by requesting an independent report on the scientific basis for establishment of the corridor (Mackey, Worboys and Watson 2010). A Memorandum of Understanding (MOU) between NSW-Victoria, NSW-Queensland, and NSW-ACT for the cooperative management and promotion of the corridor has been under discussion and negotiation for two years for possible signature and announcement in 2012.

With some exceptions, work on the corridor has made little headway in Queensland. In contrast, Victoria, NSW, and the ACT have made notable progress, where a system of interconnected protected areas in the Australian Alps is managed under a separate interstate MOU. Moreover, ACT officials have been supportive of the Initiative, and the Victorian government's *White Paper for Land and Biodiversity at a Time of Climate Change* recognizes the Initiative and extends the corridor much further west than initial descriptions (Vic DSE 2009).

The NSW DECCW identified five priority conservation areas for maintaining the integrity of the corridor. This led to the establishment of five geographically distinct partnerships with a number of respected NGOs, namely Greening Australia, the Nature Conservation Trust, National Parks Association of NSW, Bush Heritage Australia, and OzGREEN, all of which have played a lead role in facilitating the on-ground connectivity conservation work. These partnerships are also working together to identify conservation priorities, to develop conservation action plans, and to implement a range of voluntary conservation mechanisms (such as conservation covenants) and activities (such as restoration, fencing to protect habitat, weed management, and feral animal control). Land purchase for conservation is also important. With a facilitator coordinating each of the partnership groups, more than 100 other organizations have become involved. The groups ranged from eight to over thirty organizations that typically include catchment-management authorities, NGOs, local governments, government agencies, Aboriginal groups, private sector organizations, Landcare groups (a distinctly Australian conservation network), and others. With each group, these organizations work to align and integrate a range of conservation activities and mechanisms to achieve long-term conservation outcomes across all land tenure types.

In May 2010, a new partnership for the governance of the GER Initiative emerged. NGO conservation groups and NSW DECCW signed an MOU to provide ongoing management of the NSW section of the Initiative from July 2010. The partnership groups of NGOs (Greening Australia, OzGREEN, the Nature Conservation Trust, National Parks Association) and DECCW now have joint responsibility for the leadership and governance of the Initiative. A new director was also appointed by the partners, marking the beginning of a new, exciting, and challenging phase. The five partnership area facilitators have continued their operation.

The Initiative has been very well received by the citizens in the region. Voluntary involvement has been the key to this success and an essential basis for landholder and community involvement. The corridor concept just makes good sense to many people and organizations, particularly inasmuch as it pools the conservation resources to achieve success at a broad scale. Their involvement has benefited from a combination of (1) a willingness to do the right thing, (2) assistance from incentive payments, (3) advice assistance and other support services, and (4) in some cases the local social importance of being a member of the corridor "club."

Current Conservation in the Region

The 2,800 kilometer corridor includes a range of land-tenure types that interconnect many protected areas. In the ACT, the corridor extends across privately managed long-term leasehold public lands between protected areas, while it interconnects protected areas across private lands in NSW. In NSW, 38,861 square kilometers (or 37 percent) of the corridor is included in protected areas. A grand total of 55,392 square kilometers (or 52 percent) is on public land, while 48 percent is on privately owned land.

There are substantial areas on leasehold or private land in Queensland. Relatively few Queensland protected areas are interconnected by protected lands except in the Wet Tropics World Heritage Area in far north Queensland. In Victoria nearly 100 percent of the corridor is public land naturally interconnected through protected areas, state forests, or other public land.

Satellite imagery of the corridor identifies that it remains essentially interconnected by lands in a natural condition, although many "choke point" locations are problematic and require active management (fig. 16.2). Protected areas within the corridor receive the benefit of professional management in all of the states and the ACT to the level of the available resources. The majority of the protected areas are International Union for the Conservation of Nature (IUCN) Category I and II (nature reserves, flora reserves, and national parks), while in-perpetuity conservation agreements (easements), and parts of state forests apply to Category I and VI criteria, respectively. Although exploration and mining can be conducted on all lands other than protected areas, native vegetation on private land is subject to protection under individual state legislation throughout the corridor, which only allows for land clearing in limited circumstances.

Continuing land-use change, including land clearing and fragmentation of native vegetation for coal mining, industrial use, and urban development, remains the greatest threats to connectivity conservation for the corridor. These threats are greatest in the NSW Hunter Valley section and in parts of Queensland, where the protected area system is less developed and less interconnected with large gaps between reserves in many sections. Other significant threats include timber harvesting, intensification of agriculture, changing (viz., more frequent intense) fire regimes, invasive species, and other inappropriate land-management activities.

Regional Effects of Climate Change

The effect of climate change on Australia's biodiversity has been well documented (Steffen et al. 2009). The pattern of change varies across the Australian continent, with rainfall increasing in the northwest, decreasing in the southeast and eastern seaboard, and rainfall intensity and frequency increasing in the eastern tablelands

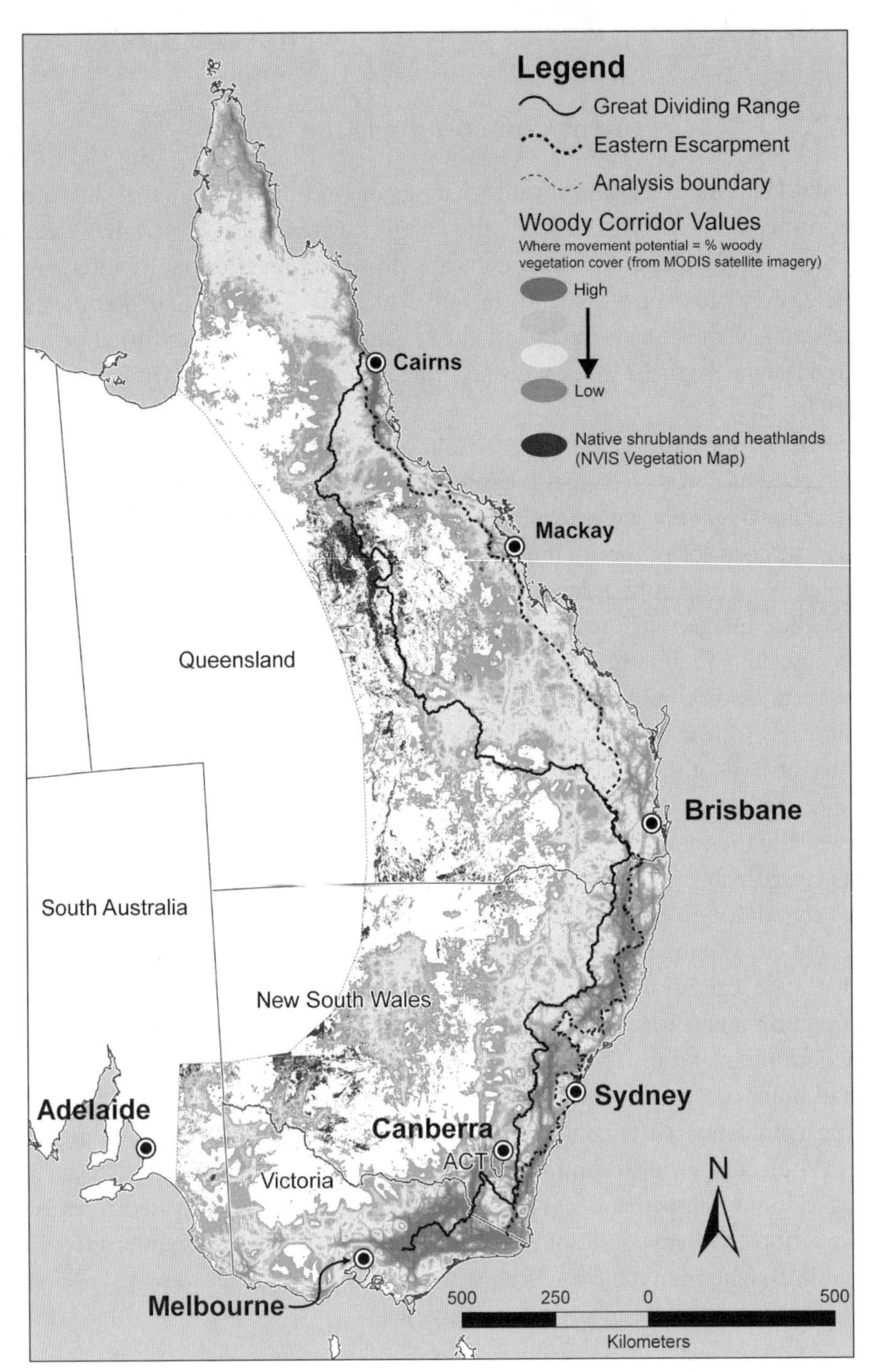

FIGURE 16.2 Connectivity of forest, woodlands, and shrublands in Eastern Australia, using a least cost-pathway analysis (Drielsma et al. 2007) based on "percent woody vegetation" measure derived from Moderate Resolution Imaging Spectroradiometer (MODIS) satellite imagery (Hansen et al. 2003). This figure also displays shrubland vegetation from the National Vegetation Information System (NVIS) Stage 1. The map illustrates that there are nonforest and woodland native-habitat connectivity values in the inland portion of the Great Dividing Range in Queensland, which was not detected by the MODIS percent woody vegetation measure. Connectivity analysis undertaken by the Department of Environment, Climate Change and Water's Landscape Modeling and Decision Support Section in 2009.

(McAlpine et al. 2007; Nicholls 2004; Steffen et al. 2009). The maximum winter snow depth in the Australian Alps section of the corridor has declined by 40 percent over the past fifty years (Green and Pickering 2002).

Scientists are forecasting continuing change, with the northern end of the corridor generally becoming warmer and wetter while the southern and alpine sections of the corridor experience longer hotter summers, including 20 percent less precipitation and the potential for more intense, frequent, and expansive wildfires. Droughts will be longer and hotter resulting in increased evaporation (Nicholls 2004). Shorter winters in the Australian Alps would mean a reduction in snow depth, duration, and cover, all of which would entail drying conditions, more frequent and extreme fires, extreme weather events, and invasive weeds (Worboys, Good, and Spate 2010; Nicholls 2004).

The corridor will require active management to deal with the climate-related increase and spread of introduced animals such as horses, deer, foxes, cats, and cane toads, as well as a large number of introduced plant species (Worboys, Good, and Spate 2010). Climate change is also likely to exacerbate the problem of disease vectors for both biodiversity (e.g., the *Phytophthora cinnamomi* fungus that can kill plants and the *Chytrid* fungus that impacts some frog species) and humans (dengue fever, Ross River virus, and Barmah Forest virus).

In response, the NSW DECCW, on behalf of an interstate agency working group, commissioned a landmark report on the scientific principles for connectivity and the Great Eastern Ranges (GER) Corridor (Mackey et al. 2010). The report identified the GER's immense importance for habitat conservation, species movement, species adaption to climate change and evolutionary development, and mitigating the effects of atmospheric carbon through native forest sequestration and the retention (nonclearing) of native forest carbon stocks. It also provided a strong scientific foundation for landscape scale connectivity conservation of Australia's unique ecosystems. Many of these principles have been incorporated into the national Caring for Our Country business plan and the draft NSW State Biodiversity Strategy. DECCW's connectivity team also spearheaded the analysis of years of accumulated data held by Birds Australia, which has demonstrated the corridor's critical role in the latitudinal, altitudinal, and seasonal migration of bird species (Smith 2010). In extreme seasons, it is a refugia site for many birds. With forecast biome shifts caused by climate change, the corridor has been shown to be a critical refugia area of habitat for species in the future (Mackey and Hugh 2010).

A close partnership between DECCW and several research organizations has generated important climate change research for the corridor using a wide variety of spatial data. Continental scale research and prediction on the possible impacts of climate change on biodiversity through modeling of the corridor is possible at coarse scales. Most of the relevant national data sets are coarse (rarely finer than 250 meter grid resolution), although finer scales are possible for sections of the corridor, such as the Border Ranges region of northern NSW. The key projects completed or under way include the following:

- ecosystem productivity and drought refugia analysis mapping to identify landscape ecosystems that are likely to function as drought refugia under climate change conditions (Australian National University);
- biodiversity and climate change modeling using Generalized Dissimilarity Modeling and analytical techniques to evaluate the expected contribution of on-ground actions to retaining compositional biodiversity under climate change (Commonwealth Scientific and Industrial Research Organisation);
- an analysis of vegetation condition and threat drivers to provide a framework for long-term forecasting and reporting of changing vegetation conditions (NSW DECCW);
- woodland management research to underpin best-practice guidelines for conserving temperate woodland in agricultural landscapes for landholders and conservation partners (Australian National University);
- prediction and decision strategies for managing extinction risks under selected climate change scenarios, including population viability models for a range of candidate plant and animal species (DECCW and Melbourne and Macquarie Universities);
- a science analysis of the corridor that identifies the opportunity for a continental scale climate change adaptation response (Australian National University); and
- a baseline condition assessment of bird movements along the corridor (Birds Australia).

Approaches to Conservation under Climate Change

In response to the global threat of climate change, both the federal and state governments established departments responsible for managing for climate change and created a draft Biodiversity Conservation Strategy that includes climate change considerations. The governments have funded a suite of expert reports on how protected areas, the Murray-Darling Basin, fire management, agriculture, fisheries, and other sectors will need to respond to climate change. The federal government's "Caring for our Country" program and investments in its protected areas continue progress toward the completion of Australia's National Reserve System and help to mobilize natural resource management with an eye to a changing climate. However, these initiatives are yet to be properly integrated into an urgently needed "National Defense Strategy for Australia." Although many regions have benefitted from the Caring for our Country program investments, national and state budgets in 2010 reduced funding levels for these programs and agencies.

Overall, in the last five years climate change has led to some realignment of governmental efforts across many sectors, including conservation. For example, climate change was a principal driver for the NSW government's facilitation of the Initiative in 2007 (viz., the allocation of A$7 million for connectivity conservation work for four years). The 2010 Federal Labor government has committed A$10 million for its Green Corridors Initiative to establish continental-scale connectivity conservation corridors as an adaptive response to climate change. For the corridor—as well as other Australian connectivity initiatives such as the Gondwana Link in Western Australia and Habitat 141 in Victoria and South Australia—this investment is a new approach, one in which NGOs have taken on a more prominent role in mobilizing on-the-ground action for large-scale connectivity conservation actions across the continent.

Within the GER, awareness of climate change has driven changing management practices for the corridor. Most important, the corridor vision has inspired people to act. It has empowered individuals to respond to climate change in an otherwise disempowering political environment. This has manifested itself in a new commitment on the part of individuals, communities, private property owners, and NGOs to volunteer their time and resources by providing valuable individual responses to the corridor. This common commitment and understanding will directly benefit a national climate change adaptation response. It is a new social phenomenon for Australia, one that will only grow.

Another broad-scale change for many protected areas embedded within the corridor has been a reconsideration of traditional protected areas management practices. Largely but not solely due to the threat of climate change, new twenty-first century approaches and tools include enhanced research on assessing changes in "base" conditions, integration of information from a range of organizations, identification of tipping points or thresholds of significant change, adaptive management, and managing for complexity and resilience (Worboys, Francis, and Lockwood 2010). Finally, for a range of sectoral land-use and management needs within the corridor, there is a new emphasis on partnerships and integrated responses to land management.

While there is some national leadership in governmental responses to climate change, far more is required if the current trends in biodiversity decline are to be arrested both nationally and within the corridor. In 2010, a majority of Australians were concerned about climate change and considered it to be a priority for government. Voting in the 2010 federal government elections reflected this when a Labor Government, which had previously shelved its promised climate change agenda, witnessed a flow of votes to the Greens. State and territory government responses have been mixed and sometimes schizophrenic with both the approval of coal mines and gas fields as well as the promotion of sustainable energy schemes. Broadly speaking, the state and federal governments of Australia are increasingly focused on reducing greenhouse gas emissions, building a prosperous low-carbon economy, and preparing for unavoidable changes in our climate.

With the renewed focus on climate change adaptation, the governments of Australia are increasingly looking to the prospects for connectivity conservation. Beyond the growing interest in the pioneering work of the corridor initiative and the Gondwana-link in southwest Western Australia, there has been increasing national acceptance of the importance of connectivity conservation by the governments of Australia. This is evinced by the following:

- the 2007 Australian Environment Protection and Heritage Council of Ministers agreement to consider the Great Eastern Ranges Corridor concept;
- the Victorian Government's White Paper on climate change adaptation that identified the importance of large scale "biolinks" (connectivity conservation areas) (Vic DSE 2009);
- the Commonwealth Government's 2010 election commitment for a National Green Corridors Plan and funding of A$10 million to implement a pilot program;
- climate change adaptation planning in 2010 by the Australian Alps Liaison Committee, which included specific connectivity conservation actions to help maintain the resilience of the Alps' catchments section of the corridor (Worboys et al. 2010); and
- the New South Wales (NSW) draft State Biodiversity Strategy's inclusion of the corridor as a backbone for a network of climate change corridors links across the coastal plain and the slopes and plains of NSW.

Conservation Planning

A number of conservation planning exercises have been completed for parts of or the entire corridor:

- Regional Forest Agreement Assessment processes for southern Australia in the 1990s that used nationally accepted criteria for establishing a comprehensive, adequate, and representative reserve system (Commonwealth of Australia 1997) in areas subject to timber harvesting on public land;
- a program to establish a national reserve system by the Federal Government based on a representative sampling of all eighty-six Australian Interim Biogeographic Regional Assessment Bioregions (Thackway and Cresswell 1995);
- a preliminary assessment of climate change impacts on the National Reserve System completed by CSIRO (Dunlop and Brown 2008); and

- Australia's first strategy addressing climate change and the National Reserve System by establishing priorities for reserve establishment (but not linkages) on a bioregional basis (NRPPC 2009).

Additionally, in NSW the Initiative has undertaken a range of spatial analyses at continental, state, and regional scales. Satellite imagery combined with vegetation mapping was used to identify connectivity (and gaps) on forest, woodland, and shrubland landscapes in eastern Australia. Species habitat corridor modeling based on the best available mapping was used to identify conservation priorities and to develop regional conservation action plans in partnership with regional community groups, agencies, catchment management authorities, and industry.

The NSW section of this corridor analysis identified five regional connectivity and buffer priority areas for more detailed research. Subject to a range of familiar land-use threats, these areas consisted of rainforest, forest, woodland, and shrubland communities mostly on private lands that interconnected the embedded protected areas along the corridor. The analysis identified strategic connectivity links that included climate and topographic gradients along the north-south corridor spine, with a network of east-west links from the coast to the woodlands on the inland slopes west of the Great Dividing Range.

Initiative Priorities

The Initiative is a large-scale climate change response, but it is yet to be matched by a supportive and coherent national policy and regulatory implementation environment. Basic operating financial resources are also a problem given that the NSW Environmental Trust funding ended during 2010. Fortunately, after a twelve months delay, an additional $4.5 million for another four years operation was announced by the NSW Government in December 2011. Other regions are yet to receive corridor-specific funding. These issues have helped to define the immediate priority actions for various actors involved in the Initiative, the most important of which include the following:

- the World Commission on Protected Areas (Oceania) of the International Union for the Conservation of Nature Resources should facilitate large-scale connectivity conservation initiatives by disseminating information about a range of supportive and incentive based instruments inspired by the GER corridor model;
- the NSW Nature Conservation Council should promote the implementation of integrated land-use policies and instruments that support the corridor Initiative;
- the Initiative-led partners in NSW should secure funding from government and philanthropic sources to maintain a small secretariat to coordinate the

Initiative and resources to continue on-ground works, especially for the established five priority areas;
- targeted research should further define priority management responses needed for maintaining the integrity of the corridor; and
- the Initiative and its integrated connectivity conservation approach should be embedded into the social and community fabric of Australia.

Roadblocks and Opportunities

For connectivity conservation, there are a number of roadblocks for including climate change in conservation planning. With the various governments supporting and implementing the corridor at very different levels, transboundary governance is challenging. For example, the particularly rapid and frequent change in environment ministers in NSW and other jurisdictions between 2007 and 2010 resulted in a loss of institutional memory and changes in priorities, all of which combined to stall effective transboundary and integrated management of the corridor.

The relative lack of tools and mechanisms makes achieving the conservation goals further challenging. The region has relatively primitive conservation and environmental planning policy and lacks support frameworks and instruments, which would otherwise help facilitate the implementation of the corridor as a climate change-adaptation response. Also, mechanisms to ensure continual and adequate financial support from governments for a governance entity and supporting GER Initiative secretariat are important. To date there has been no specific Commonwealth Government funding for connectivity conservation in the corridor. Additionally, opportunities for securing nongovernmental philanthropic support in Australia are less well developed than in, for example, the United States, making governmental funding critical. Progress is also hampered by an absence of (1) a standardized corridor-wide baseline condition data, (2) a capacity to monitor and track change in conditions, and (3) a coordinated research effort for GER (Worboys, Francis, and Lockwood 2010). Finally, inertia (or lack of leadership) by many players is problematic. This is partly due to the slow spread of information to the wider population, the impact of climate change skeptics, bureaucratic indifference, and vested interests holding back governments and agencies. There is also a natural caution and resistance to making paradigm shifts by some scientists, bureaucrats, and politicians in the development and implementation of new policy.

Conclusion and Recommendations

Connectivity conservation is a new paradigm for conservation in Australia and the world. Within conservation circles—and increasingly beyond—many have come to recognize that although protected areas are critically important as islands of conserved habitat, they will not adequately conserve biodiversity in a world subject to change. Climate change-induced biome shifts are occurring, and large-scale, interconnected protected areas provide greater opportunities for species survival through movement and adaptation. This means that large interconnected natural areas will be critical in society's response to the challenge of climate change.

The Great Eastern Ranges Connectivity Conservation Corridor Initiative has changed how many individuals, communities, planners, managers, and decision makers in the region think and act about conserving biodiversity in a climate change environment. It has helped introduce a new paradigm for conservation land use for Australia, one that has empowered individuals to make a difference when responding to climate change. It recognizes that multiple individual actions, guided by a shared vision and strategically applied, can significantly contribute to conservation, specifically, the conservation of a large-scale corridor. The Initiative has demonstrated that leadership from individuals, communities, NGOs, and others, especially when positively facilitated by governments, can provide a powerful contribution to national conservation outcomes. In the Australian conservation world, it has expanded accountability for species and habitat conservation from protected area management authorities and NGOs outward to the greater community.

Major climate change influences extend across many degrees of latitude in Australia and such big issues need big responses like the Initiative. Through enhancing species conservation, evolutionary adaptations, and movement, the corridor's protected areas and interconnections are critically important for the integrity of protected area systems of the east coast. The corridor is also helping to maintain healthy environments and healthy water supply catchments so vital for Australia's east coast communities.

Although Australians owe a great deal to a few forward-thinking leaders who have influenced land-use policy and practice, more champions from all parts of society are needed. National, state, and territory bipartisan political leadership and strategic national thinking are needed. So are progressive transboundary cooperation, greater private sector participation, partnerships, more private landowner champions, greater investments by government and industry, higher incentives, improved legislation, and more policy and supportive instruments. All of these investments are urgently needed to consolidate the Initiative for Australia's better future.

As a direct response to forecast climate change impacts to Australia, we recommend the following:

- the promulgation of new Australian legislation at national, state, and territory levels to institutionalize large-scale connectivity conservation and its effective management in the landscape (which could include concepts such as supportive instruments, development approval and conservation offset policies, incentives, private sector-government cooperation, and partnerships);

- a continued prominent leadership role for the Initiative as it continues to facilitate the management of connectivity conservation areas (while noting that the NSW Government has played an outstanding leadership role to date);
- a national assessment of strategic large-scale connectivity conservation initiatives—including the corridor—to develop a continent-wide Australian connectivity conservation response to climate change; and
- investments at national, state, and territory levels to institutionalize connectivity conservation as an Australian response to climate change.

Chapter 17

Madrean Sky Islands, North America

LAURA LÓPEZ-HOFFMAN
AND ADRIAN QUIJADA-MASCAREÑAS

The steep mountains, sheltered canyons, and grassland valleys of the Madrean Sky Island region of southwest United States and northwest Mexico is one of the world's most biologically diverse, temperate forest regions. A two-century legacy of mining and ranching has entailed significant impacts on the region's biodiversity, and today alterations due to climate change are already having visible impacts on species, ecosystems, and ecological connectivity. These changes, coupled with the global change drivers of increased urbanization and border security activities, will necessitate adaptive and forward-thinking strategies to conserve the region's biodiversity and to maintain connectivity across the border. The social, economic, and political challenges of straddling a border between two very different countries spotlight the need to foster both the institutional and scientific capacity for developing transborder conservation strategies that are adaptive to climate change.

Introduction to the Region

The term *sky islands* is widely used to describe continental mountain complexes, or inland archipelagos, comprising isolated mountain top "islands" surrounded by lower elevation valley "seas." The Madrean Sky Islands consist of roughly forty disjunct mountain complexes covering 180,000 square kilometers (fig. 17.1), with mountaintop elevations ranging from 1,800 to 3,267 meters. The mountains are separated by

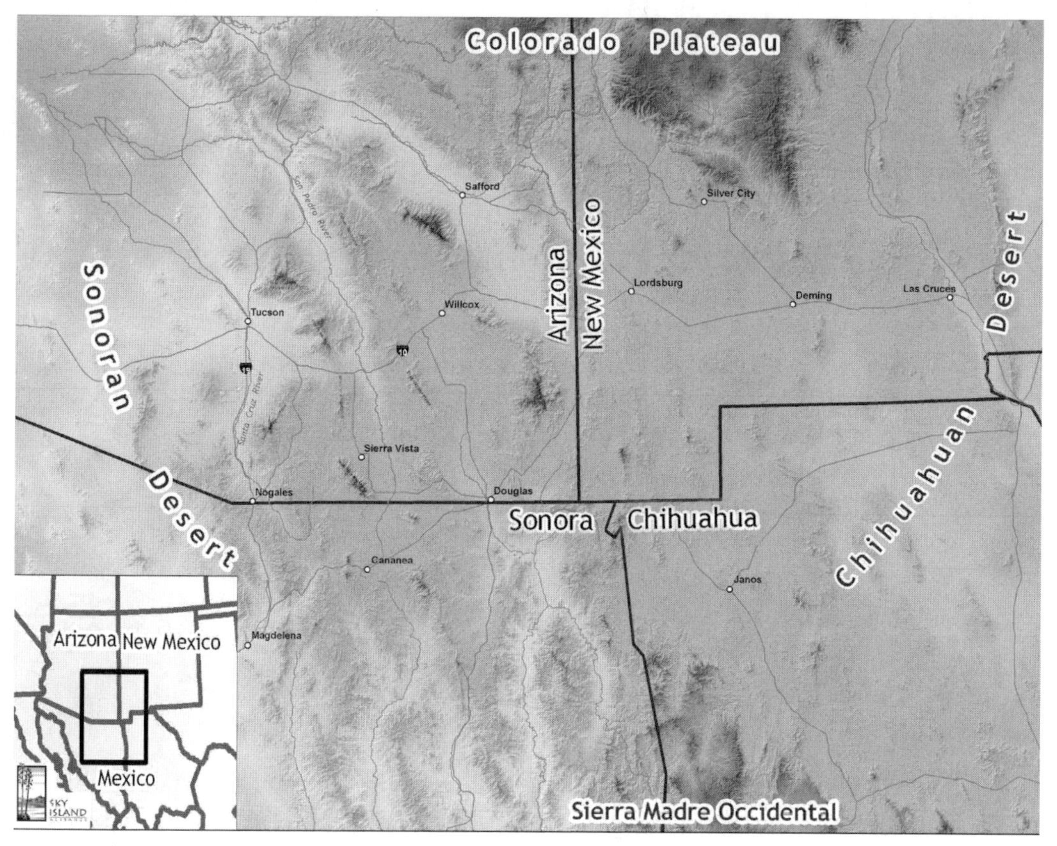

FIGURE 17.1 Map of Sky Island region (www.skyislandalliance.org).

15 to 25 kilometer-wide valleys of semidesert grassland, Sonoran desert upland, and Chihuahuan desert scrub. Valley elevations range from 750 to 1,400 meters, increasing west-to-east toward the Continental Divide.

The high degree of biodiversity in the Sky Islands region is due to two key factors: (1) steep elevation gradients and complex topography; and (2) biogeographic confluences between two climatic regions (temperate and subtropical), between two deserts (the Sonoran and Chihuahuan), and between two major cordilleras (the Rocky Mountains and Sierra Madre Occidental) (Riemann and Ezcurra 2007). Within the 250 kilometers span between the Rocky Mountains and Sierra Madre Occidental, many temperate and tropical species overlap at the edges of their distribution. The Sky Islands are the northern boundary for fourteen frost-susceptible subtropical plant family species (Felger and Wilson 1994), and eleven bird families reach either their northern or southern limits in the area (Warshall 1994).

Historical Overview of Conservation and Science Initiatives

The Sky Island landscape of today is the result of centuries, if not millennia, of interaction between people and the environment. Evidence of human activities dating back 10,000 years can be found from mountaintops to valley floors. Early inhabitants were likely seminomadic people who gathered plant resources and hunted big game, including mammoths. There is evidence of farming and substantial irrigation projects in the region by 1000 AD (Spoerl and Ravesloot 1994).

Ranching, one of the main drivers of environmental change in the region, originated with the Spanish colonial settlements and missions that began in the mid-1600s. The Spanish settlers raised livestock and spread them widely, clearing land for cattle grazing and raising wheat. The colonial period lasted until 1821, when the region became part of the new country of Mexico following its independence from Spain. Following the Treaty of Guadalupe Hidalgo in 1848 and the Gadsden Purchase in 1853, a newly drawn border divided the two countries, although both focused on exploiting the region's untapped riches. Irrigated agriculture and ranching expanded in both countries, and landscape fragmentation on both sides of the border accelerated as large tracks of land were cleared for cattle grazing. Mining, heretofore conducted on a small scale, burgeoned into an important regional industry, resulting both in the disappearance of hills and mountains into the expanding copper mine pits, and in the appearance of new mining boomtowns, such as Tombstone and Bisbee in Arizona and Cananea in Sonora. By the beginning of the 1900s, cattle barons and mine owners ruled the region (Varady and Ward 2009).

The land reform that followed the Mexican Revolution (~1910–20) entailed critical environmental consequences for the region: large ranches—many dating to Spanish colonial land grants—were partitioned into smaller, communally held lands called *ejidos*. As a result, the landscape was further fragmented and natural resources diminished by subsistence hunting, overgrazing of livestock, and timber exploitation (Culver et al. 2009). The environmental legacy of mining and livestock, including deforestation around mines, soil erosion in overgrazed areas, and diminished species populations, today remains a challenge to maintaining the region's ecological connectivity.

Between 1902 and 1907, forest reserves were declared throughout most of the Sky Islands of the United States, eventually becoming integrated into the Coronado National Forest. In 1936, Mexico designated the Sierra Los Ajos, Buenos Aires y La Púrica National Forest Reserve, and then in 1939 the Bavispe National Forest and Wildlife Reserve, the three separate areas of which are today collectively referred to as the Ajos-Bavispe Reserve (Búrquez & Martínez-Yrízar 2006). The reserve encompassed eight Sky Islands and the three most important watersheds in the state of Sonora: the Sonora, Bavispe-Yaqui, and San Pedro river basins. During the early part of the twentieth century, reserves in both countries were used to support timber and livestock production; as a result, most old-growth pine forests in the region disappeared by midcentury.

In 1994, in one of the first applied manifestations of the new "cores and corridors" approach to landscape conservation, conservationist Ton Povolitis designed a reserve plan for the US portion of the region (Povilitis 1994). The Sky Island Alliance and Wildlands Project followed in 2000 with a comprehensive framework for binational biodiversity conservation based on the protection of additional wilderness lands, the reintroduction of native species, and the conservation of landscape linkages between Sky Island ranges (Foreman et al. 2000). In 2004, The Nature Conservancy released an alternative conservation blueprint that highlighted priorities in valley locations centered on remaining grasslands and cienegas (Marshall et al. 2004). This plan in part led to the purchase of Rancho los Fresnos in northern Sonora, which is owned and comanaged by Naturalia, a Mexican nongovernmental organization (NGO) working to restore populations of the Mexican gray wolf and jaguar. In the same year, Conservation International added the Madrean Pine-Oak Woodlands region, of which the Sky Island region forms the northernmost portion, to its global list of Biodiversity Hotspots (Mittermeier et al. 2004).

Current Conservation in the Region

In the United States portion of the Sky Island region, the majority of higher elevation areas (>4,000 feet) are managed by the 1.78 million-acre Coronado National Forest, including ten wilderness areas. The US National Park Service also manages large wilderness areas within Saguaro National Park and Chiricahua National Monument. In all, 595,000 acres are managed as wilderness across different US federal jurisdictions. In Mexico, the main protected area is the Ajos-Bavispe Reserve, encompassing 445,000 acres (1,800 square kilometers) across three areas in northern Sonora. Due to its ruggedness, the Ajos-Bavispe Reserve has remained relatively protected.

In both countries, Sky Island valleys contain a mosaic of land tenures. In the United States, this includes lands held by the Bureau of Land Management, state trusts, and private owners. Within Mexico, most lands outside of the Ajos-Bavispe Reserve are privately or communally owned. Across all jurisdictions and in both countries, including privately owned land, approximately 1,500,000 acres of land are managed for conservation, representing 10 percent of the total land base.

Mining, exurbanization, and ranching remain forces of landscape change in the Sky Islands today. Mine development and expansion continues to grow due to increasing worldwide demand for copper, leading to controversy in both countries about environmental impacts (Hartman and Farr 2010; Reuters 2011). In Arizona, rapid development and exurban growth constrains conservation options for maintaining wildlife linkages between Sky Island mountain ranges. The US portion of the region is currently growing in human population at ~25 percent per decade, roughly two and a half times the national average. The growth is partly due to retirees from other parts of the country who are drawn to the region's moderate winter climate. In Mexico, growth is concentrated within existing urban areas and is not as much a threat to landscape connectivity in the Sky Islands.

Although ranching has degraded land in the region, many conservationists are examining ranching in a new light, championing the idea of "working landscapes." This growing perspective is based on the understanding that well-stewarded ranches can act as ecological buffers between protected areas and the pressures of urban areas (Knight 2009), and that the accelerated subdivision of ranch lands for exurban residential and recreation development will cause more serious and long-lasting habitat fragmentation than careful ranching (Nabhan et al., forthcoming). In the United States, a prominent example of this new approach is the Malpai Borderlands Group, a landowner-driven organization promoting ecosystem management for unfragmented landscapes in southeastern Arizona and southwestern New Mexico. In the Mexican portion of the region, a growing number of landowners, such as Carlos Robles of El Aribabi ranch in Sonora, are managing their ranches for small game hunting and bird watching, as well as cattle raising. In March 2011, the Mexican National Commission of Natural Protected Areas (CONANP) designated El Aribabi ranch as a Natural Protected Area under the Voluntary Land Conservation program. The designation protects 10,000 acres for biodiversity conservation, environmental education, and ecotourism. Such examples offer promise of ways to maintain ranching, ecological connectivity, and biodiversity in light of other threats.

The challenge of maintaining landscape and wildlife habitat permeability across the United States–Mexico border has been recently complicated by the surge in border security infrastructure and activities in the last decade (Flesch et al. 2010; Quijada-Mascareñas et al. 2011). Since 2000, pedestrian fencing (solid walls) and Normandy-style (i.e., antitank) vehicle barriers have been constructed on more than 60 percent of the United States–Mexico border. The barriers are accompanied by access roads—essentially swaths of cleared land up to sixty feet wide (Laura Lopez-Hoffman, pers. observ.). In the Sky Island region, the pedestrian fencing is mostly restricted in urban areas, with vehicle barriers in rural areas. As most areas of rugged, steep terrain do not have barriers, wildlife habitat connectivity has been mostly impaired in valleys. The lack of connectivity is particularly of concern for species such as black bear, jaguar, ocelot, and bobcats, which are more abundant on one side of the border than the other. This imbalance would put the less-abundant population at risk of extirpation if connectivity were cut off across the border (Culver et al. 2009). Because the Real ID Act of 2005 gave the secretary of the US Department of Homeland Security the right to waive laws in order to hasten border wall and road construction, environmental impact assessments on the impacts of wildlife and habitat have not been undertaken, as would normally have been the case under the US National Environmental Policy Act. An effort led by scientists from the US Geological Survey to develop a comprehensive protocol for monitoring the environmental impacts of border infrastructure and activities on wildlife, vegetation, and transboundary watershed hydrology has been languishing in review.

Regional Effects of Climate Change

The Sky Islands' climate is generally arid and warm, punctuated by rain arriving from the west in winter and cyclonic storms arriving from the south in summer (Adams and Comrie 1997; Sheppard et al. 2002). Mean annual temperatures and precipitation levels vary with elevation and range from 8 to 23 degrees Celsius and 250 to 800 millimeters, respectively. The percentage of total precipitation falling in winter decreases from west to east, whereas the percentage falling in summer decreases from south to north.

The Intergovernmental Panel on Climate Change and regional analyses suggest that the area encompassed by southwestern United States and central and western Mexico is likely to undergo significant precipitation and temperature changes throughout the twenty-first century. Mean annual precipitation is projected to decrease 5 to 10 percent, with the greatest decrease during winter months (Solomon et al. 2009; Overpeck and Udall 2010). Mean annual temperatures are increasing and projected to continue through coming decades, with consensus of a 3.0 to 3.5 degree Celsius annual mean temperature increase by 2099. This increase will not be evenly distributed throughout the year; summer high temperatures are projected to increase more than summer average temperatures, indicating greater variability in temperature extremes (IPCC-WG1 2007; Overpeck and Udall 2010).

Changing precipitation and temperature regimes are expected to affect the ecology of the Sky Islands in two main ways: (1) in the near term, fire regimes are expected to change, facilitating the spread of invasive species and eventual homogenization of the species pool; (2) over time, there will be shifts in species bioclimatic habitat envelopes—geographic spaces of climatically suitable habitat—which, within the physiography of Sky Islands, may result in "winning" species or "losing" species (i.e., local extirpations).

Fire is a fundamental ecological-organizing process in the Sky Islands. Grassland and forest fires regulate the distribution and abundance of vegetation, cycling of water and nutrients, soil formation and erosion, and carbon dynamics (DeBano and Ffolliott 1995; Falk et al. 2007). Fires influence ecosystem-level response to changing environments by resetting successional processes (Suding and Goldberg 2001). A changing climate may alter weather variables of importance to fire. For example, current projections suggest substantially decreased snowpack depth, density, and persistence, causing earlier soil drying and potentially longer fire seasons. Changes in the timing and total rainfall of the summer monsoon could also influence the prevalence of thunderstorms and, in turn, lightning ignitions of fires (Lin et al. 2007). Climate change may also alter the amplitude and duration of drought episodes: if mean temperatures increase, then the probability of extended periods above temperature and flammability thresholds will also increase (Brown et al. 2004).

The abundance of invasive, nonnative grasses at low elevations, such as buffelgrass (*Pennisetum ciliare*), may create new fire-spread pathways into higher elevation areas. As previously low-elevation fire regimes become more frequent at higher elevations, new pathways will be created for the spread of invasive species into higher elevations and the species pool of higher elevation areas might become homogenized as invasive grasses become more prevalent.

Along montane elevation gradients, area increases with ascent from the lowest elevations until reaching a centroid point (area-weighted midpoint). Above the centroid, available area decreases with ascent (Quijada-Mascareñas et al., in review). Due to this relationship between area and elevation, some Sky Island species will experience an expansion in habitat area with temperature-driven ascent under warming climate, whereas some species will experience a decrease in habitat area on a given mountain. Under ideal conditions, the latter would be able to shift to higher latitude mountain ranges with the same bioclimatic envelope conditions (Parmesan et al. 1999). However, the ability to shift latitudinally requires the ability to disperse across valleys beset by habitat loss and fragmentation due to urban growth, agriculture, mining, and ranching. Thomas et al. (2004) suggest that montane, range-restricted species that are impeded from free dispersal are at risk of extinction.

Approaches to Conservation under Climate Change

In a region such as the Sky Islands, which is divided by an international border, efforts to maintain landscape connectivity in the face of drivers of global change—climate change, exurbanization, border security, and mining—must transcend the political line. Despite the difficulties involved in developing cross-border conservation (Chester 2006; López-Hoffman et al. 2009), recent scholarship on climate change and water issues across the United States–Mexico border suggests that regional and local efforts, spearheaded by collaborations between civil society and governmental rather than purely national diplomatic initiatives, can build cross-border resilience to climate change (Wilder et al. 2010).

Three factors are critical to effective cross-border adaptive capacity (Wilder et al. 2010):

- Shared social learning is a common understanding of challenges among individuals and/or institutions. In this context, "platforms for shared social learning" refers to mechanisms by which the development of common conceptual understandings of climate change challenges can occur from the scale of individuals and NGOs to that of state and regional authorities across the international border.

- "Communities of practice" are bridges of information flow (i.e., networks) both across the border and across existing communities (i.e., associations based on shared identity, values, and practices). They are facilitated by individuals described as "network weavers" who can bring communities from both countries together (Laird-Benner and Ingram 2011), or by "boundary objects," meaning meetings or documents joining distinct communities in a common purpose (Wenger 1999).
- "Coproduction of knowledge" arises from a synergistic relationship between researchers on the one hand and stakeholders on the other to create "usable" science for framing policy (Lemos and Morehouse 2005). In a cross-border context, coproduction of knowledge requires decision makers, stakeholders, and scientists from both countries to work together to produce policy-relevant research.

Grounded in these three factors, the following discussion briefly reviews initiatives within each country to address landscape-level impacts of global change (in particular climate-related changes), and then provides a preliminary assessment of the present adaptive capacity of regional conservation initiatives and institutions. It concludes with recommendations for enhancing this capacity in the face of future changes.

Several recent initiatives by the US Department of the Interior (DOI) promise to support the development of adaptive capacity within the US portion of the Sky Island region. Under secretarial order 3289, the DOI mandated the establishment of eight Climate Science Centers around the country to promote climate change research. One of the centers will be established at the University of Arizona's School of Natural Resources and Environment, making it likely that new research will focus on the Sky Island region (DOI 2011). At the time of this writing, given that the center has yet to form, it is not clear whether the center's research will focus on basic science on climate change in the region or will engage in "coproducing" interdisciplinary knowledge that is directly applicable to policy making.

The order also mandated the establishment of Landscape Conservation Cooperatives (LCCs). The cooperatives are the applied science side of the DOI's plan to develop a coordinated, science-based response to climate change impacts on land, water, and wildlife resources. Cooperatives are intended to be communities of practice—termed "management-science" partnerships by the agency—involving scientists from academia, agencies, and NGOs that promise to support shared social learning about climate change (USFWS 2011). The Bureau of Reclamation and the US Fish and

Wildlife Service are currently in the process of defining the Desert LCC, which is to encompass portions of five US states, three deserts (Mojave, Sonoran, and Chihuahuan), and several large river systems, including the Colorado River Basin. The agency would like to eventually consider portions of the Mexican states of Sonora, Chihuahua, and Coahuila as well. Some of the stated goals of the Desert LCC are to understand the effects of long-term drought on the composition, abundance, and distribution of species; reduced water availability on vegetation, wildlife, and human populations; increased temperatures on insect outbreaks and tree mortality; soil dryness and increasing air temperature on wildfire susceptibility; and changing fire regimes caused by increased invasion of nonnative grasses. As of spring 2011, the LCC staff has held several scoping meetings with agency, conservation organizations, and university stakeholders and has formed a stakeholder-steering committee (BOR 2011). If developed as intended, the Desert LCC should create a platform for shared social learning and a community of practice around climate change adaptation within the United States. While an agency official from Mexico's Instituto Nacional de Ecologica is on the steering committee, it remains to be seen how effectively Mexican stakeholders will be incorporated.

In addition to the Department of the Interior initiatives, which are still in a formative stage, the Sky Island Alliance, a regional NGO, is taking a lead in facilitating a community of practice on adaptation to climate change. Specifically, this group is in the process of a four-year initiative to identify organizational and landscape vulnerabilities to climate change and to develop strategies for addressing vulnerabilities. It is convening a series of workshops that are serving as boundary objects for bringing together diverse stakeholders from other NGOs, agency officials, and university researchers. The group held its first workshop in late 2010 and intends to hold two more workshops over the next two years. While most of the participants to date have been from the United States, the Sky Island Alliance organizers are committed to developing a cross-border dialogue in future workshops. Several of the organization's staff and board members (e.g., Sergio Avila and one of the coauthors, A. Quijada) have deep ties with Mexican NGOs, stakeholders, and agencies, and they should be able to function as cross-border network weavers.

On the Mexican side of the border, progress is being made to develop strategies for adapting to climate change (SEMARNAT-CONANP 2010; Locatelli et al. 2011). In accordance with the strategic objectives of Mexico's National Program for Protected Areas 2007–12, the National Commission for Protected Areas has developed a Climate Change Strategy for Protected Areas. The objectives of this program include (1) increasing the adaptive capacity of ecosystems and the communities in protected areas in the face of climate change, and (2) contributing to the mitigation of greenhouse gases and enhancement of carbon stocks via carbon-capture strategies. While the plan establishes strategies and guidelines for the commission's management decisions, it recognizes the importance of incorporating key stakeholders in such processes as well as strengthening technical and institutional capacities in climate change issues. The plan will be initially implemented in the Sierra Madre Sky Islands but has the potential to be applied toward other binational efforts in the region in the coming years.

Conclusion and Recommendations

As human activities increasingly fragment the landscape and reorganize species composition and distribution in the Sky Islands, concerted action is needed to develop binational strategies for dealing with (1) changing fire regimes, (2) invasive species control, and (3) the genetic isolation and potential loss of species due to constricted ranges as a result of both climate change and border infrastructure. Government agencies, conservation leaders, and organizations in the Sky Islands are making great progress in developing communities of practice and platforms for shared learning about environmental and climate-related drivers of change within each country. A next step in these efforts is to develop mechanisms to coproduce usable scientific knowledge for framing policy. Furthermore, these organizations and initiatives must begin to create the cross-border collaborations necessary to build binational adaptive capacity and maintain future ecological connectivity across the border.

Ultimately, it is critical that the Sky Island Alliance and the Desert Landscape Conservation Cooperative live up to their stated intentions of developing the cross border communities of practice and platforms for shared social learning that will be fundamental in devising effective binational strategies for protecting transboundary connectivity. In this regard, it is incumbent upon the individuals within those organizations who have binational expertise to rise to the challenge and become border weavers, capable of fashioning cross-border, collaborative strategies for dealing with the global-change drivers—ranching, mining, border security, and climate change—that threaten the region.

Acknowledgments

The authors thank L. Misztal and M. Emerson of Sky Island Alliance for information about their climate change-adaptation program.

Chapter 18

The Northern Appalachian/Acadian Ecoregion, North America

STEPHEN C. TROMBULAK, ROBERT F. BALDWIN, JOSHUA J. LAWLER, JEFFREY HEPINSTALL-CYMERMAN, AND MARK G. ANDERSON

The Northern Appalachian/Acadian ecoregion in northeastern United States and southeastern Canada is projected to experience dramatically increased temperatures by the end of the twenty-first century, potentially driving numerous changes in species distributions throughout the region. For species to respond to such changes, landscape-scale conservation planning must result in increased levels of connectivity both within the ecoregion and with neighboring areas. Numerous initiatives have sought to promote ecological health and connectivity throughout all or a part of this ecoregion, particularly Two Countries, One Forest, a binational umbrella organization. Work in the region suggests the need for increased attention to be given to planning for linkages across landscape scales to allow for both short- and long-term movement of species, and for coupling connectivity with efforts to enhance ecosystem resilience throughout the reserve system and the surrounding matrix.

Introduction to the Region

The Northern Appalachian/Acadian ecoregion (Anderson et al. 2006) encompasses over 330,000 square kilometers in the northeastern United States and southeastern

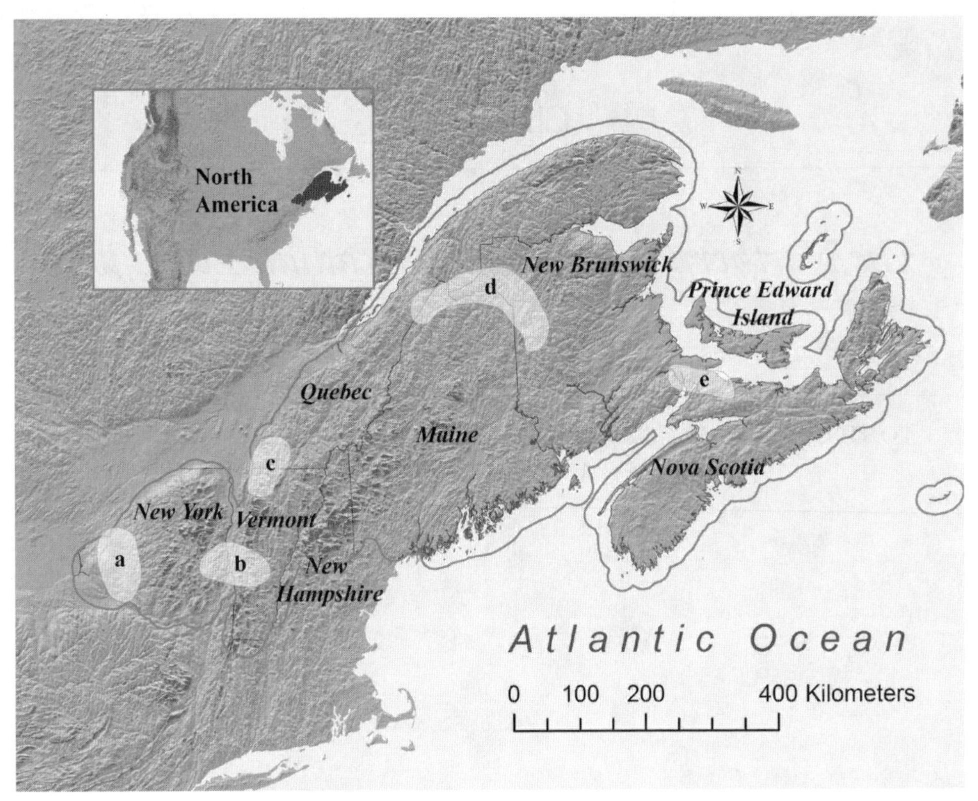

FIGURE 18.1 The Northern Appalachian/Acadian ecoregion as delineated by The Nature Conservancy and Nature Conservancy Canada. Also shown are the priority linkage areas identified by Two Countries, One Forest: (a) between the Tug Hill Plateau and the Adirondack Park, (b) between the Adirondack Park and the Green Mountains, (c) between the Green Mountains and the Sutton Mountains, (d) across the Saint John River Valley, and (e) across the Chignecto Isthmus.

Canada. It includes the Adirondack Mountains and Tug Hill Plateau in northern New York; all but the extreme southern portions of Vermont, New Hampshire, and Maine; Québec, south of the St. Lawrence River; and all of New Brunswick, Nova Scotia, and Prince Edward Island (fig. 18.1).

Both historically and currently, the ecoregion is predominantly forested with northern hardwood forests (beech-birch-maple assemblages) forming the majority of the matrix in the south and at lower elevations, and spruce-fir forests in the north and at higher elevations. It is an ecological transition zone between boreal forests to the north and warm temperate forests (e.g., oak-hickory assemblages) to the south.

The area is geologically complex, with bedrock varying dramatically in chemistry and structure across the region. Its diverse topography includes extensive coastlines, numerous inland mountain ranges (with peaks up to 1,900 meters), and glacially carved valleys. Extensive and periodic glaciations have resulted in surficial geology dominated by glacial till and outwash deposits, which give rise to thousands of swamps, bogs, lakes, and ponds, in addition to nearly 110,000 kilometers of rivers and streams. Climate in the ecoregion reflects both strong inland continental influences, especially from the Great Lakes to the west, and cool, maritime influences from the Atlantic Ocean to the east.

The ecoregion is as diverse culturally as it is biogeographically. Paleoindians colonized the land on the heels of the last retreating glacier between 15,000 and 10,000 years ago (Klyza and Trombulak 1999). Over the resulting interglacial period, a succession of cultural transitions then mirrored the ecological transitions that took place as the climate warmed and biological succession occurred: Paleoindian cultures evolved into Archaic cultures, which then evolved into Woodland cultures, which ultimately gave rise to at least seven separate First Nations, most of whom still live here today (Klyza and Trombulak 1999).

Even though evidence suggests that both Vikings and Portuguese occupied the ecoregion in pre-Columbian times, sustained European colonization did not begin until 1600s with the French in Québec and English in the United States and the other Canadian provinces. While the two countries—and even the states and provinces within each country—differ dramatically with respect to sociopolitical traditions, the entire cultural landscape is strongly dependent on the region's natural resources, particularly timber, agricultural products, and fish. In turn, the ecological landscape today is a virtual tapestry of past and current land uses (Foster and Aber 2004).These ecological and cultural traits create a shared vulnerability to future changes, particularly in a warming world.

Historical Overview of Conservation and Science Initiatives

As might be expected for a region with as many diverse political subdivisions as the Northern Appalachian/Acadian ecoregion, a plethora of conservation organizations and agencies have launched numerous landscape-scale conservation initiatives over the past several decades. Until recently, the most geographically diverse of these was the federally funded Northern Forest Lands Council process (1990–94), which sought to strengthen both conservation and the forest-products industry in the "Northern Forest" region of New York, Vermont, New Hampshire, and Maine (Trombulak 1994). The work of the Northern Forest Lands Council culminated in a suite of recommendations to foster stewardship of privately owned forestlands, to protect "exceptional" natural resources, and to strengthen the economies of rural communities in the region (NFLC 1994). Although most of the recommendations made by the Northern Forest Lands Council were either ignored or proved to be ineffectual, the council's process was largely successful in creating public and administrative awareness of the inextricable connections among the states in the region (although not with the provinces across the border with Canada) and the connections between rural economic development and ecological health. It also transformed general perceptions of what was at stake in the ecoregion and how broad the focus for achieving conservation action needed to be.

In 2001, a group of conservation advocates formed Two Countries, One Forest (2C1Forest) with the explicit mission "to protect the natural beauty, native species and ecosystems of the Northern Appalachian/Acadian ecoregion while maintaining economically healthy and culturally vibrant local communities" (Trombulak and Baldwin 2010, 12; Bateson 2005). 2C1Forest was formed on a hybrid model: on the one hand, it was a confederation of existing conservation organizations active within the eight province/state regions, including The Nature Conservancy, Nature Conservancy of Canada, Canadian Parks and Wilderness Society, Wildlife Conservation Society, Wildlands Network, and Northeast Wilderness Trust. On the other hand, 2C1Forest was also a distinct organization with its own board of directors, staff, and strategic plan. Its creation was, in part, a response to the recognition that cross-boundary coordination and cooperation are crucial to address the kinds of landscape-scale conservation challenges posed by (1) species whose survival depends on healthy ecosystems across all of the states and provinces in the ecoregion, and (2) climate change.

The primary focus of 2C1Forest's work has been to develop spatially explicit tools for strategic conservation planning throughout the ecoregion. Over a nine-year period, its Science Working Group developed (1) a high-resolution map of human transformation of the landscape, (2) forecasts of future transformation under different scenarios of human population growth and road and amenities development, (3) an assessment of the ecological importance of each location within the region, and (4) a prioritization scheme for locations based on both their current and future transformation and their ecological importance (Trombulak et al. 2008). Although it is too soon to assess the full scope of actions that will arise from these analyses, they have already provided much of the preliminary analyses used by Staying Connected, a multistate, multiorganization/agency collaborative launched in 2009 to promote wildlife connectivity across the US portion of the ecoregion, including its connections into Canada (Conservation Registry 2011).

Current Conservation in the Region

Given the diversity of governmental authorities and land trusts involved in acquiring and managing protected areas in this ecoregion, protected area coverage is shaped by numerous social and ecological factors, resulting in a diverse, haphazard assemblage of parks, managed lands, and private conservation holdings. Thus, understanding the status of protected areas in this region is difficult. Furthermore, the two countries differ in their classification systems for conservation lands: a form of International Union for the Conservation of Nature (IUCN) protected-area management categories apply in Canada, whereas US Geological Survey's Gap Analysis Program (GAP) categories apply in the United States. The absence of an unambiguous crosswalk between the two systems makes it difficult to compare levels of protection across the region (USGS 2011). However, The Nature Conservancy and Nature Conservancy of Canada have developed and regularly update a comprehensive protected areas map for the region (Anderson et al. 2006) with designations in Canada individually matched to a corresponding GAP classification in the United States. On this basis, and as of 2006, across the region approximately 7 percent of the ecoregion was exclusively devoted to biodiversity protection (GAP status 1 and 2: areas that have permanent protection from conversion of natural land cover and management plans that maintain a fully or primarily natural state of existing natural communities). Another 28 percent was secured from conversion to development (e.g., Crown or public land, privately owned conservation areas). The proportion of land secured from conversion to development (GAP status 1–3: as described, plus lands that have permanent protection from conversion but may be subject to extractive uses, and their equivalents on private land) is three times greater than that of land converted to agriculture or development. This is the only ecoregion in the eastern United States where land permanently secured from conversion to development is higher than converted lands. As is typical across the continent, most secured lands are in mountainous areas, while coastal regions and lowland valleys are the least protected.

Assessment of the status of corridors is more difficult. 2C1Forest identified five key locations within the region where conservation initiatives to promote ecological connectivity are a priority: (1) between the Tug Hill Plateau and the Adirondack Park in western New York, (2) between the Adirondack Park and the Green Mountains in central Vermont, (3) between the Green Mountains and the Sutton Mountains in southern Québec, (4) across the Saint John River Valley between northern Maine and southern New Brunswick, and (5) across the Chignecto Isthmus between New Brunswick and Nova Scotia (fig. 18.1). Various initiatives are already in progress to secure connectivity in most of these locations, including a number of local projects under the Staying Connected initiative. Also, groups such as the Canadian Parks and Wilderness Society have worked for years to improve connectivity across the Chignecto Isthmus. To date, however, none of these initiatives has been fully implemented, and connectivity in each location is at risk from continued land conversion and road construction.

Assessment of the matrix lands is equally problematic. Contiguous and ecologically complete forest ecosystems that once dominated the ecoregion are now largely young, simplified, and increasingly fragmented by timber harvesting, roads, and development (Foster et al. 2010). However, the extent of forest cover—albeit no longer in its original ecological condition—has mostly increased since the extensive deforestation throughout the region in the nineteenth century. As a result, excluding developed land, agricultural land, and roads, more than half of the region has over 80 percent natural cover. Further, Woolmer and colleagues (2008) conducted an assessment of the region's human footprint defined as a spatially explicit, multivariate measure of landscape transformation that includes assessment of human settlement, roads, and land use. This analysis showed that more than 90 percent of the ecoregion has a human footprint of less than 50 percent or, in other words, that the majority of these places are still intact. Almost 50 percent of the region has a human footprint of less than 20 percent, indicating that for the majority of the overall region, human activities are comparatively minor.

Regional Effects of Climate Change

Assessments of projected changes within the ecoregion have been conducted with respect to both climate and biological parameters. In regard to the former, 2C1Forest assessed projected changes in climate parameters with the use of Climate Wizard (Girvetz et al. 2009). Projections of changes by the end of the twenty-first century (2061–90) were made based on three General Circulation Models (CSIRO, Miroc, and HadCM3) and three emissions scenarios developed by the Intergovernmental Panel on Climate Change (A2 [high levels of emissions], A1B [intermediate], and B1 [low]) (box 2.1). Climate data were downscaled using a two-step, bias correction—statistical spatial downscaling method (Wood et al. 2004; Maurer et al. 2007).

Broad agreement in predicted climate effects across these three climate models and emission scenarios allows for a description of the effects based on an ensemble model that averages the median values for the nine separate climate model-emission scenario combinations we assessed with Climate Wizard. In general, ensemble climate projections strongly indicate that by the end of this century, the entire region will be warmer throughout the year. Both maximum and minimum temperatures are expected to increase between 2 and 4 degrees Celsius across the region. Increases in maximum temperature will be greatest in the summer (June–August) and fall (September–November) and greatest for minimum temperatures in winter (December–February), with departures from the present norm by as much as 4 to 6 degrees Celsius across the entire northern tier of the ecoregion. Not surprisingly, the magnitude of projected departures is strongly affected by whether future emissions more closely resemble the high, intermediate, or low scenarios; the higher the future emissions, the greater the projected temperature increase.

Precipitation is predicted to decline only slightly (< 10 millimeters) throughout the inland portion of the ecoregion from the current norm. Departures are predicted to vary across seasons, with increased precipitation in the winter and spring (March–May), balanced against larger decreases in summer and fall. In contrast, most of the coastline is predicted to show large decreases (≥ 50 millimeters) in precipitation across all seasons. In combination, these temperature and precipitation predictions indicate pronounced seasonal changes: winters will have much higher minimum temperatures and increased precipitation, and summers will have much higher maximum temperatures and decreased precipitation.

Numerous general projections have been made of potential biotic responses to these climatic changes, many of which focus on particular subregions such as the Adirondack Mountains (Jenkins 2010), the Champlain Basin in eastern New York and western Vermont (Stager and Thill 2010), New Hampshire (Burack 2009), Maine (Jacobson et al. 2009), Nova Scotia (NSDE 2009), and the northeastern United States (Frumhoff et al. 2007). The most detailed assessments of the potential biological effects of climate change across the Northern Appalachian/Acadian ecoregion have been taxa-specific (viz., trees, birds, amphibians, and mammals) using correlative bioclimatic models to assess changes in ranges by the end of the twenty-first century. Iverson and colleagues (Iverson and Prasad 1998; Iverson, Prasad, and Matthews 2008; Iverson, Prasad, Matthews, et al. 2008) modeled potential changes in the suitability of habitat for up to 134 tree species in the eastern United States in response to six potential future climate change scenarios. Unsurprisingly, suitable climatic conditions for many tree species in the eastern United States are projected to shift northward. For some species, the loss of climatic suitability at the southern end of their range may result in a severe contraction of the species' distribution such that they may even be extirpated from the United States. For example, balsam fir (*Abies balsamea*), paper birch (*Betula papyrifera*), northern white-cedar (*Thuja occidentalis*), and black spruce (*Picea mariana*) are all projected to experience decreases in climatic suitability over much of their US ranges and, under one or more scenarios, leave the United States altogether. Other species, whose northern range limits occur in the United States, may experience range expansions as a result of an increase in the area of suitable climatic conditions. For example, conditions for black hickory (*Carya texana*), black oak (*Quercus velutina*), and longleaf pine (*Pinus palustris*) are all projected to expand in response to climate change.

Iverson, Prasad, Matthews, and Peters (2008) also projected potential shifts in the climate space of ten different general forest types. Even the low-emissions scenario resulted in the projected, near-complete loss of the climate space for spruce-fir and white-red-jack pine forests in the eastern United States. In addition, the high-emissions scenario resulted in a mass contraction of the climates suitable for the maple-beech-birch forests and projected expansions of suitable areas for oak-hickory and oak-pine forests. That is not to say, however, that these forest types or the species that make them up will necessarily disappear from the ecoregion or further north; some species will be able to move north into Canada, and others may find climatic refugia at higher elevations or on maritime islands (Schauffler and Jacobson 2002).

Rodenhouse et al. (2008) used bioclimatic models to assess the potential effects of climate change on 150 bird species in the northeastern United States, concluding that on a species-by-species basis, the projections were highly variable—some species were projected to gain habitat and others were expected to lose habitat. Similarly, projected climate-driven shifts in the potential ranges of 2,954 bird, mammal, and amphibian species in the Western Hemisphere suggested that many species would likely experience large shifts in the distribution of their potential ranges (Lawler et al. 2009). Although many ranges have been projected to shift poleward and upslope, others have been projected to shift in less predictable ways. For example, Lawler and Hepinstall-Cymerman (2010) found that the potential ranges of the moose (*Alces alces*) and the eastern mole (*Scalopus aquaticus*) were projected to contract from and expand into, respectively, the ecoregion.

Species Changes within the Northern Appalachian/Acadian Ecoregion

Based on the projected range shifts for tree, bird, mammal, and amphibian species cited above, Lawler and Hepinstall-Cymerman (2010) examined the potential effect of climate change on the protection afforded by the current "reserve network" in the region. Of the seventeen tree species included in the analyses of Iverson, Prasad, Matthews, and Peters (2008), six were projected to experience decreases in area-weighted importance values (an index of habitat suitability) within the US portion of the ecoregion, eight were projected to experience increases, and three (red maple [*Acer rubrum*], eastern hemlock [*Tsuga canadensis*], and eastern white pine [*Pinus strobus*]) had mixed results depending on the emissions scenario. The importance values for three species (balsam fir, yellow birch [*Betula alleghaniensis*], and red spruce [*Picea rubens*]) declined by 35 to 60 percent, indicating that under a warming climate, balsam fir–red spruce, the dominant forest type in much of the northeastern United States, may occupy only 20 to 30 percent of its former range. Sugar maple (*Acer saccharum*), which has large economic value in the region, was also predicted to decline by 15 to 36 percent. Conversely, white ash (*Fraxinus americana*) and black cherry (*Prunus serotina*), both species expected to decline across the eastern United States (Iverson, Prasad, Matthews, and Peters 2008), were predicted to dramatically increase in importance within the US portion of the ecoregion and reserve network, probably due to a northward shift in their ranges. The high elevations present in the ecoregion may also provide refugia for these species as the climate warms. Quaking aspen (*Populus tremuloides*), expected to decline in the eastern United States, was projected to increase slightly within protected areas under the low-emissions scenario.

Of six bird species that Rodenhouse et al. (2008) projected to experience potential range contractions, the black-throated green warbler (*Dendroica virens*) was projected to experience the greatest decreases for both the low- and high-emissions scenarios across the ecoregion and within the reserve network. By 2100, summed incidence values for the black-throated green warbler were projected to be 38 percent and 43 percent of their current values, given a high-emissions scenario for the ecoregion and network, respectively. In contrast, the summer tanager (*Piranga rubra*) was projected to show the greatest increases across the region and reserve network (77- and 71-fold increases, respectively) given the high-emissions scenario.

Projected shifts in the potential ranges of the two other vertebrate taxa/groups (Lawler et al. 2009) resulted in turnover of 13 to 25 percent in the amphibian fauna and 7 to 21 percent in the mammal fauna, depending on the spatial scale evaluated (i.e., ecoregion, reserve network, or Adirondack Park) and whether unlimited or no dispersal was assumed. Assuming unlimited dispersal, a little more than half of the amphibian and mammal species were projected to increase their potential ranges within the ecoregion and the reserve network. Unsurprisingly, if dispersal is restricted, many more species were projected to show range contractions.

Approaches to Conservation under Climate Change

Current conservation priorities in the Northern Appalachian/Acadian ecoregion are to identify critical linkages across the landscape and work to ensure continued connectivity in order to accommodate the expected range shifts noted above. It is important, however, to recognize that these linkages are critical even if climate change were not occurring. The Northern Appalachian/Acadian ecoregion is still characterized by relatively large areas with low levels of transformation (e.g., Adirondack Park in New York, northern Maine, and both northern and southern Nova Scotia), yet they are increasingly fragmented from within and ecologically isolated from each other (Trombulak et al. 2008). Conservation planning to promote connectivity across the entire ecoregion and with neighboring ecoregions to both the north and south remains a priority.

This highlights an important point for understanding the context of climate change in conservation planning: the accelerated pace of climate change and the increased public awareness of its implications have not, on their own, made conservation planning imperative. Rather, it was already imperative before society came to appreciate more fully the magnitude of the climate crisis. Habitat destruction and fragmentation were—and continue to be—major threats to biodiversity and ecological health wherever human development is occurring (Baldwin and Trombulak 2007). However, climate change dramatically raises the stakes for what is needed in order to develop a successful conservation plan, as well as the consequences of failing to do so. Further, there is little to suggest that the results of robust planning for connectivity in the face of climate change would differ significantly from the results that emerge solely from a consideration of ameliorating the consequences of habitat destruction. While species range shifts due to climate change will involve a predominant north-south orientation, range shifts up elevational gradients and into climate refugia may occur along other directional axes as well.

The threat of climate-induced stresses has also increased awareness among conservation planners and natural resource managers of the importance of increasing the health of ecosystems—both terrestrial and aquatic—to promote their capacity to withstand environmental stresses without exhibiting change in its structure or function, also known as ecosystem resilience. Restoring natural flow regimes, minimizing air and water pollution, and controlling the spread of exotic species are all important parts of an integrated strategy to increase the ability of native biodiversity within the region to persist and function in the face of increased temperatures and altered water availability. Of course, climate projections under extreme emissions scenarios will overcome even the most resilient of landscapes. But the healthier a system is, the slower the rate of biological erosion, the slower the rate at which the community is pressured to transition to a new state with respect to species composition or abundance, and, hopefully, the greater the chance that species can reestablish themselves in more hospitable locations.

Despite the extent of modeling studies for the 2C1Forest region, incorporating climate change into ecoregional conservation planning is difficult because of the mismatch in the scales of resolution of available climate projections and conservation planning. General Circulation Models, which underlie the kinds of climate projections described earlier, are based on grid cells measured in hundreds of kilometers; conservation planning, in contrast, generally requires finer scales of discrimination. Connectivity planning, for example, requires the ability to identify landscape features of relevance to the movements of individual animals, meaning that in order to identify, say, specific locations along a road where wildlife-crossing structures would be most effective, we have to understand how particular individuals are moving across the landscape. Similarly, reserve planning requires identification of particular locations that may serve as refugia for species in a warming world because they are buffered by local topographic conditions from the overall temperature increases projected for a region. In other words, conservation planning requires the ability to identify priority locations with a resolution that often measures less than a kilometer; climate change models are not yet able to provide useful projections with that kind of resolution. As a result, some authors (Anderson and Ferree 2010; Beier and Brost 2010) argue that conservation strategies to allow adaptation to climate change require more focus on representation of a landscape's geophysical features (e.g., topography, bedrock, soil) in a reserve network simply because such strategies do not rely on future climate projections. Hence, increased attention to conserving the geophysical diversity of the region may, at this point, be a critical part of the overall strategy to address the species turnovers described previously.

Perhaps the most ambitious initiative currently active in the Northern Appalachian/Acadian ecoregion to address the needs of landscape conservation under climate change is the Staying Connected Initiative (SCI). Launched in 2009 to protect and restore connectivity for large mammals across the US portion of the region, SCI involves twenty-one organizations and state agencies across New York, Vermont, New Hampshire, and Maine. The Staying Connected Initiative primarily works on seven priority linkage areas broadly identified in the earlier work of Two Countries, One Forest: (1) Tug Hill Plateau to Adirondack Mountains, (2) Adirondacks to Green Mountains, (3) Greens to Taconic Mountains, (4) Northern Greens to the border with Québec, (5) Northern Greens to the Northeastern Highlands (i.e., "Northeast Kingdom") of Vermont, (6) Northeastern Highlands to Northern New Hampshire, and (7) to and across western Maine toward the Gaspé Peninsula.

Within each of these linkage areas, SCI focuses on five key activities:

- identifying critical locations within each linkage area, thus drawing the work down to the local level, which ultimately is necessary for implementing any site-specific conservation plan;
- providing technical assistance to local land-use planners so that local planning efforts already under way include the best available information on planning for climate change and connectivity;
- promoting local engagement in conservation planning to respect the importance of engaging local stakeholders and property owners;
- developing tools, such as conservation easements, that encourage and allow private landowners to contribute to habitat connectivity; and
- protecting land in targeted areas to increase public investment and responsibility for the goals of conservation.

Priorities and pace of work on each of these activities varies among the linkage areas in response to site-specific needs and opportunities. As of June 2011, detailed modeling and mapping for connectivity was complete in most of the linkage areas. Further, over 8,000 hectares of land have been or will soon be protected in New York and Vermont, and a project to mitigate the barrier created by a critical road that lies between the Adirondack and Green Mountains has been initiated. Much work remains to be done, but the hope is that SCI will be able to capitalize on the extensive public engagement that it has fostered and continue to increase connectivity throughout the region.

Another regional initiative, coordinated by the US Fish and Wildlife Service, is the North Atlantic Landscape Conservation Cooperative (LCC). The North Atlantic LCC partnerships, which link federal agencies, states, tribes, and environmental nongovernmental organizations, seek to identify and respond to regional science needs for achieving landscape-scale conservation goals. With respect to climate change adaptation, the North Atlantic LCC has recently launched a project to support the statewide assessments of species and habitat vulnerabilities in the face of sea-level rise, altered stream flow, and range shifts in forest ecosystems, as well as to design management strategies that will help managers prioritize conservation actions in response to these stresses.

In addition, the Wildlife Conservation Society has launched a project in the Adirondack Mountains focused on climate change adaptation in lowland boreal ecosystems, particularly its wetlands, since they are likely to be one of the ecosystems most vulnerable to climate change in the Northern Appalachian/Acadian ecoregion. This project is aimed primarily at working with landowners to set conservation priorities and implement strategies that will contribute to this overall goal: (1) development of language for conservation easements that take climate change adaptation into account, (2) establishment of monitoring sites, particularly for songbirds and plant communities, and (3) provisioning of technical support to aid local conservation planning for climate change.

Conclusion and Recommendations

Conservation planning in the face of climate change in this ecoregion must achieve the same goals as if the climate crisis were not occurring: protecting and restoring viable populations of native species and fully functional ecosystems. The fact that we are in the midst of a climate crisis does not obviate conservation planning; it simply makes the goals much harder to achieve and the stakes for failure more severe. Conservation planners had no room for complacency before; we have even less now that society is on the brink of major climate disruptions on a scale perhaps not seen in human history.

We recommend that increased attention be given to planning for linkages across landscape scales to allow for both short- and long-term movement of species. The capacity for movement is critical not only to allow populations to track changes in critical climate parameters but also to enhance effective population sizes for species already under pressure by other stresses, particularly habitat degradation and destruction. Planning for linkages need also be coupled with efforts to enhance ecosystem resilience throughout reserve systems and the surrounding matrix; healthy populations will be better able to resist climate stresses. In conclusion, we already know what we need to do: protect more lands and waters and link them together ecologically to allow for responsiveness to environmental change. The efforts now under way to conduct such planning in the Northern Appalachian/Acadian ecoregion are good beginnings; they must continue and, ultimately, be implemented successfully.

Acknowledgments

We thank Evan Girvetz for work on the Climate Wizard analyses and all the people we worked with in Two Countries, One Forest, particularly Emily Bateson, Justina Ray, Gillian Woolmer, Louise Gratton, Conrad Reining, Karen Beazley, Graham Forbes, and Jim Northup. We also thank Michale Glennon (Wildlife Conservation Society) and Mark Zankel (The Nature Conservancy and Staying Connected Initiative) for current information about their respective projects.

Chapter 19

Yellowstone to Yukon, North America

CHARLES C. CHESTER, JODI A. HILTY,
AND WENDY L. FRANCIS

The Yellowstone to Yukon (Y2Y) region represents one of the best known and most advanced large landscape conservation efforts in the world. Due to the relatively high availability of data and to the region's numerous and diverse conservation groups and agencies, the regional response to climate change has been comparatively rapid—albeit much remains to be done. Ultimately, climate change preparedness will need to occur at local and subregional scales, with these efforts scaling up to support biodiversity conservation actions and policies across the Y2Y region. The Y2Y vision and on-the-ground efforts throughout the region constitute a working hypothesis that conservation at such a continental scale will enhance ecosystem resilience and provide opportunities for adaptation during this time of climate disruption.

Introduction to the Region

Mountain ecosystems around the world are significant for their rugged beauty as well as their biological and cultural significance. Over 3,200 kilometers (2,000 miles) long and half a million square miles in area, the Y2Y region represents one of the most intact mountain systems anywhere on the planet. All of the large carnivores and ungulates that were here in 1793—when Alexander Mackenzie became the first Caucasian to cross the North American continent north of Mexico—persist in this region. Approximately 10 percent of the Y2Y region lies under some form of protected-area status, including wildlife refuges, wilderness areas, and national, state, and provincial

FIGURE 19.1 A female grizzly and her cubs wander through the southern region of Y2Y. Grizzly bears are almost completely restricted to the Y2Y region in the lower forty-eight states and exist in dangerously low densities in some parts of southern Canada. Photo courtesy of WCS Jeff Burrell.

parks. Yet even as much of the region retains its ecological intactness, human activities have entailed substantial impacts, especially in the southern third of the Y2Y region. These activities threaten a number of high-profile species such as wolverines (*Gulo gulo*), wolves (*Canis lupus*), caribou (*Rangifer tarandus*), and grizzly bears (*Ursus arctos*) (fig. 19.1). The region's 700 protected areas, including the world's first national park (Yellowstone) and Canada's first national park (Banff), are key to maintaining the region's biodiversity (fig. 19.2).

Indigenous communities, including Native Americans in the United States and First Nations in Canada, have long occupied and traveled throughout these mountains. The Blackfoot people referred to the Rocky Mountains as "miistakis," or backbone of the world. As currently defined, the Y2Y region stretches across the traditional territory of thirty-one First Nations/Native American groups (Reeves 1998). Today tribes in the United States still operate as independent nations on reservations in various parts of the region and in some cases hold ceded rights (viz., access and even management authority over wildlife and lands beyond reservations). In Canada, southernmost First Nations often govern reserves and have treaty rights to pursue traditional activities on public lands. However, First Nations in British Columbia and in parts of Yukon and the Northwest Territories are still negotiating land settlements with federal and territorial governments.

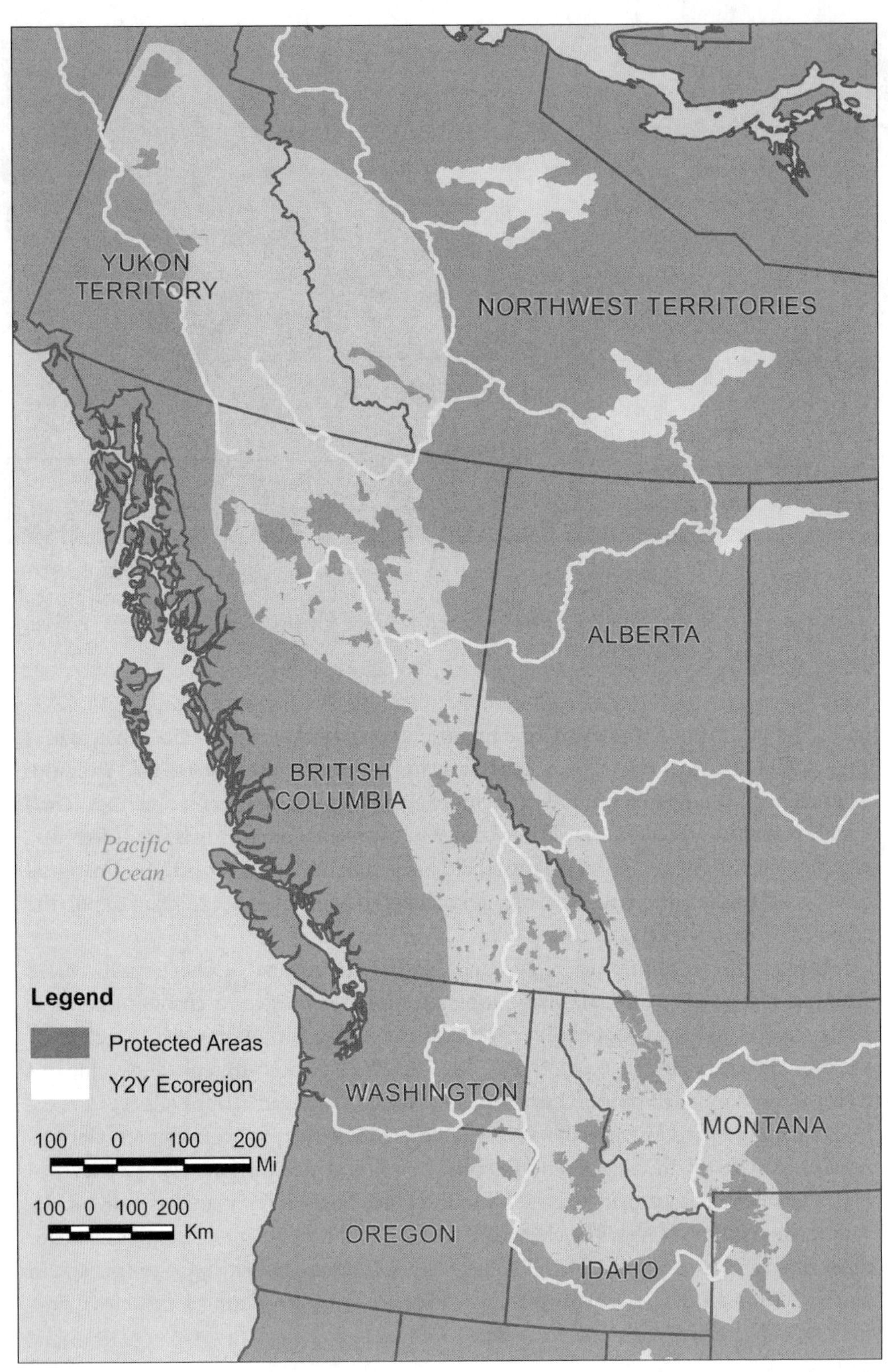

FIGURE 19.2 Major protected areas along the Yellowstone to Yukon landscape (Y2Y) in North America. Figure courtesy of WCS Andra Toivola.

With post-Columbian incursion and settlement of the Y2Y region, a new economy emerged that was largely based on various forms of natural resource extraction. These extractive industries—principally forestry, agriculture, ranching, grazing, mining, and oil and gas extraction—are extensive, meaning that they are pervasive on almost all but the most protected lands. While natural resource extraction is an increasingly significant component of the economy as one travels north, in the last few decades nonconsumptive industries have begun to dominate much of the southern extent of this region. By the 1990s natural resources industries made up less than 6 percent of all employment and less than 5 percent of all personal income in the US Rocky Mountains, down from less than 11 percent and 9.6 percent respectively in 1969 (Rasker 1994). This figure is indicative of an economy that has shifted toward tourism, outdoor recreation, and "amenity migration," the latter term being defined as "the movement of people to places, permanently or part-time, principally because of the actual or perceived higher environmental quality and/or cultural differentiation of the destination" (Glorioso and Moss 2007, 138). Other service sector businesses in the region range from recreational opportunities such as skiing and snowmobiling to restaurants and hotels. Despite the changing economic base, however, politics in the region remain largely conservative with strong allegiances toward traditional stakeholders in the extractive industries.

Although the total human population in the Y2Y region is relatively low, the region's institutional complexity is labyrinthine in its myriad federal, state, provincial, and local agencies that manage land, wildlife, minerals, and other natural resources. In addition, self-governing indigenous groups in both countries manage lands and wildlife, regulate businesses within their jurisdictions, and manage businesses in both the extractive and service-oriented industries.

Historical Overview of Conservation and Science Initiatives

The Y2Y region has long been a focal point for hundreds of environmental and conservation nongovernmental organizations (NGOs). Along with NGOs using traditional tactics of advocacy and litigation, other conservation groups include land trusts, science and research institutions, sportsmen and angler organizations, local watershed associations, and local community groups. The Y2Y region has also been the focus of academic, agency, and independent scientists who have conducted an extensive array of biological, social, economic, and political research relevant to conservation in the region. As a consequence, Y2Y is arguably one of the most studied regions in the world.

The Y2Y Conservation Initiative is an NGO whose mission is to focus on the ecological health of the whole region. Formed in the mid-1990s as a "virtual network" composed of biologists and conservationists, a core component of the Y2Y

Conservation Initiative's mission was to conceptualize and promote the larger landscape vision, as well as to prioritize activities toward critical areas of conservation concern across the region (Chester 2006). In addition to raising the profile of the Y2Y region as one of the first truly large landscape conservation efforts in the world, the organization has commissioned targeted science to identify and focus attention on a set of conservation priorities for places and issues, and has acted as a bridge helping partners share information and collaborate on shared visions or complementary efforts (for more background, see Y2Y 2011).

Since its inception, science has served as the core foundation of the Y2Y vision. Specifically, the rationale for the vision was derived from the scientific theories of island biogeography and metapopulation dynamics, both of which suggest that in the most general of terms, bigger protected areas closer together and embedded in a matrix of lands permeable to wildlife movement are more likely to increase the survival of individual species and thus to maintain the region's overall biodiversity. Research indicates that current protected areas in the Y2Y region are inadequate on their own to protect its biodiversity in the long term both because (1) they do not fully represent the region's biodiversity, and (2) they are not large enough to maintain healthy populations of wide-ranging species and the natural processes upon which they depend (Hilty et al. 2006). Such findings strongly support envisioning biodiversity conservation strategies at the scale of the Y2Y region, and of working toward both core area protection and connectivity between those core areas.

Threats to Biodiversity and the Response of Enhancing Connectivity

Threats to biodiversity in the Y2Y region run the familiar gamut, although the degree of threat shifts from south to north. Rural residential sprawl is the largest cause of land-use change in the southernmost stretches of the region, threatening many wildlife species (Hansen et al. 2002). As little as one house per section (a square mile) can turn the section from source (or good) habitat for grizzly bears to sink habitat where bears are likely to die as a result of conflict with humans (Schwartz et al., forthcoming). Farther north, areas that once experienced virtually no permanent human footprint are increasingly being explored and developed for various forms of natural resource extraction. Other threats range from roads and unsustainable forestry practices to intolerant attitudes toward some wildlife species. Some new categories of threat are looming, including the placement and extent of renewable energy projects and infrastructure, which represent a largely unknown threat. And of course, climate-related changes also threaten biodiversity.

The conservation community collectively uses a range of tools including private land conservation, wildlife restoration efforts, backcountry access management, various types of incentive programs, living with wildlife education and conflict reduction programs, and coordinated strategies that integrate core habitat protection campaigns. A major focus of conservation activities in the region is to target key public and private lands to maintain and enhance core areas and connectivity between core areas. Connectivity work in the Y2Y landscape has focused on mitigating the impacts of roadways such as the Trans-Canada Highway through Banff National Park and Highway 3 through the Crowsnest Pass area of southern Alberta (Ford et al. 2010; T. Lee et al. 2010) (fig. 19.3). Notably, the Y2Y region boasts the first federally designated migration corridor in the United States, the Path of the Pronghorn (Cohn 2010; ENS 2008). First documented by biologists from Grand Teton National Park and the Wildlife Conservation Society, the discovery of this tenuous pathway inspired a collaborative effort by land management agencies and local conservation groups to secure protection for this migratory route.

Collectively, the connectivity work at each of these scales supports the broader vision of the entire region functioning as an enormous corridor composed of protected core areas and linkages that ensure the long-term viability of all of the region's native species. While the Path of the Pronghorn represents an important step forward for corridor conservation, much work remains to be done in the broader Y2Y region.

Regional Effects of Climate Change

The Y2Y region currently hosts myriad research initiatives on the effects of climate change that range from studies on particular species in site-scale habitats to extensive large landscape analyses. A few examples of broad-scale research initiatives include the following:

- The Pacific Climate Impacts Consortium, which focuses on climate change impacts in the Pacific Northwest, including the portion of British Columbia that lies within the Y2Y region (http://pacificclimate.org);
- The Climate Adaptation Secretariat of the Province of British Columbia, which provides advice to government and reports on actions to implement a climate action plan (http://www.env.gov.bc.ca/cas);
- The US Forest Service Rocky Mountain Research Station, which is examining the effects of climate change on water supply and quality, wildland fire, and terrestrial ecosystems (RMRS 2009);
- The Northwest and North Central Climate Science Centers, established by the US Geological Survey, which "synthesize existing climate-change-impact data and management strategies, help resource managers put them into action on the ground, and engage the public through education initiatives" (NCCSC 2010; USDOI 2010); and

FIGURE 19.3 Across the Trans-Canada Highways bisecting Banff National Park, more than twenty-five wildlife underpasses and overpasses have been built to enhance wildlife connectivity, including this overpass. Photo courtesy of WCS Jodi Hilty.

- The Department of the Interior's Great Northern Landscape Conservation Cooperative, a goal of which is to foster conservation science-management partnerships to provide scientific and technical support for conservation at landscape scales under conditions of climate change and other landscape stressors (http://www.nrmsc.usgs.gov/gnlcc).

Such efforts demonstrate that both academic scientists and government officials have begun to take steps to understand the impacts of climate change across the Rocky Mountains and adjacent ecoregions. These and other science efforts were summarized in *Moving toward Climate Change Adaptation: The Promise of the Yellowstone to Yukon Conservation Initiative for Addressing the Region's Vulnerabilities* (Graumlich and Francis 2010). Much of the following discussion summarizes this report's assessment and review of climate impacts in the Y2Y region.

Scientists have long recognized the potential for rapid climate change in the Rocky Mountains (Luckman 1990). In an extensive study on climate change in the US portion of the Rocky Mountains, Reiners et al. (2003, 178) noted that "[s]hifts in environmental gradients across Rocky Mountain landscapes are likely to lead to rapid growth in the numbers of recognizably threatened species." Today, a rapidly expanding body of scientific evidence indicates that the Y2Y region is experiencing climate change impacts and that such continued changes will significantly alter the native biota. Mean annual temperatures have been increasing throughout the region, and scientists have documented changing precipitation patterns and an increased frequency of extremes in comparison with historical records. Although average temperature changes are relatively small, it is the increase in extreme temperatures that could have the greatest impact. This is because many species exist within relatively narrow climate envelopes, and extreme temperatures can lead to stress, reproductive failure, and even death. As Graumlich and Francis (2010, 28) put it, there is strong evidence that:

> … the climates of the Y2Y region have already changed beyond the limits of historic variation; that these climatic changes are having ecological impacts; that continued changes, especially warming, will have long-term, unprecedented future impacts; and that landscape-scale conservation is a central element of limiting and adapting to such inevitable changes.

Winters are shorter and warmer in the twenty-first century than in previous decades, and precipitation is increasingly falling as rain rather than snow, the effects of which include changes in seasonal water availability and glacier endurance. Glaciers are literally disappearing, and Glacier National Park could soon be a misnomer, a fact that has made the park a prominent "poster-child" of climate change (Hall and Fagre 2009). Those species that depend on the presence of snow and ice on a year-round basis will experience the greatest impacts. For example, because wolverines typically depend on cool places both to cache their food where it will not rot, and to den in areas where deep snow provides insulation, the loss of permanent snow could be problematic for this species (Chadwick 2010; Copeland et al. 2010; McKelvey et al. 2011; Peacock 2011). In addition to changes in climatic and physical variables, scientists are also documenting phenological alterations (such as changes in hibernation patterns) and shifts in species distributions. Ecosystem function can also depend on phenological timing between codependent species, which can decouple as such species seek to adapt to changing conditions (Graumlich and Francis 2010).

Some species may benefit from climate change. Yellow-bellied marmots (*Marmota flaviventris*) in the Rockies are gaining prehibernation weights much earlier in the summer and then exiting hibernation earlier in the spring, leading to their population expanding beyond historical numbers. However, researchers have posited that such benefits may be short lived due to the long-term probability of extended drought (Ozgul et al. 2010). More broadly, both observational data and modeling indicate that species' distributions are changing and are likely to further change over the next century, and that those species that cannot move to areas with desirable climate and habitat conditions are likely to perish (Graumlich and Francis 2010).

Arguably the most visible impacts of climate change in the Y2Y region are the widespread outbreaks of pine beetles (mostly the mountain pine beetle, *Dendroctonus ponderosae*). Although large infestations have occurred in the past, researchers have characterized the extent and longevity of recent infestations as unprecedented, particularly inasmuch as warmer conditions at higher elevations have brought the beetles into contact with five-needle pine species, such as whitebark pine (*Pinus albicaulis*) (Carroll et al. 2006; Logan et al. 2010; Robertson et al. 2009; Taylor and Carroll 2004). Cold winters have historically limited pine beetle infestations, and the extensive infestations seen on the landscape today result from milder winters and, at high altitudes, the pine beetle's corresponding ability to shorten its life cycle and thus increase its reproductive rate (Bentz et al. 2010). Extensive and rapidly expanding stands of dead and dying trees are increasingly ubiquitous throughout much of the Y2Y region.

Approaches to Conservation under Climate Change

Conservationists in the Y2Y region have long recognized the threat of climate change. For example, when the Y2Y Conservation Initiative was founded in 1993, one of the rationales for envisioning conservation across such a large scale was the need to conserve landscapes at sufficient scale to be robust to climate change (Harvey 1998). Yet as evinced by the large-scale regional climate prediction and adaptation efforts covering large portions of the Y2Y region discussed earlier, climate change is emerging as a new focus for many conservationists in the region.

Not surprisingly, government agencies, communities, conservation NGOs, and other entities in the region vary widely in the terms of their engagement on climate change. Some have continued their current line of work and have not added any emphasis or focus to climate change. Many local land trusts, for example, have a fairly narrow remit, and with the exception of a few instances where water conservation is beginning to take a higher profile, their business strategy and programmatic approach is often relatively unaffected by climate change. Other entities indicate that their conservation priorities and actions already *are* a response to climate change, inasmuch as their focus on protecting core areas, maintaining connectivity, and removing or minimizing nonclimate stressors constitute their priorities for ensuring that ecosystems are resilient to and/or are able to adapt to climate change (Graumlich and Francis 2010; Hansen et al. 2010). For example, groups collaborating to restore and maintain the last tendrils of connectivity in the transboundary Cabinet-Purcell region describe their work as both a strategy to conserve the region's grizzly bears and a tool for enhancing the region's resilience to climate change (Y2Y 2011).

Other entities have reevaluated their biodiversity conservation or management goals in light of climate change. For the Crown of the Continent Ecosystem, the Wildlife Conservation Society (WCS) analyzed roadless areas for their value to wildlife, a component of which focused on connectivity and climate change to guide protected area design (Weaver 2011). This work addresses the difficult question: if we cannot protect all of the current roadless areas, which ones might be most important to conserve, given climate change and connectivity needs? Likewise, targeted beaver (*Castor canadensis*) restoration efforts in Alberta by a local watershed group are part of a larger effort to raise the profile of water under climate disruption and of beaver as a component of free ecosystem services that have the capacity to help retain mountain water runoff for longer periods, cool in-stream water, and raise ground water levels (WCS Canada 2009).

The reality is that many institutions throughout the Y2Y region are still working to incorporate climate change considerations at the planning level. Management plans for the US Forest Service, for example, are supposed to consider the impacts of climate change, but how this translates to on-the-ground natural resource decisions is less clear. Similarly, western states are working to incorporate connectivity and climate change into their state wildlife action plans to help them map out their priorities into the future (WGA 2008; AFWA 2009). Other efforts focus on bringing climate science information to local stakeholders, helping them make robust decisions under uncertain climatic conditions. For example, for both the Greater Yellowstone Ecosystem and the transboundary US–Canada region, the US Fish and Wildlife Service and WCS collaborated on two workshops to discuss climate change impacts on grizzly bears and wolverines, to identify strategies to conserve these species given climate change, and to enhance communication and collaborative action among regional stakeholders

(Cross and Servheen 2009; Servheen and Cross 2010). Farther north, the Yukon Conservation Society has an Energy and Climate Change Program that has developed a proposal for a "Yukon Carbon Fund" (Taggart 2009). On-the-ground examples of management decisions that consider the impacts of climate change are limited, but if trends continue, this will likely become a common land management practice throughout the Y2Y region.

Roadblocks and Opportunities

Perhaps the most difficult challenge facing conservationists in the Y2Y region is their own history—specifically, their longstanding tradition of protecting biodiversity by staking out specific areas for protection of *in situ* resources, a pattern originating with the 1872 establishment of Yellowstone National Park. Although Yellowstone was originally intended to protect its rare geological features, it soon became as much a sanctuary for wildlife that was rapidly disappearing from a developing continent. The working assumption was that if humankind were to place a sufficient portion of land under conservation status, protected areas could play a key role in conserving a region's biodiversity. Although this is still the case, that role is shifting; climate change now demands that conservationists work across much larger landscapes, both to protect the ability of species to move across those landscapes and to preserve ecosystem functions rather than just their constituent components. Precedent, not to mention mechanisms, are lacking for effective and coordinated multijurisdictional planning and implementation.

Another major challenge is that many citizens reject the premise of anthropogenic climate change. Despite overwhelming scientific evidence, this backlash influences particular sectors in the Y2Y region and presents significant challenges in incorporating climate change considerations into policy and management decisions. Also, the region lacks new resources to retain or expand climate expertise either on the science or policy side. This is particularly true in government agencies, where a mandate for climate change adaptation can translate into additional work without new resources. Further, efforts to reduce greenhouse gas emissions mean an increased demand for renewable energy sources, and growing wind and hydrogenerated electricity are already showing an increasing footprint in the Y2Y region. As these expand, they may to come into conflict with conservation priorities and could decrease the region's intactness and even its resilience to climate change. Examples are the proposed Site C hydroelectric dam on the Peace River in northern BC and the hundreds of independent power projects under consideration for streams and rivers throughout the BC portion of the Y2Y region (Evenden 2009).

Finally, stakeholders have not yet fully conceptualized the problem of rapid climate change or its solutions. There is a lack of agreement around whether the goal is to resist the impacts of climate change or to enable transformative ecological changes that sustain changing configurations of biodiversity. For example, should a waterfowl reserve with drying wetlands begin to pump water to maintain the same conditions? If so, under what circumstances? When will pumping become too costly or ineffective? When should the focus instead be on managing for change?

Conclusion and Recommendations

Given the significant and unpredictable ecosystem changes that will inevitably accompany alterations in temperature and precipitation, a business-as-usual conservation model will not protect the biodiversity of the Y2Y region. For example, the availability of water is already a contentious issue in much of the region. As water reserves and the timing of peak season flows change, securing adequate supplies of water for all users is likely to become one of the most contentious conservation issues in the twenty-first century. Changes in precipitation quantity and type (viz., rain or snow) in the Y2Y region mean that certain seasons may be more arid, resulting in higher agricultural and municipal demand during certain times of year. Pressure to construct dams as a means to prepare for times of drought is likely to build. At the same time, species dependent on free-flowing and cool streams will almost certainly be at odds with such demands. Conservationists in the Y2Y region need to prioritize this issue and design proactive solutions before it becomes too late. Likewise, ensuring connectivity for those species that have the capacity to move will be increasingly important, and consequently it will be essential to understand what kinds of corridors or linkages may be robust over time (and for which species) given climate change (Cross et al. in press).

In addition, it may be useful to identify the likely locations of refugia from climate change and to prioritize these areas for protection (Graumlich and Francis 2010). During the last ice age, the Y2Y region contained several refugias where glaciation did not occur. Although refugias in a warming future will likely be quite different than those of a glaciated past, the varied terrain of the Y2Y landscape harbors conditions that could provide species with refuge from the current and looming changes.

The Y2Y region is fortunate inasmuch as it has attracted many scientists over the past century, and thus we have a significant amount of historic and current science to guide biodiversity conservation efforts under conditions of climate change. Yet even with this knowledge, considerable uncertainty remains in terms of predicting a climate change trajectory within any subregion of the landscape. Despite such uncertainties, conservation activities must ensure that the Y2Y region can be a place where plants and animals can shift to find appropriate niche space as habitats change.

The Y2Y region arguably represents one of the planet's best opportunities to work at a large landscape scale. As one of the world's most intact mountain ecosystems, the Y2Y region's ecological integrity makes it more resilient to climate change than many other regions. As conditions change, dispersal-limited species are likely to find more of an opportunity to move to appropriate climatic spaces due to the diversity of climate niches within the vast Y2Y region—niches that result from the region's relative intactness, its wide latitudinal expanse, and its mountainous terrain that incorporates both aspect (north-east-southwest) and elevational gradients (Chen et al. 2011). Given the inherent resilience and adaptability of the Y2Y region, it is all the more important to implement sound conservation actions now to ensure that this region can be robust for species conservation over the coming centuries.

PART 6

Polar Land to Seascapes

Although least impacted by non-climate related human activities now, accelerated warming trends in the polar regions may offer increased access to the region and thereby open the door for new human-caused stressors. At the same time, there is no farther north or south for species currently living in these regions to go, so identifying and protecting those places that are most likely to retain resources and habitat for polar species may constitute a critical conservation strategy for the future.

Chapter 20

Arctic Alaska, USA

STEVE ZACK AND JOE LIEBEZEIT

Although Arctic Alaska contains some of the last remaining wild lands and intact wildlife populations in North America, the region is challenged by increasing energy development pressures and a climate that is changing more rapidly than anywhere else on Earth. The inevitability of transformative change in Arctic ecological systems suggests that there are few options available to offset impacts to wildlife and landscapes. Therefore, the main opportunities for addressing climate change in Arctic Alaska is the protection of large areas important to wildlife in advance of energy development and climate change planning, which takes into account interactions with energy development and seeks to identify climate refugia where arctic systems stand the best chances of persisting.

Introduction to the Region

Arctic Alaska encompasses much of the northern portion of the state, including the Brooks Range, associated foothills, and the entire Arctic Coastal Plain (fig. 20.1). The northern boundary of Arctic Alaska abuts the Beaufort Sea to the north and the Chuckchi Sea to the west, with the region's largest town, Barrow, at the coastal junction of these seas.

Arctic Alaska landforms are shaped by permafrost, frozen soil that may reach to depths of up to 650 meters below the surface (Davis 2001). Permafrost plays an important role in shaping the unique topography of the arctic landscape, helping form

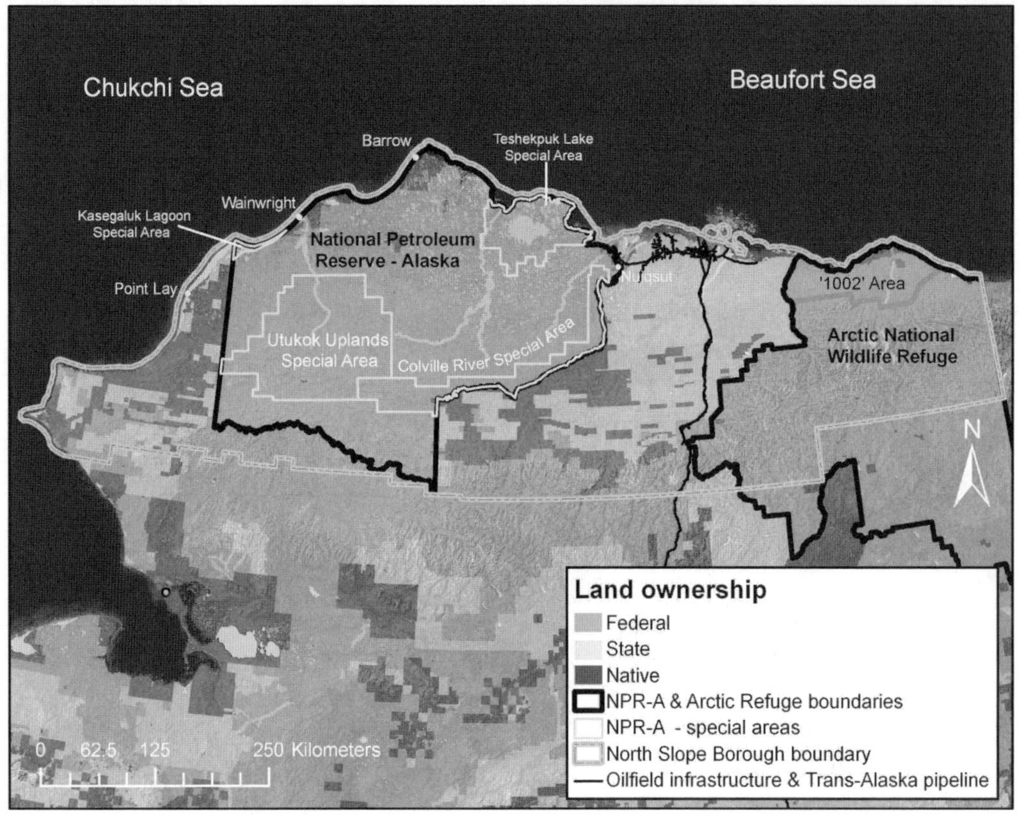

FIGURE 20.1 Land ownership in the coastal plains of Arctic Alaska. Figure courtesy of WCS Joe Liebezeit.

the myriad tundra polygons, strangmoors (low ridges), and other unique land features. These surface features, in turn, influence water flow and thus the entire surface hydrology. Arctic tundra, dominated by sedges in the coastal plain and low shrubs in the foothills, characterize the vegetation in the northern two-thirds of this region. The immense Arctic Coastal Plain is dominated, primarily in the west, by the largest wet meadow-tundra complex in the circumpolar Arctic (CAVM Team 2003). Boreal forest makes up most of the remaining habitat in the Brooks Range and southern foothills.

These environs encompass a diversity of wildlife (fig. 20.2). The wetlands of Arctic Alaska attract millions of migratory birds during the brief breeding season, most conspicuously, important waterfowl and shorebird populations (Johnson et al. 2007; Larned et al. 2010). All told, over fifty bird species migrate here from all the world's continents and oceans to nest and rear young. Four large caribou (*Rangifer tarandus*) herds migrate to and within Arctic Alaska (Valkenburg 1998) to calve young on the coastal plain, including the Western Arctic herd (ca. 300,000–500,000). Resident

FIGURE 20.2 (a) Resident musk ox face effects of a warming Arctic; (b) Snowy owls depend on lemming abundance; (c) Arctic caribou herds are in decline worldwide; (d) Red-necked phalaropes at risk with development; (e) Increasing red fox are a threat to arctic fox; (f) Spectacled eider winter in Bering Sea polynyas. Photos courtesy of WCS Steve Zack.

wildlife species that brave the bitter winter include several birds, such as ptarmigan (*Lagopus leucura*) and snowy owl (*Bubo scandiacus*); and mammals such as musk ox (*Ovibos moschatus*), polar bear (*Ursus maritimus*), and arctic fox (*Vulpes lagopus*). The seas surrounding Arctic Alaska are, likewise, rich in wildlife, including bowhead whales (*Balaena mysticetus*), beluga whales (*Delphinapterus leucas*), walrus (*Odobenus rosmarus*), bearded seals (*Erignathus barbatus*), spotted seals (*Phoca largha*), and some of the most productive fisheries in Alaska.

Despite its immense size (~231,000 square kilometers—larger than the state of Minnesota, USA), this region is sparsely populated by some 6,752 residents (0.021 people per square kilometer: U.S. Census Bureau 2010). Most residents are Inupiat Eskimos (68 percent), with several thousand transient oil-industry workers in the region for two-week stints centered in the oil fields in Prudhoe Bay, Kuparuk, and the supply town of Deadhorse. Two pieces of federal land dominate the landscape: the National Petroleum Reserve–Alaska (NPR–A: 95,000 square kilometers), and the Arctic National Wildlife Refuge (ANWAR: 78,050 square kilometers).

Historical Overview of Conservation and Science Initiatives

With much of the landscape in Arctic Alaska in federal and, to a lesser degree, state holdings (fig. 20.1), the principal actors of Arctic Alaskan conservation are the federal, state, and local government agencies that administer these lands—including, respectively, the Department of the Interior (most prominently), the State of Alaska, and the North Slope Borough (the most important, local political constituency). Conservation nongovernmental organizations (NGOs) are active in Arctic Alaska, including Audubon Alaska, The Wilderness Society, The Nature Conservancy, the Wildlife Conservation Society, and others. Oil and gas companies (prominently BP [Exploration]; Alaska, Inc.; ConocoPhillips Alaska, Inc.; and Shell) are significant actors in the sense that the dominant development activity in this region is defined by oil (and in the future, gas) extraction. Mining companies and interests are also a growing presence. The Inupiat community, which relies strongly on subsistence hunting, mostly for whales, seal, caribou, and waterfowl, plays a role in conservation. Although much of their subsistence activities occur across public land, they own several smaller but significant land parcels in the region (fig. 20.1).

Conservation in the Alaskan Arctic has entailed intense debate at the federal level largely because of (1) the significant natural resources in Arctic terrestrial and oceanic regions, (2) important wildlife populations and intact wildlands, and (3) to a growing degree, accelerated climate change impacts in the region. The conservation issues of the Arctic are thus conspicuously affected by extrinsic forces such as political election cycles, ongoing debate over balancing domestic and foreign oil use, commercial interests such as the opening of *bona fide* shipping lanes through the Northwest Passage, and others (e.g., the 2010 oil spill in the Gulf of Mexico). While all stakeholders profess an interest in balanced development with wildlife protection, the expressed parameters of what constitutes such a balance can and do vary tremendously.

Arctic Alaska has a rich, complex, and contentious history of conservation and landscape protection (Borneman 2003). Controversies over federal land designations in northern Alaska persist today for many Alaskans. The far-reaching Alaska Native Claims Settlement Act of 1971 resolved land and financial claims made by the Alaska Natives, including those in Arctic Alaska. This settlement set the stage for the 1980 Alaska National Interest Lands Conservation Act, which provided for the creation or revision of fifteen National Park Service properties and set aside other public lands for the US Forest Service and the US Fish and Wildlife Service. In all, the act provided for the designation of 321,900 square kilometers of public lands, fully a third of which were set aside as wilderness area. While the Department of the Interior had established the Arctic National Wildlife Refuge in 1960, the 1980 act set aside the coastal plain region of the refuge (the "1002 Area") for potential development—yet only through specific congressional action. Since that time, debate over drilling in this region has accompanied virtually each election cycle, becoming one of the signature conservation battles of our time and setting the seed for today's controversies and debates over conservation in Arctic Alaska.

Although there are significant protected areas in northern Alaska, less than 5 percent of the wildlife-rich coastal plain is protected. Development on the coastal plain is currently restricted to the Prudhoe Bay and associated oil fields with infrastructure spanning to the borders of the NPR–A to the west and the Arctic Refuge to the east (fig. 20.1). Decisions in the coming decades will determine the establishment of any new protected areas and the extent of energy and mining development on both the land and in the nearby seas.

The tremendous importance of Arctic Alaska for wildlife is well documented, and the impacts of oil development and its associated infrastructure on Arctic wildlife have been relatively well studied (Ballard et al. 2000; Liebezeit et al. 2009, Liebezeit and Zack 2008; NAS 2003; Troy 2000). Such research has highlighted the effects of development on caribou movements and calving, of road dust on nearby tundra and wildlife habitats, and of "subsidized" predators (via garbage and oil field structures) on nesting birds. However, to date, no cumulative effects analysis has assessed the multiple direct and indirect effects of development on Arctic wildlife or, moreover, scaled such effects to the current and proposed level of development.

Current Conservation in the Region

There is very little development in the totality of Arctic Alaska—the existing oil fields and a few dispersed settlements (mostly along coastlines) comprise the only development in this immense region. Outside of the oil field network, there is little actual or possible "habitat management" in the Arctic; there are no forests to thin or burn, no invasive weeds to manage and control, no altered watersheds to mitigate (with the exception of altered water flow in the oil fields). Tundra environs are slow growing, a process measured in centuries, with low vegetation and extreme vulnerability to any disturbance (Emers et al. 1995). Because efforts to rehabilitate tundra habitat have to date not matched adjacent habitat, restoration of habitats cannot be viewed as viable in the conservation toolbox for Arctic Alaska.

Our ongoing national dependence on fossil fuels will continue to press for development in Arctic Alaska landscapes and seascapes. Whenever gas prices spike, the political focus inevitably heads north to Arctic Alaska. More than any other force, the economic potential of Arctic Alaska for oil and gas development will constantly press and threaten the wildlife and landscapes there that have been historically remote from development pressures.

The major gap to future conservation in Arctic Alaska is the absence of an Arctic-wide comprehensive conservation plan. Each federal, state, and tribal jurisdictional unit is administered as an island unto itself. This issue is important because the wildlife—particularly the caribou, but also other species—move broadly across this immense region without regard to political boundaries. An associated need is an integrated analysis of the synergistic effects of climate change and potential energy development in the Arctic. As those are the major threats, their intersection in relation to future conservation planning is clearly needed.

Finally, there is tremendous uncertainty about how species in the Arctic will respond to a much warmer year-round landscape. For resident terrestrial species (e.g., musk ox or ptarmigan) warmer springs and summers may cause considerable physiological stress. For the many migratory birds that breed in the summer months of Arctic Alaska, the threat would appear to be less about the warming per se, and much more about the change in habitat (e.g., predicted tundra conversion to taiga) and in timing (e.g., asynchrony between insect emergence and chick hatch).

Regional Effects of Climate Change

The Arctic is warming almost twice as fast as the global average (IPCC-WG2 2007; Martin et al. 2009). Regional warming in the near future will be transformative (Lenton et al. 2008), particularly in regard to the two conspicuous threshold changes of dramatic, seasonal sea-ice recession and permafrost melting (IPCC-WG2 2007). Under such conditions, Arctic Alaska will move from tundra-like conditions to more boreal-like conditions, with cascading changes to the region's wildlife and human communities (Both et al. 2004; Euskirchen et al. 2009; Høye et al. 2007; Ims and Fuglei 2005; IPCC 2007; Martin et al. 2009; Post et al. 2009).

The most important trends and concerns that have emerged for wildlife conservation in the Arctic include the following (these are not mutually exclusive, and synergistic effects may exacerbate or temper outcomes):

- Continued warming. Despite the likelihood of increased precipitation, the prospect of continued warming temperatures—particularly in the foothills interior of Arctic Alaska—nonetheless means more evapotranspiration (Martin et al. 2009). The trend toward warmer and earlier springs, milder winters, and lingering falls is projected to continue.
- Drying. Some parts of the Arctic have experienced dramatic and recent drying, prominently evidenced by disappearing lakes (Smith et al. 2005; Smol and Douglas 2007). The immense and productive wetland complex of western Arctic Alaska is wetter and is underlain by deep permafrost, and consequently it remains unclear if widespread drying is a feasible scenario in the near future, as the interactions between surface water, eroding permafrost, and summer drainage patterns are unclear in the increasingly warmer Arctic (Martin et al. 2009).
- Phenological mismatches. Arctic fauna and flora are initiating their phenologies (e.g., flowering, nesting) earlier and earlier (Both et al. 2004; Euskirchen et al. 2009; Høye et al. 2007; Martin et al. 2009; Post et al. 2009). It is not yet known how this will affect health, reproduction, and survivorship of wildlife (Troy 2000; Tulp and Schekkerman 2008).
- Vegetation shifts. The changing climate is enabling more vegetation from further south to spread northward into the Arctic, with the dominant trend of more woody vegetation displacing sedge tundra environs (Euskirchen et al. 2009; Hudson and Henry 2009; Tape et al. 2006). Woody vegetation changes soil characteristics, which in turn alters productivity patterns (Martin et al. 2009). Moreover, tundra fires are increasing in frequency and in scale of effect (Racine and Jandt 2008), jump-starting the invasion of woody vegetation recruitment (Martin et al. 2009).
- Shoreline erosion. Significant shoreline erosion is occurring on the Arctic Ocean shorelines of Arctic Alaska (Jones et al. 2008; Mars and Houseknecht 2007) as a result of recession of shore ice, increased severity of summer storms, and melting of shoreline permafrost. The erosion, and the associated inundation of salt water with storms, is leading to alteration of shoreline habitats to more salt tolerant vegetation (Martin et al. 2009). These effects have the potential to disrupt the post-breeding feeding biology of shorebirds and many waterfowl species that depend on shoreline habitats to gain energy in advance of their southern migrations (e.g., Derksen et al. 1979; Taylor et al. 2010).
- Boreal forest invasion. Finally, there is increasing evidence of the "invasion" of boreal fauna and flora into the Arctic environs (Euskirchen et al. 2009; Martin et al. 2009; Post et al. 2009). For Arctic wildlife, boreal species could mean increased competition, change in predator-prey dynamics, and possibly heightened susceptibility to disease (see Martin et al. 2009). Arctic fox have

> been replaced by red fox (*Vulpes vulpes*) elsewhere in the Arctic (Hersteinsson and MacDonald 1992), and anecdotal evidence (Pamperin et al. 2006; pers. observ.) suggests that it is happening at least in the oil fields of Prudhoe Bay where red fox are preying on Arctic fox, with the red fox seeming to have increased in recent years. There is also anecdotal (unpublished) evidence that increasing numbers of brown bear (*Ursus arctos*) are in Arctic environs. Some scientists speculate that brown bear increases had a role in the deaths and disappearance of musk ox in the coastal plain of the Arctic National Wildlife Refuge.

Efforts to link climate science to the Arctic are on the rise. For instance, The Wilderness Society and The Nature Conservancy led the development of the first generation of downscaled General Circulation Models (see chap. 2) of climate change in Arctic Alaska (Martin et al. 2009). The Wildlife Conservation Society is evaluating Arctic bird vulnerability (using a modified version of the NatureServe Climate Change Vulnerability Index), with an effort to understand both wintering ground and Arctic breeding ground vulnerabilities for a subset of species. Similarly, investigations into landscape change, as detected by satellite imagery, are really just getting under way. Such investigations are just now being coupled with the long-term monitoring of wildlife populations to create, for the first time, a spatial and temporal sense of how wildlife and vegetation communities are changing with the changing Arctic. The University of Alaska–Fairbanks has an active climate change group, the Scenarios Network for Alaska Planning (SNAP 2011), which is beginning to bring forth major syntheses.

The major impediment to regional modeling efforts in Arctic Alaska is the paucity of weather stations (Martin et al. 2009). Although the coast, with its dispersed Inupiat communities, has reasonable weather station capacity, the interior of Arctic Alaska is virtually devoid of such stations. This means poorer estimates of regional variation in key weather variables and the consequent uncertainty in modeling regional-scale future climate projections. Nonetheless, best available modeling efforts do predict something of a differentially changing Arctic Alaska, with the interior warming more than the nearshore coastal plain in the near future (Martin et al. 2009). The spatial and temporal resolution is not as great as that of other regions.

Approaches to Conservation under Climate Change

Climate change has become an important organizing factor in conservation—but only recently. Due to the politically charged nature of climate change, state and federal agencies historically averted the topic. This has changed dramatically over the past five years, with the number of discussions, workshops, and actions relevant to climate change amplifying rapidly. The first major conference was held by the US Fish and Wildlife Service in 2008 and involved representatives of state and federal agencies, universities, and NGOs. The conference report laid out the primary drivers of change in the warming Arctic and some tentative discussions of potential ecological interactions and species effects of such changes (Martin et al. 2009). A follow-up workshop focused on identifying species at risk due to several scenarios of future climate change impacts (WCS 2010). Additional climate change science has been spurred on by the creation of the Arctic Landscape Conservation Cooperative (see chap. 2) in 2009, and in 2010 the US Geological Survey placed a regional Climate Science Center at the University of Alaska–Fairbanks. This recent and growing federal activity has positioned Alaska at the forefront of climate change research and understanding—appropriately so, given how quickly this region is changing due to climate change.

Although there has been considerable effort (1) to describe the changes in the Arctic and their effects on wildlife, and (2) to attempt to predict and project near-future changes, there is little to no capacity to manage or offset such changes that are resulting from the changing climate. Despite our understanding of how climate change is transforming the Arctic, it is unclear if and how the practice of conservation can help wildlife or landscapes adapt to the pending transformation. That is, there are no evident adaptation tools to offset the dramatic changes afoot and the unclear responses of wildlife to such changes in the Arctic. Moreover, the Arctic is changing so quickly that species are unlikely to adapt in the evolutionary sense (Berteaux et al. 2004).

One hypothetical avenue of applying a conservation solution to the changing climate in the Arctic is to identify relative *climate refugia*—regions that, by projection, may change comparatively less climatologically in the near future so as to maintain Arctic tundra and weather conditions. For example, projections of climate change make clear that habitats near the Arctic Ocean are to some degree buffered against spring and summer warming due to maritime conditions and frequent fog. However, it is entirely unclear if the scale of that effect is large enough geographically, or moderates temperature change sufficiently, to make a tangible difference for wildlife (Martin et al. 2009).

Despite earlier recommendations (NAS 2003), there has been no landscape conservation-needs assessment to date for Arctic Alaska. Although the elements of such an analysis have been proposed (Martin et al. 2009; WCS 2010), they consist largely of efforts to identify research and information gaps across Arctic geographies, rather than of any concerted, region-wide effort. The Bureau of Land Management, which administers the National Petroleum Reserve–Alaska, is required to produce a cumulative effects analysis of proposed development patterns in that immense region, and the pending release of new development plans would also require such an analysis and include consideration of the effects of the changing climate.

Roadblocks and Opportunities

The most likely roadblock to any kind of conservation planning in the Arctic—regardless of whether it incorporates the threat of climate change—will be the political backing for extensive energy and mining development. The push for Congress to open up the 1002 Area of the coastal plain in the Arctic National Wildlife Refuge and the recent efforts to develop oil and gas operations in the Teshekpuk Lake Special Area of the National Petroleum Reserve–Alaska are ongoing. The State of Alaska, the North Slope Borough, and the oil companies interested in the continued viability of the Trans-Alaska Pipeline System all signal a strong interest in growing the development footprint in Arctic Alaska. Such growth would coincide and, indeed, significantly interact with the push for offshore oil development in the Chukchi and Beaufort Seas. Despite these pressures, many of the stakeholders do have a strong interest in maintaining viable wildlife populations, particularly for those species important for subsistence.

A second important roadblock is the availability of funding for future research, modeling, and land-use planning. The scale of work needed to accurately predict impacts to wildlife from climate change—not to mention developing a conservation response—in such a remote region is daunting. At the same time, funding opportunities appear to be diminishing as the federal budget debt increases and research program funding remains on the chopping block. Ironically, industry funds could, and in some cases do, provide support for much needed climate change research and planning.

Conservation is, by one definition, the planned management of natural resources to prevent overexploitation. The long and contentious conservation battles in Arctic Alaska make clear that there is no consensus on finding a balance between use and protection. It remains an open question as to how and whether the emerging plans resulting from climate change analyses will affect these long-standing debates.

The main opportunity for conservation in Arctic Alaska is in the protection of large areas important to wildlife in advance of development. Despite the uncertainties of exactly how Arctic Alaska will change in the next several decades (Martin et al. 2009), it is important to recognize that it will change dramatically (ACIA 2004; IPCC-WG2 2007; Lenton et al. 2008). Thus, the best option for conservation is to protect large areas in advance of such changes with the idea that the protection of large areas is the most conservative and best tool available now for conservation in a changing world (Halpin 1997; McCarty 2001; Heller and Zavaleta 2009).

The main opportunities for protection of large landscapes in Arctic Alaska under climate change align with the current high-priority areas of concern by conservation groups, viz., the 1002 Area (6,070 square kilometers) of the coastal plain of the Arctic National Wildlife Refuge, the three large special areas in the NPR–A (Teshekpuk Lake, 7,082 square kilometers; Utukok River Uplands, 15,620 square kilometers;

and the Colville River, 9,752 square kilometers) and the coastal plain regions west of Teshekpuk Lake that have no particular designation or planned consideration for protection by the Department of the Interior. Each of these areas is very large and contains distinct and important wildlife attributes. As a whole, protection from development of each of the areas is the best available buffer against the manifold changes coming to this most important part of the world for wildlife.

Conclusion and Recommendations

With a rate of change double that of the global average, climate change in Arctic Alaska promises dramatic alterations to the region's wildlife and landscapes. Despite political impediments in initiating research on the particular effects of climate change in the region, the recent significant infusion of federal funding dollars into state agencies and the University of Alaska–Fairbanks has quickly made climate change the focus of planning and wildlife conservation.

There is a clear and urgent need for climate change planning for the entirety of Arctic Alaska, planning that would cross agency and political landscapes. Wildlife does not respect those arbitrary boundaries, and most species range widely across landscapes that are expected to differentially change with the warming climate in the near future. The US Fish and Wildlife Service-led Arctic Landscape Conservation Cooperative effort is a significant step in this direction, but it will take partnership and collaboration between agency, NGO, and industry stakeholders to develop an effective plan.

Such planning should also account for the potential interaction of energy development, the changing climate, and wildlife concerns, both on and offshore. The development footprint will be larger, as winter ice road-use diminishes with the warming Arctic and as Arctic ice cap melting allows increased shipping traffic. As for wildlife in the region, our current understanding of climate change impacts with the synergistic impacts from development is limited. We have little knowledge about how caribou movement patterns are changing with the changing climate, and only a few studies of the displacement effects on caribou by development. Further, we know that development incurs increased pressures on nesting birds by nest predators, yet we do not know how expansion of development into areas of higher diversity and abundance of breeding birds will affect nest productivity and how the additional stressor of climate change will influence these vital population parameters. The pending Bureau of Land Management assessment of potential development patterns in the NPR–A should address some of these issues.

Both research and planning efforts should emphasize ecosystem processes as well as an interdisciplinary approach layering wildlife and relevant climate data with habitat, hydrology, and other drivers. As an example, tens of thousands of migratory caribou and millions of nesting birds migrate to the coastal plain with the expectation of high productivity; if Arctic drying is a key climate-related stressor on wildlife (since wildlife abundance and diversity mirror the location of lakes and wet tundra), then the associated changes in primary productivity with warming demands close evaluation.

We need to understand the complex changes affecting Arctic food webs and trophic levels brought on by species that win or lose in the face of climate-mediated habitat changes and by apparent boreal wildlife invasions into the Arctic (e.g., red fox and grizzly bear). For example, are the recent observations of hybridization between grizzlies and polar bear just noise (Kelly et al. 2010), or do they portend a real change for top predators in the terrestrial Arctic?

Finally, we need to know if there are regions within the Arctic that will change less with the warming climate and thus be capable of acting as climate refugia for at least some species. Identifying such areas is critical, as the future of energy extraction for much of the Arctic will be decided in the coming years and decades. The special areas in the NPR–A, and the 1002 coastal plain region of the Arctic Refuge, appear to be the current, viable options for wildlife protection in Arctic Alaska. Such protections would be, at best, placeholders, as we begin to witness a region transforming and understand, the hard way, how wildlife in this once remote region is to be affected.

Acknowledgments

Support from the Liz Claiborne Art Ortenberg Foundation, British Petroleum, the Disney Foundation, the Kresge Foundation, the Doris Duke Charitable Foundation, The U.S. Fish and Wildlife Service (Neotropical Migratory Bird Conservation Act, Atctic Landscape Conservation Cooperative), the Bureau of Land Management, the National Fish and Wildlife Foundation, the state of Alaska, private funders, and the Wildlife Conservation Society has been essential to this work.

Chapter 21

Antarctica

DAVID AINLEY AND TINA TIN

With parts of Antarctica warming at seven times the rate of the rest of the globe, climate change stands to significantly affect terrestrial and aquatic species and ecosystems. Disappearing sea ice, declines in Antarctic krill stocks, and expanding plant communities over the last several decades portend even more notable changes in the future. While many of the forty-nine signatory nations to the Antarctic Treaty engage in extensive climate change research, climate change has only recently become part of the treaty agenda. Greenhouse gas emissions from human activities in Antarctica are small within the global context, yet the symbolic value of reducing emissions from human activities in one of the world's most vulnerable environments cannot be underestimated. Visionary thinking on climate change-adaptation, including strategic planning for large-scale protected or specially managed areas, will be required to achieve conservation goals into the future.

Introduction to the Region

Antarctica is the fifth largest continent and the Southern Ocean, the sixth largest ocean, within which lies the Antarctic Circumpolar Current, the largest current system on Earth. In total, this south polar region covers 14 million square kilometers of land and 32 million square kilometers of sea (fig. 21.1). Zoogeographically, the cold Antarctic Circumpolar Current has cut off this region from the rest of the planet for millions of years. Endemism is high.

Antarctic terrestrial biodiversity is low. At the species level, richness is low, while at higher taxonomic levels, many groups are missing (Convey and McInnes 2005). The terrestrial fauna is composed of invertebrates, predominantly nematodes and microarthropods, with only two higher insect species present (Block 1984). Plant communities are largely cryptogamic (e.g., algal mats in ponds and streams, and mosses, liverworts, and lichens not associated with streams) with only two higher plants present on the Antarctic continent, both of which occur only at the northerly tip of the Antarctic Peninsula (Convey et al., in review).

In accord with global zoogeographic "rules," Southern Ocean biodiversity is also relatively low, swamped by a huge biomass of a number of ecologically dominant species. Waters south of the Antarctic Polar Front, and particularly the southern boundary of the Antarctic Circumpolar Current, are the destination for numerous whales and seabirds, which forage there during summer but calve in warm waters or breed on islands at sub-Antarctic or temperate latitudes. Nowhere else in the planet's oceans do numbers of such cetacean or avian populations compare with those of the Southern Ocean (fig. 21.1).

The sea ice-covered area of the Southern Ocean is also home to an endemic seal and bird fauna. The fish fauna is dominated by one unique family, the *Nototheniidae*, which has radiated extensively. The invertebrate fauna of midwaters is dominated by several copepod and euphausiid species, including Antarctic krill. Over slope and deeper waters, the Antarctic krill assumes the role of the anchovy or sardine in temperate, upwelling-rich waters. Over shelves, it is replaced by the crystal krill. In some southern areas, especially where the colonial alga *Phaeocystis* predominates, pteropods (small pelagic molluscs) are especially abundant (W. O. Smith et al. 2010).

Governance

To understand conservation planning in the Antarctic, it is necessary to understand its particular governance structure. The original 1959 Antarctic Treaty arose at the end of the International Geophysical Year 1957–58 in response to conflicting territorial claims and the need to ensure continued freedom for scientific research. The treaty effectively "froze" territorial claims—parties did not endorse or revoke existing claims but agreed to not lay new claims or enlarge old ones. The treaty also prohibits military activities and nuclear testing in the treaty area and requires parties to take measures regarding the preservation and conservation of living resources in Antarctica. Over the next fifty years, the Antarctic Treaty System expanded to include other legal instruments, twenty-eight consultative and twenty-one nonconsultative parties, and a number of official observers. The 1964 Agreed Measures on the Conservation of Flora and Fauna provided protection for native mammals and birds. The 1972 Convention for the Conservation of Antarctic Seals set harvesting limits and seal reserves, should sealing recommence. The 1980 Convention for the Conservation of Antarctic Marine

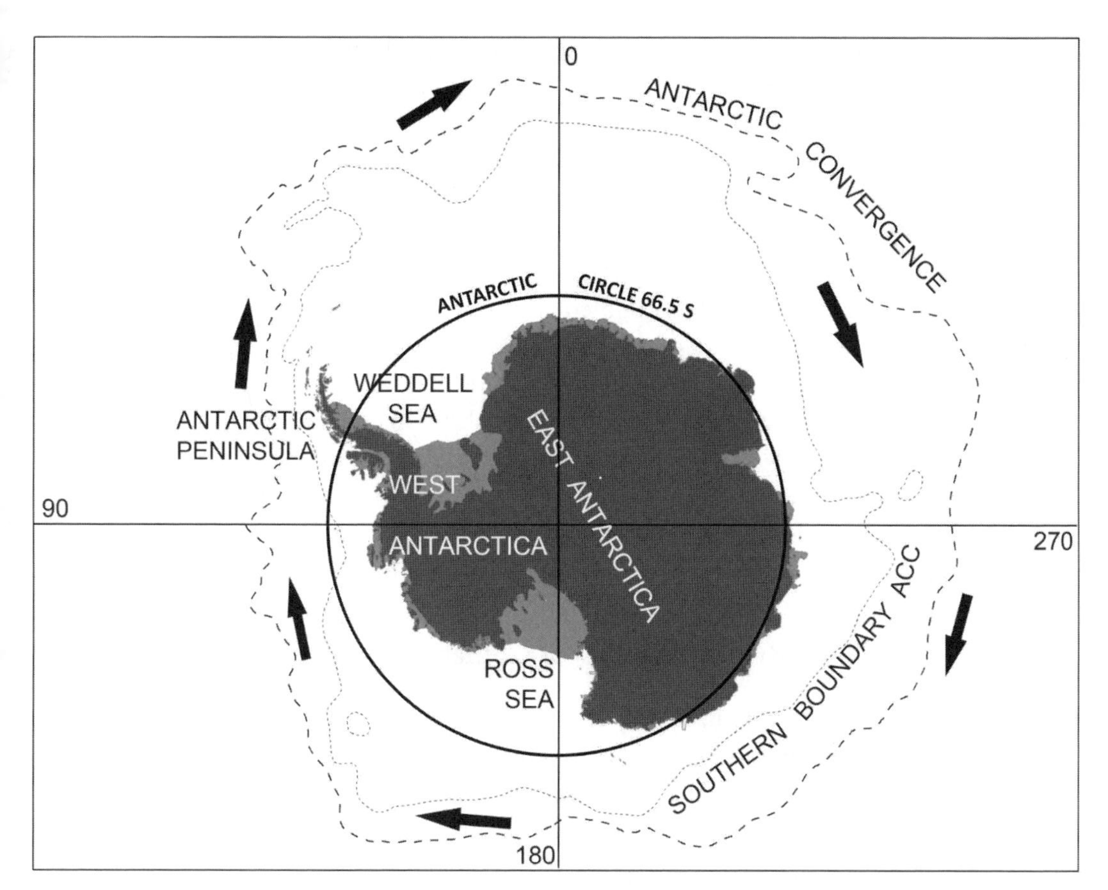

FIGURE 21.1 The Southern Ocean is bounded on the north by the Antarctic Polar Front (APF), known in political documents as the Antarctic Convergence. The westward-moving circumpolar winds drive the Antarctic Circumpolar Current, the Southern Boundary (sbACC) of which is an important front, to the south of which ocean productivity is exemplary. The sbACC is also usually the edge of the sea-ice zone. These features (APF, sbACC) have isolated the high-latitude portion of the Southern Ocean from zoogeographic communities to the north. The Ross Sea is exemplary as having, by far, the widest continental shelf of Antarctica, as well as having a portion free of continental ice during glacial maxima; thus, it has a unique fauna. East Antarctica holds the Antarctic plateau, otherwise a 3 km thick sheet of glacial ice; it is separated from West Antarctica, which is mostly ice grounded on the ocean floor, by the Trans-Antarctic Mountains, which run along the western coast of the Ross Sea to eventually pass up the spine of the Antarctic Peninsula.

Living Resources (CCAMLR) was developed to ensure the conservation (including rational use) of all Antarctic marine living resources. The 1991 Protocol on Environmental Protection to the Antarctic Treaty (Environmental Protocol) banned mining and aims to provide "comprehensive protection of the Antarctic environment and dependent and associated ecosystems."

Antarctic Treaty parties and parties to CCAMLR meet independently on an annual basis, and not all states are parties to both. Agreements are by consensus. Official observers, including the Council of Managers of National Antarctic Programmes, the Scientific Committee on Antarctic Research, the Antarctic and Southern Ocean Coalition, and the International Association of Antarctica Tour Operators are invited to attend these meetings. In particular, the Antarctic and Southern Ocean Coalition occupies the only environmental nongovernmental organization (NGO) observer seat and represents over 100 NGOs worldwide having interests in Antarctica's future.

The Antarctic Treaty System designated the entire treaty area as a special conservation area. In addition, the Environmental Protocol and CCAMLR contain mechanisms to allow the designation of specific protected areas within the bigger treaty and convention areas. Currently, seventy-one Antarctic Specially Protected Areas protect 0.011 percent and 0.008 percent of the terrestrial and marine portions, respectively, of the treaty area (New Zealand 2005; New Zealand 2009a). Approximately 1 percent of the convention area occurs within Marine Protected Areas (WWF 2008; Toropova et al. 2010). It is clear that the size of existing protected areas is inadequate to address climate adaptation and protect Antarctic marine and terrestrial biodiversity.

Regional Effects of Climate Change

Scientists have long recognized that climate change can severely alter high-latitude ecosystems (Roots 1989; IPCC-WGII 1990). However, for its size and vulnerability to climate change, and despite extensive research, the Antarctic has received relatively little attention in the various assessment reports of the UN Intergovernmental Panel on Climate Change. In 2009, the report, "Antarctic Climate Change and the Environment" (Turner et al. 2009), provided the most in-depth summary of our knowledge of climate change in Antarctica and its likely impacts:

- Since the 1950s, significant warming has occurred across the Antarctic Peninsula, and to a lesser extent West Antarctica, with little change across the rest of the continent. Parts of western and northern Antarctic Peninsula are among areas that are warming the fastest on Earth—seven times faster than the average global rate (IPCC 2007; Turner 2010).
- The waters of the Antarctic Circumpolar Current, particularly Circumpolar Deep Water, have warmed more rapidly than the global ocean as a whole.
- From 1991–2007 the concentration of carbon dioxide in the ocean increased south of 20 degrees South, especially in the Southern Indian Ocean, leading to progressive ocean acidification.
- Net sea ice extent has changed little around the Antarctic over the last thirty years. It has decreased by 40 percent in the western Antarctic Peninsula, compensated by increases in the Ross Sea sector. Where sea ice has been lost, changes in marine algal growth have occurred. Consequently, stocks of Antarctic krill—which feed on algae that grow under sea ice—have declined signi-

ficantly. The distribution of Adelie penguins has also changed, with many populations on the northern Antarctic Peninsula disappearing due to a reduced period of sea ice, whereas Ross Sea and East Antarctica populations are generally stable or increasing (Ainley, Russell, et al. 2010; Ducklow et al. 2007) (fig. 21.2).

- Plant communities have expanded across the Antarctic Peninsula. Along with higher temperatures, this area has experienced a marked switch from snowfall to rain during the summer, as well as a retreat of ice fields. These linked changes have led to the colonization of newly available land by plants and animals.
- Many ice shelves, which originally form on land and can be the size of small nations, have broken off and melted into the ocean with the total ice shelf area on the Antarctic Peninsula having been reduced by over 27,000 square kilometers in the last fifty years. Ice shelves bordering the Amundsen Sea could be entering a phase of collapse, which could amplify global sea-level rise.

Climate models project significant surface warming over Antarctica by the year 2100. Annual average total sea ice area is projected to decrease by a third (Turner et al. 2009), though dramatic effects will be seen well before then (Ainley, Russell, et al. 2010). Consequently, all marine ecosystem components closely tied to sea ice can be expected to show significant changes in their ecological performance. In areas where large Antarctic krill populations currently exist, population size is likely to stabilize at a lower level as sea ice disappears (Atkinson et al. 2004). The minke whale, currently the most abundant cetacean, could lose 5 to 30 percent of its ice-associated habitat, while blue, humpback, and fin whales could encounter a compression of foraging habitat in the highly productive zone south of the Southern Boundary of the Antarctic Circumpolar Current (Tynan and Russell 2008). The consequences of historical harvesting reduce our ability to differentiate between the impacts of overharvesting and climate change, particularly in regard to fish, seals, and whales (Ainley and Blight 2009; Ainley et al. 2007; Ainley, Ballard, et al. 2010; Nicol et al. 2010). While some whale species are now recovering from whaling, low krill availability may compromise further recovery. The ice-obligate emperor and Adelie penguins, as well as crabeater, Weddell, leopard, and Ross seals, will likely experience local extinctions due to changes in both habitat and food-web dynamics (Siniff et al. 2008; Jenouvrier et al. 2009; Ainley, Russell, et al. 2010). Species of any systematic group with a sufficient initial population size and circumpolar distribution are expected to survive at least in the Pacific sector, where according to predictions the sea ice is likely to remain relatively stable (Turner et al. 2009) or to continue to grow in the next few decades (Ainley, Russell, et al. 2010). If surface ocean pH levels become more acid by 0.2 to 0.3 units, this may lead to the thinning of the aragonite skeletons of pteropods and have disproportionate and negative consequences on ecologically key species.

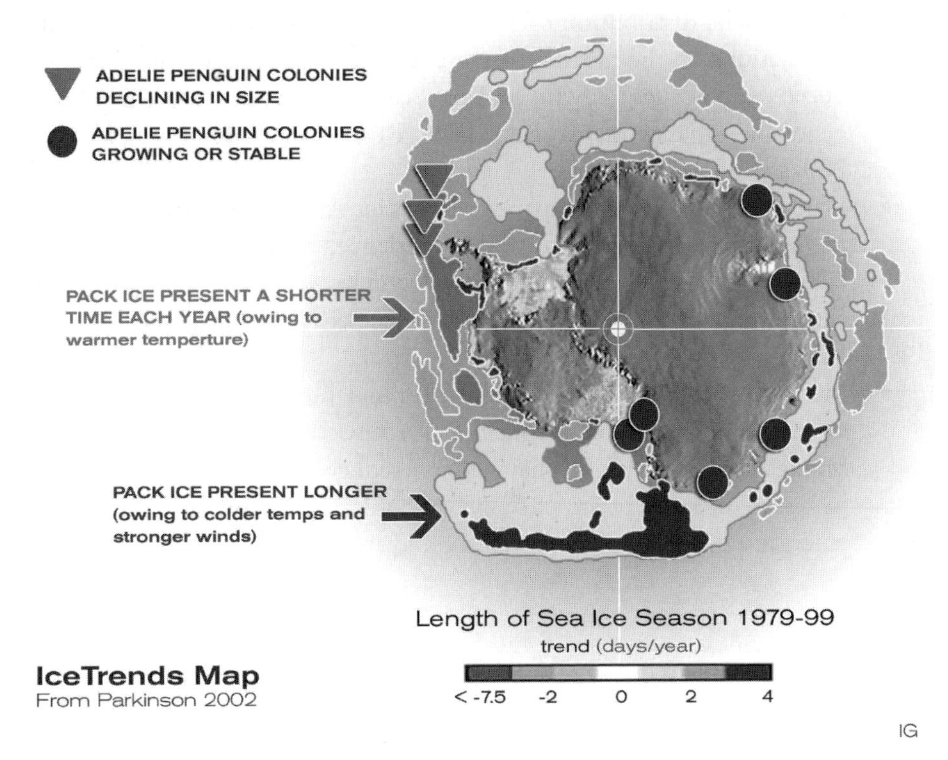

FIGURE 21.2 Changes in the length of the ice season in the Southern Ocean, 1979–99, and how Adelie penguin colonies have responded. Ice map redrawn from Parkinson (2002), labels added; Adelie trends, from colonies having long data time series, from Ainley, Russell, et al. (2010).

Regarding terrestrial ecosystems, the frequency of freeze-thaw events and extreme low temperatures is predicted to increase under climate change. The tolerance limits of arthropods and continental bryophytes could readily be exceeded, although lower lethal temperatures show substantial capacity for both phenotypic plasticity and evolutionary change. High among future scenarios is the likelihood of invasion by more competitive nonnative species, as warmer temperatures and increased human visitation combine to make it easier for nonnative species to colonize (Turner et al. 2009). The cumulative effects of an increased human presence and disturbance, the higher probability of nonnative species becoming established, and climate change (e.g., changes in precipitation, wind, and temperature) are not well understood.

Approaches to Conservation under Climate Change

Climate change appeared late within the formal agendas of the Antarctic Treaty Consultative Meeting and Parties to CCAMLR. It appeared in 2008 due to repeated efforts of the Antarctic and Southern Ocean Coalition, the Scientific Committee on Antarctic Research, and some treaty parties to highlight the urgency of the climate change threat. In particular, the Antarctic and Southern Ocean Coalition has been the lead advocate for the need for climate adaptation planning (ASOC 2010a–d). Although the term *climate adaptation* is rarely used by the Antarctic conservation community, ideas on incorporating the effects of climate change into conservation practices in the Antarctic region are emerging.

The CCAMLR recognizes the importance of understanding the potential impact of climate change on krill recruitment variability (CCAMLR 2010). The Convention's Working Group on Ecosystem Monitoring and Management has recently constructed ecosystem models that attempt to take into account trends in the populations of Antarctic predators and prey (Trathan and Agnew 2010). CCAMLR also recognizes the need for well-designed surveys that could provide time series data of relative recruitments, and that such data are particularly important for monitoring future effects of climate change and as input into stock assessment models (CCAMLR 2010). CCAMLR is considering the need for a review of its Ecosystem Monitoring Program in order that the effects of krill fishing may be monitored in the face of rapid climate change (CCAMLR 2009).

Nonnative species have rapidly risen to the fore as one of the major concerns facing conservation in the Antarctic Treaty area. Antarctic terrestrial biota evolved to cope well with the harsh environment but this has come at the expense of reduced competitive ability, leaving them vulnerable to the impact of colonization by competitors that may have more advantage under changed climatic conditions (Turner et al. 2009). Treaty parties are in the process of developing a quarantine manual and a set of measures to minimize the risk of nonnative species introduction (France 2010). Ideas include procedures for cleaning vehicles to prevent the transport of nonnative species into and within the Antarctic Treaty area; guidance to personnel following the discovery of a suspected nonnative species (United Kingdom 2010a, b); and a framework for analyzing nonnative species risks for national Antarctica programs (New Zealand 2009b).

Outside the relationship between climate change and nonnative species, the consideration of climate change in the conservation of terrestrial environment is developing slowly. Proposals include the following (Norway and United Kingdom 2010; United Kingdom 2010c):

- a classification of existing protected areas according to climate change vulnerability;
- the need for more sophisticated and coordinated ecosystem monitoring;
- a review of existing management tools to assess their continuing suitability in a climate change context; and

- a more strategic approach to the selection and designation of protected areas that is dynamic and flexible enough to fast-track the protection of important new sites and vulnerable species at marginal locations and facilitate the delisting of sites for which the principal values no longer exist.

Strategic Planning and Protected Areas

Including climate change in the conservation of the Antarctic implies dealing with large-scale and long-term issues such as inter- and intracontinental transfer of non-native species, gradual or catastrophic disappearance or appearance of habitat, shifts in species distribution, and possible trophic cascades throughout the entire ecosystem. It requires (1) holistic governance that cuts across multiple sectors and issues, and (2) proactive visions for different temporal and spatial scales that can account for complex scenarios and interactions that go beyond the direct effects of climate change (Hansen and Hoffman 2011).

The CCAMLR and the Environmental Protocol are examples of how the Antarctic Treaty System have acted proactively, in anticipation of newly developed human activities and their environmental impacts (Roura and Hemmings, forthcoming). The CCAMLR was set up in advance of the development of any significant commercial krill fishery and the Environmental Protocol was negotiated before the start of any mining activity. This strategic thinking needs to be expanded and filtered through to the day-to-day governance of the Antarctic. Key issues that require strong consideration include the extent to which biological redistribution in Antarctica can and should be minimized, and how much time and resources the Antarctic community is willing to dedicate to reducing the rate at which unique biological assemblages are compromised and eventually lost (Hughes and Convey 2010).

A coherent vision, one that includes elements such as the desirable evolution of the footprint, type, and scale of human activities in the Antarctic, needs to be articulated and used to guide management decisions that are more proactive and encompass longer-term and larger-scale considerations (Convey et al., in review). Tools such as *strategic environmental assessment* can be used as guiding frameworks (Roura and Hemmings, forthcoming). Climate change is then incorporated as one of the environmental stresses that act across multiple sectors. Such strategic planning exercises can first be tested at a regional level, in areas where different actors and activities are already engaged. These represent microcosms of the treaty area and lessons learned may be scaled up or applied to other locations.

Protected areas are another tool that can be used to support the strategic planning process proposed above. They can be used to set aside climate refugia, that is, areas less likely to change. For example, the Ross Sea is an area where sea ice is likely to persist the longest, perhaps of any place on Earth. Its protection will ensure a refuge for sea ice-dependent species as sea ice continues to disappear from other areas

(Jenouvrier et al. 2009; Ainley, Russell, et al. 2010). Protected areas free from the pressures of resource extraction will also be important as climate change-reference areas. Establishment of inviolate, no-go zones can further help to protect the wilderness and scientific values of key high-risk areas for the future (ASOC 2010a). Sea ice is a dynamic habitat, and accordingly its associated species have far-ranging natural history patterns. Therefore, large-scale thinking and planning is required if representative areas are to be protected.

The Precautionary Approach, Scenario Planning, and Feedback Management

Data and scientific understanding of ecosystems will never be complete, and this extant uncertainty in Antarctic conservation will only be exacerbated by climate change. The precautionary principle adopted by the CCAMLR is particularly applicable to the consideration of reducing nonclimate stresses arising from human activities. Fishing has been shown to magnify the sensitivity of food webs to climate change (e.g., Österblom et al. 2007; Watermeyer et al. 2008). Studies have shown that under some scenarios of climate change, even moderate levels of fishing in the Southern Ocean may exacerbate any declines, and populations may not be able to recover within two to three decades in the absence of fishing (Trathan and Agnew 2010). Therefore, it is necessary that CCAMLR reviews current fishing strategies and quotas now, as it is likely that those currently construed as precautionary will be inadequate in the future (ASOC 2010b). As sea ice disappears, some areas will become more accessible, and CCAMLR needs to exercise extreme precaution in deciding whether new and exploratory fisheries should be allowed in these areas where data are extremely limited (Trathan and Agnew 2010). In general, there may be a need to reduce human access to certain vulnerable areas in order to allow ecosystems to have the space to cope with the stresses arising from climate change.

Uncertainty can also be addressed through scenario planning and adaptive management, often referred to as *feedback* management among the community of managers and researchers in Antarctica (see chap. 2). The need for a feedback management approach for Antarctic fish and krill stocks becomes even more pressing in the face of climate change (Trathan and Agnew 2010; ASOC 2010c). Such a management system depends on a good operational monitoring program, and the existing CCAMLR Ecosystem Monitoring Program needs to be significantly reformed and expanded in its coverage to provide the needed information (Trathan and Agnew 2010; ASOC 2010c).

Reducing the Rate and Extent of Climate Change

There is a limit to what can be accomplished by adaptation. For many polar species that depend on snow and ice and cold temperatures, limiting the rate and extent of climate change may be the only adaptation option (Hansen and Hoffman 2011). Within the global context, energy consumption and emissions from human activities in Antarctica are small, but activities in and travel to Antarctica are energy intensive and emissions of greenhouse gases, relative to the number of people involved, are proportionally high (e.g., Shirsat and Graf 2009; Lamers and Amelung 2007). The use of renewable energy and energy efficiency applications even in the harsh Antarctic environment has proven to lead to reductions in greenhouse gas emissions and cost savings (Tin et al. 2010). Additional measures such as strategic planning for new facilities as well as reduced travel and presence in the region will certainly lead to increased local and global environmental benefits at a low cost. As the only continent collectively managed internationally, and one that plays an important role in the study of global climate change, there is an ethical importance to reducing emissions and leading by example using best practices (ATCM 2008).

Conclusion and Recommendations

Because there are no permanent inhabitants in Antarctica, and in light of the unique legal status of Antarctica and the Southern Ocean, the South Polar region has largely fallen outside the global deliberations on climate adaptation under the Kyoto Adaptation Fund, Global Environment Facility, and others. The Antarctic Treaty System, the main governing mechanism for the region, has only very recently begun to consider climate change in its official deliberations. Despite a late start, ideas on incorporating the effects of climate change into conservation practices in the Antarctic region are emerging, and climate change is poised to rise as an important part of the work of the parties to the Antarctic Treaty and the Commission for the Conservation of Antarctic Marine Living Resources in coming years.

Many of the potential roadblocks to including climate change in conservation planning and action are shared among most other conservation issues, with challenges arising from the particular governance structure and geopolitics of the region. Creative thinking will certainly be essential for exchanging ideas with adaptation practitioners outside Antarctica and in making current conservation practices within Antarctica evolve. Within the Antarctic, a strategic planning process that includes a coherent vision and a strategic approach to selection of protected areas could help to bring about holistic governance that can account for complex scenarios and interactions that go beyond the direct effects of climate change. Uncertainty can be managed through the precautionary approach, scenario planning, and feedback management. Yet there is a limit to what can be accomplished by adaptation. While greenhouse gas emissions from human activities in Antarctica are small within the global context, the global symbolic value of reducing emissions from human activities in one of the world's most logistically challenging environments should not be underestimated.

Acknowledgments

This chapter grew out from an earlier contribution to an IUCN volume: T. Tin, D. Ainley, J. N. Barnes, and S. Hajost. 2010. "Impacts and Ecosystem-based Adaptation in the Antarctic and Southern Ocean." In *Building Resilience to Climate Change: Ecosystem-based Adaptation and Lessons from the Field*, A. Andrade Pérez, B. Herrera Fernandez, and R. Cazzolla Gatti, eds., Gland, Switzerland: IUCN. Thanks to the editors and reviewers of both volumes for the opportunity to express our views and improve our text.

PART 7

Lessons Learned

Based on this volume's nineteen case studies from around the world, an examination of how climate change is being incorporated into landscape and seascape science, planning, and action in these scapes offers valuable insights. Guidance comes both from seeing how some tools and approaches may be robust across scapes, and from understanding how others necessarily differ in response to local ecological, political, and socioeconomic circumstances. The case studies represent an ongoing set of experiments as to how best to conserve biodiversity during this time of rapid climate change.

Chapter 22

Moving Forward on Climate Change Science, Planning, and Action

JODI A. HILTY, MOLLY S. CROSS,
AND CHARLES C. CHESTER

Working in more than seventy landscapes and seascapes around the world, the Wildlife Conservation Society is but one example of many institutions working to bring about conservation at the large *scape* level (see extended definition of scape in chap. 1). Many of these locations face serious and imminent threats that, if left unresolved, will lead to the loss of the wild areas and wildlife within them forever. For example, the Western Ghats region of India—essentially, the mountain range running parallel to the country's southwestern coast—is one of those scapes where rising human population pressures translate into expanding impacts including agricultural expansion, road development, livestock grazing, and habitat fragmentation. It is an all too familiar panoply of threats, but one that conservationists have no choice but to face head on. Quite simply, if conservationists are to ensure that populations of globally imperiled tigers (*Panthera tigris*), Asian elephants (*Elaphus maximus*), and other rare species maintain their stronghold in this region, they must bring immediate attention and redress to these threats.

Biodiversity in the Western Ghats faces all of these challenges with or without the manifestation of regional climate change. We do not want to understate this critical point, particularly in a volume that focuses on "scape-level responses to climate change." At the same time, we are impelled to juxtapose—not to contradict, but to juxtapose—the equally critical point that climate change will likely entail serious ramifications in the Western Ghats not only as a direct threat to biodiversity in and of itself, but also as (1) an indirect threat due to climate change's exacerbating effects on current threats such as habitat change and invasive species, as well as (2) an indirect threat via interference with traditional conservation responses to the familiar listing of the drivers of biodiversity loss.

Admittedly, these may at first seem to be fine distinctions, but they are not—far from it. Looking backwards in time, we face the prospect that our failure to address the already evident and increasing impacts of climate change could ultimately imperil many of our myriad conservation efforts to date. Even more ominously, those conservation solutions being proposed today for the Western Ghats may or may not work under conditions of climate change tomorrow. Consequently, it has become important to take into account a wide range of questions that climate change will likely entail:

- How might climate change directly affect tiger populations?
- How will climate change affect the prey populations upon which the already-threatened tigers depend?
- What type of vegetation shifts could occur that would affect elephant diets?
- How will humans respond to vegetation shifts associated with climate change?
- Can the current forms of local and regional governance within the Western Ghats effectively respond to these kinds of shifts?
- Will historical and ongoing conservation activities continue to serve their stated purpose as climate changes?
- What new conservation options or altered management practices might be needed to achieve conservation goals given the impacts of climate change?

These questions become all the more pertinent in light of the expectation that the Western Ghats will be one of the most vulnerable areas to projected impacts of climate change in India (Chaturvedi et al. 2011). They are also examples of questions that may vary in degree and emphasis from scape to scape but exist everywhere. In essence, failure to address climate change questions in the context of conservation may well result in long-term conservation failure. For this reason, we need to begin incorporating climate considerations into conservation around the world and share the lessons as we learn.

Of course, be it within the Western Ghats or at the global level, conservation is fraught with complexities and challenges even without considering climate change. With the global expansion of threats ranging from habitat destruction and invasive species to acid rain and endocrine disruptors, the vast majority of the Earth's surface has felt the impacts of human activities (Sanderson et al. 2002). Conservationists have achieved a number of key successes, including increased protected areas, new regulations, halting or slowing human impacts in globally important ecosystems, and many others. Still, conservation remains a low priority to society, posing persistent challenges to the mission of protecting biodiversity in its many manifestations. As but one example, the portion of private charitable donations directed toward environmental causes constitutes about 2.3 percent of all giving in the United States, of which only a small portion is targeted for biodiversity conservation (GUSAF 2011).

So why should climate change be added as another factor when we have not been able to eliminate other threats and when resources to do so are scarce? We argue that failure to consider climate change in protecting biodiversity would be failure to recognize a new and permanent organizing factor in conservation. As the nineteen scapes in this book demonstrate, today's changing climate is already affecting biophysical factors such as the relative distribution and abundance of many species. Predictions of potential future impacts are even more dire. For example, in 2004 the journal *Nature* published an article covered by newspapers worldwide suggesting that of the approximately five million terrestrial species that currently exist, 18 to 34 percent of them are likely to go extinct, given current climate projections (Thomas et al. 2004). This would amount to somewhere around a million species lost forever. The article did not pass by without a good deal of critical review from within the scientific community. But as one critique noted:

> Whether or not we can improve our ability to forecast the magnitude of impending species losses, however, one generalization from available knowledge seems robust. All estimates using available methods point to a huge impending extinction episode from unmitigated anthropogenic climate warming and that the uncertainty in these methods are likely to underestimate the extent of extinctions in the coming centuries. The data suggest that life on earth, as we have known it, is going to change dramatically over the coming century. (Harte and Kitzes 2011)

In other words, regardless of any specific estimate, it is widely acknowledged that a substantial number of species are likely to experience extinction due to climate change. For that reason alone, natural resource managers, conservationists, and others interested in maintaining the diversity of life on earth need to pay attention to climate change.

The chapters in this book offer cautious hope of how climate change can be and is being incorporated into scape-scale conservation. These case studies touch on three basic tenets of incorporating projections into management: science, planning, and action. While not new to conservation, these three tenets will be vital to successful climate change adaptation. Although the degree to which science, planning, and action occur in each scape varies across the case studies, these examples illustrate how the conservation of large scapes is moving forward since at least some stakeholders located in each geography are exploring ways to incorporate climate change.

The take home messages from these chapters are nuanced, revealing similarities and differences across the case studies. Many scapes would benefit from expanded basic science and most are trying a variety of tools to identify and implement actions that address a changing climate. They also are situated in different political contexts. Below, we distill key messages, highlight unique approaches, and lastly close with a list of recommendations for others who may be embarking on efforts to address climate change adaptation at a scape scale.

Reasons for Addressing Climate Change

Given that biodiversity in each of the case study scapes is imperiled by numerous threats independent of climate change, we want to understand why the stakeholders in these nineteen scapes have decided to incorporate climate change into their thinking and action. The answer is varied. In the Yellowstone to Yukon landscape, the founding conservation vision was formed with climate change in mind, although it played a more subdued role compared with general concerns over genetic connectivity (chap. 19). Along with the Yellowstone to Yukon Conservation initiative, the Sundarbans mangrove forest in Asia (chap. 12) and the northern Appalachian/acadian ecoregion (chap. 18) are all focused on existing protected area networks. In recent years, these scape conservation efforts have more formally incorporated climate change as an explicit programmatic priority, reassessing whether the envisioned protected area networks will be adequate to achieve their original intent of preserving target species and ecosystems as the climate changes.

Other entities are following international or national political imperatives. In the case of freshwater conservation in the Alps (chap. 11), European Union-wide agreements and resulting directives strongly influence the direction of conservation efforts. In Australia, the federal government made climate change adaptation a continental priority, resulting in the Great Eastern Ranges Corridor becoming a federal priority for conserving biodiversity from the threat of climate change (chap. 16). In the state of Washington, United States, the state government mandated a closer examination of and planning for climate change (chap. 10). One of the challenges with jurisdiction-by-jurisdiction efforts is that they often do not include consistent planning across state, provincial, or other political boundaries, which is necessary for planning to be ecologically meaningful. Although Washington State has a strong plan, ensuring compatibility with activities in neighboring jurisdictions remains an ongoing challenge for which strong cross-jurisdictional collaboration is essential (chap. 10).

For other scapes, conservation efforts have begun addressing climate change as a result of pressure from international entities or funders. In the Albertine Rift in Africa, funders largely drove the push for integrating climate change into conservation planning and actions (chap. 3). This is similar for the Mesoamerican Biological Corridor in Latin America, which offers strong cautions on donor-driven agendas (chap. 5). The history of conservation in Central America has long been fraught with donor-influenced priorities, and chapter 5 offers a refreshingly frank assessment of how donor-driven agendas can derail conservation priorities and how the region could recover using climate change adaptation as a way to refocus on biological conservation.

Relative Priority of Climate Change

The understanding that climate change and biodiversity are interrelated is generally accepted across all of the case studies. Most case studies suggest that current efforts to protect biodiversity can make the impacts of climate change and subsequent biodiversity loss less profound. Similarly, all the case studies infer that incorporating climate change adaptation into conservation will be important for conservation to be successful over the long term. Despite these similarities, the relative priority of addressing climate change adaptation varies significantly across the different case studies.

As previously mentioned, the creation of the Great Eastern Ranges Corridor in Australia (chap. 16) was driven by a plan to conserve biodiversity during a changing climate. There, the federal prioritization of climate change as an issue drove the announcement of this continental conservation effort. The authors go so far as to say that climate change "has helped introduce a new paradigm for conservation land use for Australia" (chap. 16, p. 214).

For many scapes, climate change is increasingly recognized as a dominant organizing factor for conservation, particularly those more isolated from human activities (e.g., Ramirez-Llodra et al. 2011). In Antarctica (chap. 21), this recognition is driven by the fact that climate change may have more sizeable impacts on species and ecosystems in the near term than will human activities since that area currently has a relatively low human footprint. Similarly, in Arctic Alaska (chap. 20) where the polar scape presently benefits from being less affected by human activities, one of the key threats of a changing climate is that it enhances the potential expansion of the human footprint. As both Arctic Alaska and Antarctica become more accessible with climate change, the consequences of increased visitation, introduction of invasive species, and various types of natural resource extraction grow. Also, the fate of species currently thriving in polar regions is unclear, and conservation efforts are focused on identifying those places that are most likely to serve as refugia or functional habitat in this time of rapid change.

For other regions where human activity levels are already high, the presence of a notable human footprint is part of the motivation for focusing increased attention on climate change. In low-lying coastal areas such as Fiji (chap. 13), rising sea levels could have significant effects on both biodiversity and the human communities living along the coast. For this reason, the urgency and acceptance with which climate change is being addressed is arguably stronger than in many other regions.

In contrast, for the majority of case studies, climate change has not risen as a singular priority, but rather has come to modify how conservation strategies are being prioritized and how they are being conducted. For example, for places such as the Northern Appalachian/Acadian Ecoregion (chap. 18), climate change is not what made connectivity or the strategic design of a protected area network imperative. Instead, the destruction and fragmentation of species' habitats are the major threats to biodiversity. Placing climate change on top of those other threats does, however, raise the level of urgency associated with these actions, and can offer better direction as to where to focus connectivity and land conservation activities within the landscape. In this way, climate change-adaptation planning can influence the priority of myriad conservation needs, and is being incorporated as one factor in existing conservation planning and decision-making processes (IPCC-WG2 2007). For other scapes such as Africa's Albertine Rift (chap. 3), although climate change is recognized as a necessary component to consider in a long-term vision for conservation success, it is also seen as a low immediate priority despite donor attention to the matter. In the case of the Amazon (chap. 4), authors of the case study make a compelling argument that current local land-use change is the driving factor affecting local Amazon climate and that this land-use conversion, exacerbated by global climate change, will transform the system if it continues to expand. Further, the major priority for climate change adaptation is to stop this land-use change although the authors recognize that analyses and planning around climate change adaptation are lacking in the region.

If the impacts of climate change become more profound in these and other scapes, it is likely that the priority of addressing it will also rise—hopefully not too late to be able to respond effectively. While it is commonly recognized that failure to address immediate threats will render moot the consideration of climate change in long-term biodiversity conservation, various stakeholders in each case study are emphasizing the need to factor climate change into longer-term planning. Indeed, if the nineteen case studies share a common trait, it is the recognition that waiting for reactive reprioritization is counterproductive and that incorporating climate factors into conservation planning *now* is the right approach. Moreover, what is especially promising is that many of these scapes have demonstrated ways of incorporating climate change adaptation into ongoing conservation efforts. In other words, the land, sea, and wildlife management decisions and priorities in these scapes address both ongoing, immediate threats as well as climate change. Whether or not their efforts will be adequate is highly uncertain, but we suggest that important steps, albeit initial steps, are being taken in all of the scapes discussed in this volume.

Importance of Collaboration

A striking similarity across the nineteen scapes is that virtually all are engaged in complex partnerships including local, subnational, and national government agencies;

regional and/or international nonprofit conservation groups; businesses; academics; local communities; and/or individuals. Given the nature of large scape conservation, the success of a broad geographic vision arguably will require the engagement of a large subset, if not all, of the interested entities in a region. Where the science, planning, and implementation were the strongest, the products or outcomes appeared to be strongly driven by collaborative efforts, such as efforts to conserve marine turtles in the Wider Caribbean Region (chap. 14). The Madrean Sky Island landscape (chap. 17) details how success in working across the United States–Mexico border has been based on shared learning, networking, and building capacity across the region—all necessary prerequisites for collaborative adaptation projects to move forward. As the Madrean Sky Island example indicates, collaborations do not just appear, but rather the right circumstances need to be created and fostered for successful partnerships.

Failure to create these circumstances can hamper projects; lack of collective action across whole scapes meant that in several case studies implementation occurred in only some parts of large scapes. The Northern Appalachian/Acadian Ecoregion (chap. 18) is but one case where a strong international vision and scientific foundation exists to enact climate adaptation actions on the ground, but funding for implementation is lacking on the Canadian side such that the United States partners are moving forward ahead of their Canadian counterparts. This theme exists in other scapes as well, including the Altai-Sayan landscape (chap. 15). For the Mesoamerican Biological Corridor (chap. 5), actions are dependent on players within individual countries such that Costa Rica and Mexico moved further ahead on implementing the vision than most of the other partner countries. Decisions for Antarctica and surrounding waters (chap. 21) are dominated by consensus processes, which require collective action and are more challenging politically to move forward quickly. Even in-country political lines can mean that visions are acted upon differently. In Australia, Queensland's efforts were reported to remain weak compared to those of Victoria, New South Wales, and Australian Capital Territory (chap. 16). While there are many factors underlying the circumstances in each particular case, in general understanding what drives these differences should be examined more closely. Even as it is all too evident that politics and financial resources play important roles, it is likely that biophysical factors, land ownership regimes, and other issues also come into play.

While many scapes are challenged by multijurisdictional contexts, others seem to be engaging across borders fairly well. Although the United States and Canadian governments do not formally recognize a shared agreement about the whole region, the Yellowstone to Yukon Conservation Initiative and vision has now been operating for over fifteen years (chap. 19). The efforts of many entities, from governments and local communities to nonprofits and academics, have been steadily driving forward policies and land practices to advance the vision. In the Wider Caribbean Region, an NGO has led efforts to build regional capacity and reduce climate change threats to coastal ecosystems that support marine turtles and human communities; this effort has successfully engaged existing regional collaborations and a vast network of science experts to advance adaptation planning and action (chap. 14).

Although some transboundary mechanisms exist through international law, such as the 1916 Migratory Bird Treaty or the North American Agreement on Environmental Cooperation under NAFTA (North American Free Trade Agreement), the formalization of shared biodiversity priorities at a scape scale across borders is lacking in most international conservation efforts. One example of where such agreements exist is in Europe where shared directives assist in pushing forward a European environmental network including a focus on freshwater conservation in the Alps (chap. 11). For existing and new transboundary scape-scale projects designed to address climate change adaptation to be successful, a careful evaluation of what mechanisms and partnerships need to exist for the region to progress as a whole will be important.

Role of Science

In all scapes examined in this volume, science played a role in guiding climate change-adaptation priorities. But it did so at significantly varying levels and in different ways. For some scapes, conservation actions have been broadly informed by syntheses of climate change impact research. In the Yellowstone to Yukon landscape (chap. 19), a climate change-informed conservation agenda was laid out in response to a comprehensive overview of relevant research in the region (Graumlich and Francis 2010). Climate change impacts reports for other scapes, such as the Altai-Sayan region (chap. 15), have clearly played an important role in amassing information to motivate and inform conservation actions. In other scapes, conservation planning is being driven by targeted climate change analyses and/or focused engagement of experts. For example, in the eastern Mongolian grassland steppe, The Nature Conservancy has developed new models of climate change effects on net primary productivity to prioritize areas for protection and for inclusion in regional "grass banks" to support domestic grazing needs during drought periods (chap. 8). In the Northern Great Plains (chap. 9), regional experts considered downscaled climate data and associated vegetation model outputs during a scenario planning workshop, and developed recommendations on managing grasslands to maintain a diversity of habitats for birds and providing incentives for maintaining wildlife habitat on private lands. In the Wider Caribbean Region (chap. 14), marine turtle scientists and managers have been engaged in region-wide climate change adaptation planning since 2007 through workshops and self-administered surveys. These examples, and others covered in this book, illustrate the ways that climate science is being integrated into conservation planning at the scape scale.

Incorporating climate science into conservation planning has yet to become an integral and automatic part of any planning process, and such will never happen unless and until scientists, managers, and conservation practitioners ensure that it does. In the Madrean Sky Island landscape (chap. 17), a priority has been placed on the "co-production of knowledge" by having decision makers, stakeholders, and scientists from both the United States and Mexico work together to create usable science for framing climate change-adaptation policy. Even when climate science is successfully integrated into planning, the challenge remains on how to move that science and planning into action. In the Cape Floristic Region of South Africa (chap. 7), conservation scientists met with officials from a public institution with the statutory responsibility for biodiversity conservation in the region to discuss their use of models to identify priority areas for conservation given climate change. While it remains to be seen just how much of that modeling will inform land protection efforts in the Cape, this kind of effort to link the science to relevant decision makers is necessary for adaptation planning to be meaningful and result in successful implementation.

Several case studies noted that a paucity of data and uncertain forecasts for the future pose significant hurdles to integrating climate change into conservation. The lack of adequate weather stations, such as in the Albertine Rift and Arctic Alaska (chaps. 3 and 20), was cited as a limitation to understanding historic climate trends and correlations between climate and ecological conditions, thereby limiting the capacity to generate effective and useful models of future conditions. Even in regions with more extensive history of and capacity for ecological and climate science, uncertainties in modeling future conditions presents substantial difficulties. As is the case with much science related to natural resources, there is often a desire for more precise data and analysis, and yet also a pressing need to take action under imperfect information. In response to this common dilemma, several scapes presented scenario planning and adaptive management as tools for moving forward without perfect information (chap. 2). During the scenario planning workshops in the Northern Great Plains (chap. 9), managers were able to identify a range of conservation actions that are relatively robust to uncertainty since they would apply under differing yet plausible scenarios of the future. Scenario planning is also recommended for Antarctica (chap. 21), along with adaptive management of Antarctic fish and krill stocks to adjust actions and priorities in response to monitored changes.

The continuing need for science to support climate change adaptation in these scapes is clear. First, we need science to detect and understand the changes happening at a species and ecosystem level. Second, we need to integrate the climate sciences and conservation sciences to understand impacts and management options at the scape scale, which in turn will guide our decision-making and adaptive management processes. In the Canadian boreal forest (chap. 6), the Conservation Matrix Model demonstrates a way to prioritize protection of large functioning ecosystems, conserve special elements, and design a human land-use matrix that complements conservation in the region. Third, interdisciplinary science that connects across social, political, economic, and biological sciences will help us understand the links between these different fields and find solutions that may be beneficial for both ecosystems and humans, a concept utilized in the Madrean Sky Island case study (chap. 17). Finally, since information is often, although not always, inversely proportional to acrimony, the creation of shared information at the scapes level can drive different stakeholders to work together toward common solutions.

End Goals

The goals for climate change adaptation can range from resisting change, increasing resilience to change, and facilitating transformation (see chap. 2 for a review of concepts). In most scape case studies, transformation is not overtly mentioned as a goal and at most vaguely referenced. The Northern Great Plains and Antarctic landscapes (chaps. 9 and 21) do recognize that current species composition may not be possible to maintain in the future, but none of the scapes set an explicit way forward to plan or prepare for such a transformation. Some species may benefit from climate change and others may perish, and such a transformational tenet has not yet deeply penetrated the natural resource management or conservation psyche. Perhaps part of this is embedded in the history of conservation that has focused on referencing back to historical baselines (chap. 1). This might also be due to the uncertainty of future scenarios and/or humanity's general denial of what the future may look like.

Overall, the emphasis on resistance and resilience-building strategies over transformation in the case studies covered here is not unusual. As but one example, when The Nature Conservancy integrated climate change into twenty conservation projects from around the globe, project teams identified a total of forty-two adaptation strategies—only two of which addressed transformation (Poiani et al. 2011). Such a limited focus on transformation is probably shortsighted. Managers in these and other scapes considering climate change adaptation should take pause and explore what the scape could look like, to ensure that near-term actions are conducive to plausible longer-term future scenarios. That said, there are valid reasons to focus on building both resilience and resistance in the near term, and even if ultimately unsuccessful, those strategies may help a scape be amenable to transformation as systems inevitably change.

Whereas the term *resistance* indicates to resist changes or maintain status quo, resilience and adaptation are used and interpreted a number of ways. For example, when stakeholders in the Yellowstone to Yukon landscape suggest that the purpose of incorporating knowledge of climate change is to enhance ecosystem resilience (chap. 19), it is difficult to deduce from such general statements if the entity or entities laying out such a vision expect ecosystems to be able to persist as climate changes. In other words the goal of resilience could be for ecosystems and the components to more or less remain where they are currently located, continue to be located somewhere within the scape, or to be functionally replaced. Another general concept that appears in a number of the case studies that leaves much room for interpretation concerns providing opportunities for adaptation. This could mean different endpoints as well. It could mean allowing or facilitating transformative change, or it could suggest adaptation toward the end of maintaining current systems, or some combination of both.

Because these are relatively new terms for conservation in the context of climate change, we suggest that visions and goals that use these terms should be as explicit as possible about what they mean with respect to what kinds of changes might be inevitable, and which changes we are trying to reduce or prevent. And just as we must be careful about language and definitions, we must also be mindful of the possibility that we find ourselves in the midst of a dramatic shift in conservation philosophy. For example, in the United States where national parks were first established, society has not really grappled with what it means to "protect and preserve" if fifty years from now most of the country's national parks look very different from what they look like today. There is no easy solution to reconciling our traditional outlook with looming changes, but the sooner we recognize the significance of these changes and begin grappling with them, the sooner our management of national parks and other such important conservation places will allow for "best-possible" protection of a changing ecosystem.

Connectivity

Establishing, maintaining, and restoring connectivity as a scape-scale adaptation strategy is a recurring theme through many of the chapters. This includes the marine-terrestrial connections inherent in the Sundarbans mangrove forest and the "ridge to reef" continuum in Fiji (chaps. 12 and 13). Similarly, conservation practitioners in a number of terrestrial landscapes are adopting various approaches to protecting connectivity as a climate change-adaptation tool (box 2.2). In some cases, such as the Altai-Sayan landscape (chap. 15) and the Y2Y landscape (chap. 19), connectivity efforts are focused on linking large and sometimes near-contiguous protected areas that span north-south latitude and low-high elevation gradients. In other scapes, conservation scientists are explicitly incorporating climate and/or species modeling tools into establishing connectivity priorities. For example, researchers in the Cape Floristic Region (chap. 7) are coupling future climate models, species distribution models, and reserve selection tools to examine the connectivity needs of species as they track shifts in suitable habitat through time and space. Given the caveats associated with downscaling future climate model outputs to fine resolutions (box 2.1), connectivity modelers in Washington State (chap. 10) have developed a new technique for identifying what they call "climate-smart connectivity" that relies on information about current climatic gradients, land-use patterns, and topography.

Given that connectivity is one of the most frequently recommended tools for climate change adaptation in the scientific literature (Heller and Zavaleta 2009), it is hardly surprising to find connectivity as a common theme across many scapes. As with many of the proposed climate adaptation principles (chap. 1), the protection and restoration of connectivity specifically for the purpose of enabling species responses to climate change remains largely untested, and there are numerous uncertainties and complicating factors in assessing the effectiveness of maintaining connectivity. Yet, as we suggested elsewhere:

> Despite these unknowns, it is hard to think of any alternative to connectivity conservation, reasonable or unreasonable. Translocation? Genetic manipulation? Controlling greenhouse gas emissions to a safe level? Connectivity conservation areas appear to be our best comparatively reasonable hope for protecting biodiversity in the long term. (Chester and Hilty 2010, 31)

As mentioned in chapter 1, one alternative to enhancing connectivity is translocation of species (also described as "assisted colonization," "assisted migration," and "managed relocation"; see Loss et al. 2011). Connectivity and translocating species may be seen as opposite ends of a movement spectrum—at one end of which connectivity allows for species to move on their own, while at the other end humans move species from one locality to another. Although translocating species likely to perish because of climate change has gained significant traction in the literature as an option that should be on the table (Hoegh-Guldberg et al. 2008), it is notable that none of the case studies entertained translocation of species as a tool under consideration. We do, however, feel safe in asserting that this generally results from a long-standing preference on the part of conservationists and biologists in favor of enhancing connectivity over translocation as an adaptation strategy, a preference grounded in the assumptions that managed relocation: (1) is generally costly and inefficient, (2) may be ineffective in terms of long-term survival, and (3) will increase the possible spread of what could inadvertently become invasive species (Ricciardi and Simberloff 2009). The long list of invasive species that were intentionally introduced by humans make this last issue particularly relevant because it is extremely difficult to predict how a new species will "fit" in the context of other species in a new ecosystem. Although the resistance to considering translocation as a viable adaptation strategy is unlikely to change in the near term, the door has been opened to discussion of it as a possible conservation strategy. If worst-case climate change impacts become a reality and we see more species teetering on the edge of survival, we might expect discussions about translocations to escalate, necessitating the need for detailed cost/benefit assessments. For now, it seems that for most scapes such measures are not being seriously considered.

Bridging Adaptation and Mitigation

Although not the intended focus of this book, many chapters compellingly argue that addressing mitigation—the reduction of greenhouse gas concentrations in the atmosphere—has to be part of a comprehensive adaptation strategy. Failure to address greenhouse gas emissions will ultimately doom adaptation efforts as the magnitude of climate changes reaches a dire level. In this context, the irony of increasing pressure for oil drilling in the Arctic (chap. 20) deserves particular notice given that burning of such oil will only contribute to already strong climate change challenges in the region.

While reducing greenhouse gas emissions to the atmosphere is the most important approach to altering our current trajectory, the sequestration and storage of carbon such as in soils, peatlands, and vegetation is also an important component of mitigation. Several chapters described how conservation actions can simultaneously contribute to climate change adaptation and mitigation if projects are designed with both goals in mind. In the case of the Amazon (chap. 4), the government pledge around climate change is focused solely on mitigation and not adaptation. Since mitigation-related funding is currently higher than that for adaptation efforts, identifying projects that achieve mitigation and adaptation goals can essentially increase available resources for adaptation. Currently within the international arena, the primary mechanism for funding carbon storage projects is through Reduced Emissions from Deforestation and Forest Degradation (REDD), a controversial and evolving program that offers forest owners (e.g., governments, private companies, or local communities) rewards for keeping their forests intact rather than cutting them down (Harvey et al. 2010; chap. 1). REDD had evolved to REDD+, incorporating conservation, sustainable management, and increasing forest carbon storage as well. In the case of the Mesoamerican Biological Corridor (chap. 5), the opportunity exists to store carbon through reduced deforestation in key protected areas and corridors. However, the selection of areas for carbon sequestration and storage projects in Central America is not closely tied to biodiversity conservation priorities. If the international community follows through on its commitments and the implementation of REDD and similar types of "payment for ecosystem services" become more widespread, conservationists must take advantage of and build on this exceptional opportunity—indeed, one of the only such opportunities that seem practicable—to enhance the overlap between such mitigation and adaptation activities.

Actions in the Alps were viewed both as adaptation and mitigation measures (chap. 11), such as the example of marshland and forest restoration to enhance carbon dioxide storage. They also point to the restoration of rivers for mitigation, pointing out that doing so can reduce natural hazards such as flooding and landslides from increasingly common extreme rainfall events. The boreal forest in Canada (chap. 6) offers another example for aligning mitigation and adaptation priorities. Linear fragmentation from roads, pipelines, and habitat loss due to natural resource extraction not only diminishes the integrity and related resilience and resistance of the boreal system to climate change, but also leads to enormous carbon release into the atmosphere. Forests and underlying peat at the growing edge in boreal systems are more likely to dry out, experience changes in species composition, and be susceptible to fire. Prioritization of carbon storage in key intact boreal forests could both diminish carbon release and provide incentives to maintain large blocks of boreal forest as part of a climate change-adaptation strategy.

Adaptation for Biodiversity and Human Systems

Similar to finding win-win scenarios for mitigation and adaptation, several case studies discussed the value of connecting adaptation strategies for biodiversity conservation with maintaining or improving human livelihoods. The Mongolian grassland steppe landscape (chap. 8) insisted that, "overall, the climate adaptation strategies employed must consider the natural landscapes as well as the social, cultural, and political contexts in Mongolia." Conservation practitioners in Mongolia are explicitly integrating human needs into climate change-informed biodiversity conservation by identifying areas that can largely be set aside as protected areas, but in which cattle grazing is allowed during particularly bad droughts (i.e., grass banks). In the Vatu-i-Ra seascape in Fiji (chap. 13), practitioners are focusing on maintaining and restoring ecosystems for the role they play in increasing the ability of humans to adapt to climate change, and on building the social resilience of local human communities to allow for flexible and adaptive management of natural resources. Freshwater conservation in the Alps (chap. 11) provides yet another example where the current and future outlook for both biodiversity and humans can be improved through maintaining and restoring key waterways. In the case of the Amazon (chap. 4), deforestation for crops and pasture must be stopped before the Amazon reaches a tipping point such that local climate moves from a moist recirculating system to a hotter drier system. Continued agricultural expansion will imperil the Amazon as we know it, while simultaneously negatively affecting human livelihoods as neither system benefits from hotter and drier conditions. By recognizing the mutual benefits to biodiversity and human communities afforded by particular conservation and management actions, there is an increased likelihood that those actions will be embraced and implemented.

Recommendations

The issues discussed in this and the preceding chapters lead to some clear recommendations for strengthening current and future efforts to incorporate climate change adaptation into scape-scale conservation (table 22.1).

The first and most obvious step is to evaluate conservation priorities in light of climate change. We need to step back and evaluate what conservation means in this

TABLE 22.1 Summary recommendations for strengthening efforts to incorporate climate change adaptation into scape-scale conservation

Conduct a critical examination of what conservation looks like, given increasing climate changes.
Define and be explicit about goals from resistance and resilience to transformation.
Create broad and inclusive collaborations at all levels within scapes.
Work at the appropriate scape scale to address conservation needs in the face of climate change.
Ensure that federal and sub-federal governments as well as international agreements prioritize climate change adaptation.
Advocate for and enable international entities to form agreements or other binding mechanisms that support transboundary scape conservation and climate adaptation.
Establish mechanisms to fund both national and transboundary collaborations.
Prioritize science that informs climate-relevant decision making and actions, including adaptive management.
Reduce atmospheric greenhouse gases.
Link relevant mitigation efforts in scapes to climate change-adaptation priorities.
Practice ecosystem-based adaptation: connect conserving and managing scapes to maintaining and improving livelihoods.

rapidly changing world. Doing so means formally letting go of the perception that ecosystems are static, and balancing our human tendency to look backwards at past conditions with the difficult challenge of looking forward to likely future changes in order to understand what actions are needed to maximize biodiversity conservation over the long term. This returns us to the strategies of *resistance*, *resilience*, and *transformation*, and the critical importance of being clear about our goals. While human-caused climate change impacts have been increasing for many years, we are more or less just beginning to see confirmed signals and more extreme impacts are predicted to come such that we will be forced to make hard decisions about what to do with scarce conservation resources. Establishing clear priorities, to the degree possible given inevitable uncertainties about the future, will help ensure that we ultimately maximize our ability to conserve biodiversity.

Building a suite of collaborators even at the early stage of goal setting helps move climate change-adaptation priorities forward over the long term. Both setting of goals and building a team of collaborators must occur at the appropriate scale. We argue that a large scape scale is necessary given that some parts of such a larger scape may offer refugia, and that looking at too small an area may lead to the unneccessary prioritization of species or ecosystems that are expected to fare well in other parts of a scape.

With goals articulated and collaborators on board, it is advantageous to develop a policy framework under which climate adaptation priorities can move forward. Ensuring that climate change adaptation and mitigation become federal or other government-level priorities is a game changer in accomplishing scape-scale conservation; indeed, it appears to be one of the strongest ways to obtain recognition, funding, and prioritization. Similarly, a lack of clear international agreements can hamper progress across international scapes, leading to unevenness in funding levels and differential priorities along political boundaries. Mechanisms to formalize large scape-scale conservation that incorporates climate change are needed. Such mechanisms could be in the form of legislation, or establishing formalized but voluntary collaborations and incentive systems. The challenge of securing funding to work across large scapes and especially across political boundaries is ongoing, so developing funding streams in parallel with formal mechanisms is crucial.

As with collaborative mechanisms and funding, sound science that addresses applied conservation questions also is needed to drive priorities forward. The reality is that the field of climate science is new and evolving. New information can change priorities, and yet the threat of climate change to biodiversity conservation is so evident that we cannot afford to wait until the climate science is clearer. Given this, science is key for guiding decisions and informing adaptive management into the future. Moveover, to assist with decision making in light of the uncertainties associated with climate change, we suggest that climate change planning and action prioritization consider multiple plausible scenarios for the future.

The use of mitigation funding mechanisms to address not only mitigation of climate change but also climate change-adaptation priorities makes intuitive sense. Stronger links between mitigation and adaptation priorities should be sought across scapes as well as at global, country, and even sub-regional levels.

Similar to finding win-win scenarios with mitigation and climate change, connecting biodiversity conservation adaptation strategies to maintaining or improving human livelihoods also make sense. That said, if biodiversity conservation receives lower priority compared to improving livelihoods, then conservation will not succeed. It may be that some conservation adaptation priorities are not associated with human livelihoods but should still be supported and implemented.

Conclusion

Because of the scale at which many species and ecosystems need to be conserved, and because of the scale at which climate change is causing impacts, conservation today requires working across large scapes. Failure to do so will shortchange conservation significantly. We cannot afford to continue to consider conservation only within the jurisdiction of a single national park, a single forest, a single stretch of river, or a single community, but rather we must understand how these entities contribute to conservation across a broader scape. The case studies in this book give compelling examples of both why and how to do so. These scapes are at different stages and engaging at different levels in the science, planning, and implementation of climate change-adaptation strategies. At the same time, both individually and as a whole, they offer important jewels of information as to how to incorporate climate change adaptation into conservation. It is not that the case studies are necessarily adding major innovations to the conservation toolbox. Rather, they illustrate shifting priorities that range from minor adjustments in the scape's conservation agenda to profound alterations in stakeholder priorities. We are early in incorporating climate change into scape conservation. Watching these "experiments" going on around the world will be important to building on our understanding and honing our tools and approaches. We hope this book will serve as inspiration and guide for others to now engage in climate change adaptation at the landscape or seascape scale.

REFERENCES

Abell, R., J. D. Allan, and B. Lehner. 2007. "Unlocking the Potential of Protected Areas for Freshwaters." *Biological Conservation* 134:48–63.

ACIA (Arctic Climate Impact Assessment). 2004. *Scientific Report.* Cambridge, UK: Cambridge University Press.

Adams, D. K. and A. C. Comrie. 1997. "The North American Monsoon." *Bulletin of the American Meteorological Society* 78:2197–2213.

Adams, V. M., M. Mills, S. D. Jupiter, and R. L. Pressey. 2011. "Improving Social Acceptability of Marine Protected Area Networks: A Method for Estimating Opportunity Costs to Multiple Gear Types in Both Fished and Currently Unfished Areas." *Biological Conservation* 144:1, 350–61.

Adger, W. N., N. W. Arnell, and E. L. Tompkins. 2005. "Successful Adaptation to Climate Change across Scales." *Global Environmental Change* 15:77–86.

AFWA (Association of Fish and Wildlife Agencies). 2009. *Voluntary Guidance for States to Incorporate Climate Change into State Wildlife Action Plans and Other Management Plans*. Washington, DC: Climate Change Wildlife Action Plan Work Group, AFWA.

Ahmad, I. U., C. J. Greenwood, A. C. D. Barlow, M. A. Islam, A. N. M. Hossain, M. M. H. Khan, and J. L. D. Smith. 2009. *Bangladesh Tiger Action Plan 2009–2017*. Dhaka, Bangladesh: Bangladesh Forest Department, Ministry of Environment and Forests, Government of the People's Republic of Bangladesh. www.bforest.gov.bd/tiger_plan.php (accessed January 11, 2012).

Ainley, D. G., G. Ballard, S. Ackley, L. K. Blight, J. T. Eastman, S. D. Emslie, and A. Lescroël. 2007. "Paradigm Lost, or Is Top-down Forcing No Longer Significant in the Antarctic Marine Ecosystem?" *Antarctic Science* 19:283.

Ainley, D. G., G. Ballard, L. K. Blight, S. Ackley, S. D. Emslie, A. Lescroël, S. Olmastroni, et al. 2010. "Impacts of Cetaceans on the Structure of Southern Ocean Food Webs." *Marine Mammal Science* 26:482–89.

Ainley, D. G. and L. K. Blight. 2009. "Ecological Repercussions of Historical Fish Extraction from the Southern Ocean." *Fish and Fisheries* 10:13–38.

Ainley, D. G., J. Russell, S. Jenouvrier, E. Woehler, P. O. Lyver, W. R. Fraser, and G. L. Kooyman. 2010. "Antarctic Penguin Response to Habitat Change as Earth's Troposphere Reaches 2°C above Pre-industrial Levels." *Ecology* 80:49–66.

Alam, K., M. Shamsuddoha, T. Tanner, M. Sultana, M. J. Huq, and S. S. Kabir. 2011. "Political Economy of Climate Resilient Development Planning in Bangladesh." *IDS Bulletin* 42: 52–61.

ALPARC (Alpine Network of Protected Areas). 2004. *Réseau écologique transfrontalier. Collection Signaux Alpins n°3, Secrétariat permanent de la Convention alpine, Innsbruck.*

Alpine Convention. 1994. Protocol *Conservation of Nature and the Countryside*. http://www.alpconv.org/theconvention/conv02_en.htm (accessed January 7, 2012).

Anderson, E. R. E., E. A. Cherrington, and A. I. Flores. 2008. *Potential Impacts of Climate Change on Biodiversity*. Panama City, Panama: Water Center for the Humid Tropics of Latin America and the Caribbean and U.S. Agency for International Development.

Anderson, M. G. and C. E. Ferree. 2010. "Conserving the Stage: Climate Change and the Geophysical Underpinnings of Species Diversity." *PLoS ONE* 5:e11554.

Anderson, M. G., B. Vickery, M. Gorman, L. Gratton, M. Morrison, J. Maillet, A. Olivero, C. Ferree, D. Morse, G. Kehm, et al. 2006. *The Northern Appalachian/Acadian Ecoregion: Ecoregional Assessment, Conservation Status and Resource CD.* The Nature Conservancy, Eastern Conservation Science and The Nature Conservancy of Canada: Atlantic and Quebec regions. conserveonline.org/workspaces/ecs/napaj/nap (accessed January 11, 2012).

Andréfouët, S., F. E. Muller-Karger, J. A. Robinson, C. J. Kranenburg, D. Torres-Pulliza, S. A. Spraggins, and B. Murch. 2006. "Global Assessment of Modern Coral Reef Extent and Diversity for Regional Science and Management Applications: A View from Space." *Proceedings of the 10th International Coral Reef Symposium*, Okinawa, Japan, June 28–July 4, 2004, 1732–45.

Angerer, J., G. Han, I. Fujisaki, and K. Havstad. 2008. "Climate Change and Ecoystems of Asia with Emphasis on Inner Mongolia and Mongolia." *Rangelands* 30:46–51.

Araújo, M. B., M. Cabeza, W. Thuiller, L. Hannah, and P. H. Williams. 2004. "Would Climate Change Drive Species Out of Reserves? An Assessment of Existing Reserve-selection Methods." *Global Change Biology* 10:1618–26.

ARCOS (Albertine Rift Conservation Society). 2004. *A Framework for Conservation in the Albertine Rift: 2004–2030*. Report of Strategic Planning Process for Albertine Rift, ARCOS.

Ashcroft, M. B. 2010. "Identifying Refugia from Climate Change." *Journal of Biogeography* 37:1407–13.

Asner, G. P., J. Elmore, L. P. Olander, R. E. Martin, and A. T. Harris. 2004. "Grazing Systems, Ecosystem Responses, and Global Change." *Annual Review of Environment and Resources* 29:261–99.

ASOC (Antarctic and Southern Ocean Coalition). 2010a. *Key Climate Change Actions in Antarctica: Emissions Reduction, Adaptation and Science*. IP 73. 32nd ATCM, May 3–14, Punta del Este, Uruguay.

———. 2010b. *Climate Change and the Role of CCAMLR*. CCAMLR-29/BG/19.

———. 2010c. *Antarctic Krill Fisheries and Rapid Ecosystem Change: The Need for Adaptive Management*. Antarctic and Southern Ocean Coalition. IP 10. ATME on Climate Change, April 6–9, 2010. Svolvær, Norway.

———. 2010d. *Environmental and Economic Benefits of Climate Change Mitigation and Adaptation in Antarctica*. Antarctic and Southern Ocean Coalition. IP 9. ATME on Climate Change, April 6–9, 2010. Svolvær, Norway.

ATCM (Antarctic Treaty Consultative Meeting). 2008. "Final Report of the ATCM 31." June 2–13, 2008. Kyiv, Ukraine.

Atkinson, A., V. Siegel, E. Pakhomov, and P. Rothery. 2004. "Long-term Decline in Krill Stock and Increase in Salps within the Southern Ocean." *Nature* 432:100–103.

Auer, I., R. Bühm, A. Jurkovic, W. Lipa, A. Orlik, R. Potzmann, and W. Schüner, et al. 2007. "Histalp: Historical Instrumental Climatological Surface Time Series of the Greater Alpine Region." *International Journal of Climatology* 27:17–46.

AWF (African Wildlife Foundation), IGCP (International Gorilla Conservation Programme), and EcoAdapt. 2011. *The Implications of Global Climate Change for Mountain Gorilla Conservation in the Albertine Rift*. www.igcp.org/wp-content/themes/igcp/docs/pdf/The-Implications-of-Global-Climate-Change-for-Mountain-Gorilla-Conservation-in-Albertine-Rift.pdf (accessed January 11, 2012).

Bachelet, D., R. P. Neilson, T. Hickler, R. J. Drapek, J. M. Lenihan, M. T. Sykes, B. Smith, S. Sitch, and K. Thonicke. 2003. "Simulating Past and Future Dynamics of Natural Ecosystems in the United States." *Global Biogeochemical Cycles* 17:14-1–14-21.

Baldwin, R. F. and S. C. Trombulak. 2007. "Losing the Dark: A Case for a National Policy on Land Conservation." *Conservation Biology* 21:1133–34.

Ballard, W. B., M. A. Cronin, and H. A. Whitlaw. 2000. "Carobou and Oil Firlds," in *The Natural History of an Arctic Oil Field*, J. C. Truett and S. R. Johnson, eds. San Diego, Academic Press, 85–104.

Balshi, M. S., A. D. McGuire, P. Duffy, M. D. Flannigan, J. Walsh, and J. Melillo. 2009. "Assessing the Response of Area Burned to Changing Climate in Western Boreal North America Using a Multivariate Adaptive Regression Suplines (MARS) Approach." *Global Change Biology* 15:578–600.

Bannayan, M., S. Sanjani, A. Alizadeh, S. S. Lotfabadi, and A. Mohamadian. 2010. "Association between Climate Indices, Aridity Index, and Rainfed Crop Yield in Northeast of Iran." *Field Crops Research* 118:105–14.

Barnett, J. 2001. "Adapting to Climate Change in Pacific Island Countries: The Problem of Uncertainty." *World Development* 29:977–93.

Barnett, J. and J. Campbell. 2010. *Climate Change and Small Island States: Power, Knowledge and the South Pacific*. London: Earthscan.

Bateson, E. M. 2005. "Two Countries, One Forest—*Deux Pays, Une Forêt*: Launching a Landscape-scale Conservation Collaborative in the Northern Appalachian Region of the United States and Canada." *George Wright Forum* 22:35–45.

Batt, B. D. 1996. "Prairie Ecology/Prairie Wetlands." In *Prairie Conservation: Preserving North America's Most Endangered Ecosystem,* F. B. Samson and F. L. Knopf, eds. Washington, DC: Island Press, 77–90.

Battin, J., M. W. Wiley, M. H. Ruckelshaus, R. N. Palmer, E. Korb, K. K. Bartz, and H. Imaki. 2007. "Projected Impacts of Climate Change on Salmon Habitat Restoration." *Proceedings of the National Academy of Sciences of the United States of America* 104:6720–25.

Beever, E. A., C. Ray, J. L. Wilkening, P. F. Brussard, and P. W. Mote. 2011. "Contemporary Climate Change Alters the Pace and Drivers of Extinction." *Global Change Biology* 17:2054–70.

Beier, P. and B. Brost. 2010. "Use of Land Facets to Plan for Climate Change: Conserving the Arena, Not the Actors." *Conservation Biology* 24:701–10.

Bentz, B. J., J. Régnière, C. J. Fettig, E. M. Hansen, J. L. Hayes, J. A. Hicke, R. G. Kelsey, J. F. Negrón, and S. J. Seybold. 2010. "Climate Change and Bark Beetles of the Western United States and Canada: Direct and Indirect Effects." *BioScience* 60:602–13.

Berkes, F. and C. Folke. 1998. *Linking Social and Ecological Systems: Management Practices and Social Mechanisms for Building Resilience*. Cambridge, UK: Cambridge University Press.

Berkes, F., M. Kislalioglu, C. Folke, and M. Gadgil. 1998. "Exploring the Basic Ecological Unit: Ecosystem-like Concepts in Traditional Societies." *Ecosystems* 1:409–15.

Berteaux, D., D. Reale, A. G. McAdam, and S. Boutin. 2004. "Keeping Pace with Fast Climate Change: Can Arctic Life Count on Evolution?" *Integrative and Comparative Biology* 44:140–51.

Betts, R. A., Y. Malhi, and J. T. Roberts. 2008. "The Future of the Amazon: New Perspectives from Climate, Ecosystem and Social Sciences." *Philosophical Transactions of the Royal Society B: Biological Sciences* 363:1729–35.

Biswas, A. K. 2008. "Management of Ganges-Brahmaputra-Meghna System: Way Forward." In *Management of Transboundary Rivers and Lakes*, O. Varis, C. Tortajada, and A. K. Biswas, eds. Berlin: Springer, 143–64.

Blancher, P. 2003. *Importance of Canada's Boreal Forest to Landbirds*. Ottawa: Canadian Boreal Initiative and the Boreal Songbird Initiative.

BLC (Boreal Leadership Council). 2003. "Boreal Forest Conservation Framework." Ottawa, Ontario: BLC. www.borealcanada.ca/framework-e.php (accessed January 11, 2012).

Block, W. 1984. "Terrestrial Microbiology, Invertebrates and Ecosystems." In *Antarctic Ecology*, R. M. Laws, ed. London: Academic Press, 163–236.

Bomhard, B., M. Richardson, J. S. Donaldson, G. O. Hughes, G. F. Midgley, D. C. Raimondo, A. G. Rebelo, M. Rouget, and W. Thuiller. 2005. "Potential Impacts of Future Land Use and Climate Change on the Red List Status of the Proteaceae in the Cape Floristic Region, South Africa." *Global Change Biology* 11:1452–68.

BOR (Bureau of Reclamation). 2011. *Landscape Conservation Cooperatives (LCCs): The Desert and Southern Rockies (LCC)*. Washington, DC: US BOR.

Borneman, W. R. 2003. *Alaska: Saga of a Bold Land*. New York: Harper Collins.

Boruff, B. J. and S. L. Cutter. 2007. "The Environmental Vulnerability of Caribbean Island Nations." *Geographical Review* 97:24–45.

Both, C. A. V., A. V. Artemyev, B. Blaauw, R. J. Cowie, A. J. Dekhuijzen, T. Eeva, and A. Enemar, et al. 2004. "Large-scale Geographical Variation Confirms that Climate Change Causes Birds to Lay Earlier." *Proceedings of the Royal Society London B* 271:1657–62.

Botkin, D. B., H. Saxe, M. B. Araujo, R. Betts, R. H. W. Bradshaw, T. Cedhagen, and P. Chesson, et al. 2007. "Forecasting the Effects of Global Warming on Biodiversity." *BioScience* 57:227–36.

Bradley, B. A. 2009. "Regional Analysis of the Impacts of Climate Change on Cheatgrass Invasion Shows Potential Risk and Opportunity." *Global Change Biology* 15:196–208.

Bradley, B. A., L. D. Estés, D. G. Hole, S. Holness, M. Oppenheimer, W. R. Turner, H. Beukes, R. E. Schulze, M. A. Tadross, and D. S. Wilcove. 2012. "Predicting how adapation to climate change could affect ecological conservation: secondary impacts of shifting agricultural suitability." *Diversity and Distributions*. Early view online.

Bradshaw, C. J. A., I. G. Warkentin, and N. S. Sodhi. 2009. "Urgent Preservation of Boreal Carbon Stocks and Biodiversity." *Trends in Ecology and Evolution* 24:541–48.

Bragg, T. B. and A. A. Steuter. 1996. "Prairie Ecology: The Mixed Prairie." In *Prairie Conservation: Preserving North America's Most Endangered Ecosystem*, F. B. Samson, and F. L. Knopf, eds. Washington, DC: Island Press, 53–63.

Brown, T. J., B. L. Hall, and A. L. Westerling. 2004. "The Impact of Twenty-first Century Climate Change on Wildland Fire Danger in the Western United States: An Applications Perspective." *Climatic Change* 62:365–388.

Bryant, D., D. Nielsen, and L. Tangley. 1997. *The Last Frontier Forests: Ecosystems and Economies on the Edge*. Washington, DC: World Resources Institute.

Burack, T. S. 2009. New Hampshire Climate Action Plan. des.nh.gov/organization/divisions/air/tsb/tps/climate/action_plan/nh_climate_action_plan.htm (accessed January 11, 2012).

Burgess, N. D., A. Balmford, N. J. Cordeiro, J. Fjeldså, W. Küper, C. Rahbek, E. W. Sanderson, J. P. W. Scharlemann, J. H. Sommer, and P. H. Williams. 2007. "Correlations among Species Distributions, Human Density and Human Infrastructure across the High Biodiversity Tropical Mountains of Africa." *Biological Conservation* 134:164–77.

Burke, E. J., S. J. Brown, and N. Christidis. 2006. "Modeling the Recent Evolution of Global Drought and Projections for the Twenty-first Century with the Hadley Centre Climate Model." *Journal of Hydrometeorology* 7:1113–25.

Burke, L. and J. Maidens. 2006. *Reefs at Risk in the Caribbean*. Washington, DC: World Resources Institute.

Burke, L., K. Reytar, M. Spalding, and A. Perry. 2011. *Reefs at Risk Revisited*. Washington, DC: World Resources Institute.

Burkett, Virginia R., D. A. Wilcox, R. Stottlemyer, W. Barrow, D. Gagre, J. Baron, and J. L. Nielsen. 2005. "Nonlinear Dynamics in Ecosystem Response to Climate Change: Case Studies and Policy Implications." *Ecological Complexity* 2:357–94.

Burns, C. E., K. M. Johnston, and O. J. Schmitz. 2003. "Global Climate Change and Mammalian Species Diversity in U.S. National Parks." *Proceedings of the National Academy of Sciences of the United States of America* 100:1147–77.

Búrquez, A. and A. Martínez-Yrízar. 2006. "Conservation and Landscape Transformation in Sonora, Mexico." In *Dry Borders: Great Natural Reserves of the Sonoran Desert*, R. S. Felger and B. Broyles, eds. Salt Lake City: University of Utah Press, 537–47.

Bush, G. K., S. Nampindo, C. Aguti, and A. J. Plumptre. 2004. "*Valuing Uganda's Forests: A Livelihood and Ecosystems Approach*." Unpublished report to National Forest Authority, Uganda. www.albertinerift.org (accessed January 9, 2012).

Butorin, A. and T. Yashina. 2009. *Golden Mountains of Altai, World Heritage Property*. Krasnoyask, Russia: UN Development Program and Global Environment Facility Altai-Sayan Ecoregion Project.

Cambers, G. 2009. "Caribbean Beach Changes and Climate Change Adaptation." *Aquatic Ecosystem Health and Management* 12:168–76.

Campbell, J. 2006. *Traditional Disaster Reduction in Pacific Island Communities*. GNS Science Report 2006/38. Wellington, New Zealand: GNS Science.

CAN (Climate Action Network). 2010. *Bangladesh Launches Climate Change Resilient Fund.* Washington, DC: CAN-International. www.climatenetwork.org/blog/banglades-launches-climate-change-resilient-fund (accessed January 11, 2012).

Cardillo, M., G. M. Mace, J. L. Gittleman, and A. Purvis. 2006. "Latent Extinction Risk and the Future Battlegrounds of Mammal Conservation." *PNAS* 103:4157–61.

Carr, A., III. 2004. "Utopian Bubbles: What Are Central America's Parks For?" *Wild Earth* Spring/Summer: 34–39.

Carroll, A. L., J. Régnière, J. A. Logan, S. W. Taylor, B. J. Bentz, and J. A. Powell. 2006. *Impacts of Climate Change on Range Expansion by the Mountain Pine Beetle.* Victoria, British Columbia: Pacific Forestry Centre, Canadian Forest Service, Natural Resources Canada. Mountain Pine Beetle Initiative Working Paper, 2006–14.

Carroll, C., J. R. Dunk, and A. Moilanen. 2010. "Optimizing Resiliency of Reserve Networks to Climate Change: Multispecies Conservation Planning in the Pacific Northwest, USA." *Global Change Biology* 16:891–904.

CAVM (Circumpolar Arctic Vegetation Mapping) Team. 2003. *Circumpolar Arctic Vegetation Map [Scale 1:7,500,000].* Conservation of Arctic Flora and Fauna (CAFF) Map No. 1. Anchorage, Alaska: U.S. Fish and Wildlife Service. www.geobotany.uaf.edu/cavm (accessed January 11, 2012).

CBFA (Canadian Boreal Forest Agreement). 2010. *Canadian Boreal Forest Agreement.* www.canadianborealforestagreement.ca (accessed January 11, 2012).

CBI (Canadian Boreal Initiative). 2009. *The Carbon the World Forgot: Conserving the Capacity of Canada's Boreal Region to Mitigate and Adapt to Climate Change.* Ottawa: CBI.

CBP (Canadian BEACONS Project). 2011. *Conservation Matrix Model.* www.beaconsproject.ca/cmm (accessed January 11, 2012).

CCAD (Central American Commission on Environment and Development). 2002. *Mesoamerican Biological Corridor: A Platform for Sustainable Development.* Technical Series 02, Proyecto Corredor Biologico Mesoamericano, Managua, Nicaragua.

———. 2003. *Estado del Sistema Centroamericano de Areas Protegidas: Informe de Sintesis Regional.* Comision Centroamericano del Ambiente y Desarrollo, Managua, Nicaragua.

CCAMLR (Commission for the Conservation of Antarctic Marine Living Resources). 2009. *Report of the Twenty-eighth Meeting of the Commission.*

———. 2010. "Report of Scientific Committee Chair to the Commission 2." CCAMLR-XIX/BG/50.

CCCCC (Caribbean Community Climate Change Centre). 2009. *Climate Change in the Caribbean: A Regional Framework for Achieving Development Resilient to Climate Change.* Belmopan, Belize.

CCEA (Canadian Council on Ecological Areas). 2011. *Conservation Areas Reporting and Tracking System (CARTS).* CCEA. www.ccea.org/en_carts.html (accessed January 9, 2012).

CDERA (Caribean Disaster Emergency Response Agency). 2003. *Adaptation to Climate Change and Managing Disaster Risk in the Caribbean and South-East Asia.* Report of a seminar, Barbados, July 24–25, 2003.

CEC (Commission of the European Communities). 2009. *Adapting to Climate Change: Towards a European Framework for Action.* Brussels: CEC.

——. 2010. *Terrestrial Protected Areas of North America*. Montreal: CEC. www.cec.org/naatlas (accessed January 7, 2012).

Chadwick, D. H. 2010. *The Wolverine Way*. Ventura, CA: Patagonia.

Chapin, F. S., III, A. L. Lovecraft, E. S. Zavaleta, J. Nelson, M. D. Robards, G. P. Kofinas, S. F. Trainor, G. D. Peterson, H. P. Huntington, and R. L. Naylor. 2006. "Policy Strategies to Address Sustainability of Alaskan Boreal Forests in Response to a Directionally Changing Climate." *Proceedings of the National Academy of Sciences* 103:16637–43.

Chapman, A. D. 2005. *Numbers of Living Species in Australia and the World*. Canberra: Department of Environment and Heritage.

Chaturvedi, R., R. Gopalakrishnan, M. Jayaraman, G. Bala, N. Joshi, R. Sukumar, and N. Ravindranath. 2011. "Impact of Climate Change on Indian Forests: A Dynamic Vegetation Modeling Approach." *Mitigation and Adaptation Strategies for Global Change* 16:2, 119–42.

Chen, C., J. K. Hill, R. Ohlemüller, D. B. Roy, and C. D. Thomas. 2011. "Rapid Range Shifts of Species Associated with High Levels of Climate Warming." *Science* 333:1024–26.

Chester, C. C. 2006. *Conservation across Borders: Biodiversity in an Interdependent World*. Washington, DC: Island Press.

Chester, C. C. and J. A. Hilty. 2010. "Connectivity Science." In *Connectivity Conservation Management: A Global Guide*, G. L. Worboys, W. L. Francis, and M. Lockwood, eds. London: Earthscan, 22–31.

Chuine, I. and E. Beaubien. 2001. "Phenology Is a Major Determinant of Tree Species Range." *Ecology Letters* 4:500–510.

Church, J. A., N. J. White, and J. R. Hunter. 2006. "Sea-level Rise at Tropical Pacific and Indian Ocean Islands." *Global and Planetary Change* 53:155–68.

CI (Conservation International). 2004. *Conserving Earth's Living Heritage: A Proposed Framework for Designing Biodiversity Conservation Strategies*. Washington, DC: CI. library .conservation.org/Published%20Documents/2004/Conserving%20Earth%27s%20 Living%20Heritage.pdf (accessed January 10, 2012).

——. 2007. *Proceedings of the Annual Seascapes Strategy Workshop*. August 13–18, 2007, Kri Island, Papua, Indonesia. Arlington, VA: CI. www.theseascapes.com/Seascapes/workshop _files/seascape.proceedings.workshop.printlayoutpy.pdf (accessed January 10, 2012).

CIG (Climate Impacts Group). 2004. *Overview of Climate Change Impacts in the U.S. Pacific Northwest. Climate Impacts Group*. Background paper prepared for the West Coast Governors' Climate Change Initiative.

CIPRA (Commission Internationale pour la Protection des Alpes). 2010. *Relevant Instruments in the Field of Ecological Networks in the Alpine Region*. CIPRA alpmedia background report. www.cipra.org/alpmedia (accessed January 7, 2012).

Clark, J. S., E. C. Grimm, J. J. Donovan, S. C. Fritz, D. R. Engstrom, and J. E. Almendinger. 2002. "Drought Cycles and Landscape Responses to Past Aridity on Prairies of the Northern Great Plains, USA." *Ecology* 83:595–601.

Clarke, P. and S. D. Jupiter. 2010a. "Law, Custom and Community-based Natural Resource Management in Kubulau District (Fiji)." *Environmental Conservation* 37:98–106.

——. *Principles and Practice of Ecosystem-based Management: A Guide for Conservation Practitioners in the Tropical Western Pacific*. Bronx, NY: Wildlife Conservation Society.

Coates, A. G. 1997. *Central America: A Natural and Cultural History*. New Haven, CT: Yale University Press.

Cody, M. L. 1986. "Diversity, Rarity and Conservation in Mediterranean-climate Regions." In *Conservation Biology*, M. Soulé, ed. Sunderland, MA: Sinauer Press, 122–52.

Coe, M. T., M. H. Costa, and E. A. Howard. 2008. "Simulating the Surface Waters of the Amazon River Basin: Impacts of New River Geomorphic and Dynamic Flow Parameterizations." *Hydrological Processes* 22:2542–53.

Coe, M., M. H. Costa, and B. S. Soares-Filho. 2009. "The Influence of Historical and Potential Future Deforestation on the Stream Flow of the Amazon River: Land Surface Processes and Atmospheric Feedbacks." *Journal of Hydrology* 369:165–74.

Cohn, J. P. 2010. "A Narrow Path for Pronghorns." *BioScience* 60:480.

Collinge, S. K., W. C. Johnson, C. Ray, R. Matchett, J. Grensten, and J. F. Cully Jr. 2005. "Testing the Generality of a Trophic-cascade Model for Plague." *EcoHealth* 2:102–12.

Commonwealth of Australia (CoA). 1997. *Nationally Agreed Criteria for the Establishment of a Comprehensive, Adequate and Representative Reserve System for Forests in Australia*. Joint ANZECC/MCFFA National Forest Policy Statement Implementation Subcommittee.

Conservation Registry. 2011. *Staying Connected in the Northern Appalachians: Mitigating Fragmentation and Climate Impacts on Wildlife through Functional Habitat Linkages*. Wildlife Conservation Society. Wildlife Action Opportunities Fund Portal. wcs.conservation registry.org/projects/3837 (accessed January 11, 2012).

Convey, P., K. A. Hughes, M. S. de Villiers, D. Haase Liggett, N. Holmes, and T. Tin. In review. "Conservation of Terrestrial Flora and Fauna in the Antarctic and Sub-Antarctic in the Face of Increasing Human Activity."

Convey, P. and S. J. McInnes. 2005. "Exceptional, Tardigrade Dominated, Ecosystems from Ellsworth Land, Antarctica." *Ecology* 86:519–27.

Copeland, J. P., K. S. McKelvey, K. B. Aubry, A. Landa, J. Persson, R. M. Inman, and J. Krebs. 2010. "The Bioclimatic Envelope of the Wolverine (*Gulo gulo*): Do Climatic Constraints Limit Its Geographic Distribution?" *Canadian Journal of Zoology* 88:233–46.

Coops, N. C., M. A. Wulder, D. C. Duro, T. Han, and S. Berry. 2008. "The Development of a Canadian Dynamic Habitat Index Using Multitemporal Satellite Estimates of Canopy Light Absorbance." *Ecological Indicators* 8:754–66.

Costa, M. H., A. Botta, and J. A. Cardille. 2003. "Effects of Large-scale Changes in Land Cover on the Discharge of the Tocantins River, Southeastern Amazonia." *Journal of Hydrology* 283:12.

Costa, M. H. and J. A. Foley. 2000. "Combined Effects of Deforestation and Doubled Atmospheric CO_2 Concentrations on the Climate of Amazonia." *Journal of Climate* 13:18–34.

Covich, A. P., S. C. Fritz, P. J. Lamb, R. D. Marzolf, W. J. Matthews, K. A. Poiani, et al. 1997. "Potential Effects of Climate Change on Aquatic Ecosystems of the Great Plains of North America." *Hydrological Processes* 11:993–1021.

Cowling, R. M. 1992. *The Ecology of Fynbos: Nutrients, Fire and Diversity*. Cape Town: Oxford University Press.

Cowling, R. M. and C. Hilton-Taylor. 1999. "Plant Biogeography, Endemism and Diversity." In *The Karoo: Ecological Patterns and Processes*, W. R. J. Dean and S. J. Milton, eds. Cambridge, UK: Cambridge University Press, 42–56.

Cowling, R. M. and P. M. Holmes. 1992. "Endemism and Speciation in a Lowland Flora from the Cape Floristic Region." *Biological Journal of the Linnean Society* 47:367–83.

Cowling, R. M., I. A. W. MacDonald, and M. T. Simmons. 1996. "The Cape Peninsula, South Africa: Physiographical, Biological and Historical Background to an Extraordinary Hotspot of Biodiversity." *Biodiversity and Conservation* 5:527–50.

Cowling, R. M. and P. J. Mustart. 1994. *Vegetation and Conservation*, vol. 2, append. 5. Southern Overberg Sub-regional Structure Plan. MLH Architects and Planners, Cape Town.

Cowling, R. M. and R. L. Pressey. 2001. "Rapid Plant Diversification: Planning for an Evolutionary Future." *Proceedings of the National Academy of Sciences of the United States of America* 98:5452–57.

———. 2003. "Introduction to Systematic Conservation Planning in the Cape Floristic Region." *Biological Conservation* 122:1–13.

Cowling, R. M., R. L. Pressey, A. T. Lombard, P. G. Desmet, and A. G. Ellis. 1999. "From Representation to Persistence: Requirements for a Sustainable System of Conservation Areas in the Species-rich Mediterranean-climate Desert of Southern Africa." *Diversity and Distributions* 5:51–71.

Cowling, R. M., R. L. Pressey, M. Rouget, and A. T. Lombard. 2003. "A Conservation Plan for a Global Biodiversity Hotspot: The Cape Floristic Region, South Africa." *Biological Conservation* 112:191–216.

Cowling, R. M., P. W. Rundel, P. G. Desnet, and K. J. Esler. 1998. "Extraordinarily High Regional Scale Plant Diversity in Southern African Arid Lands: Subcontinental and Global Comparisons." *Diversity and Distributions* 4:27–36.

Cox, P. M., R. A. Betts, C. D. Jones, S. A. Spall, and I. J. Totterdell. 2000. "Acceleration of Global Warming Due to Carbon-cycle Feedbacks in a Coupled Climate Model." *Nature* 408:184–87.

Crain, C., B. S. Halpern, M. W. Beck, and C. V. Kappel. 2009. "Understanding and Managing Human Threats to the Coastal Marine Environment." *Annals of the New York Academy of Sciences* 1162:39–62.

Craine, J. M., E. G. Towne, A. Joern, and R. G. Hamilton. 2009. "Consequences of Climate Variability for the Performance of Bison in Tallgrass Prairie." *Global Change Biology* 15: 772–79.

Cramer, W., A. Bondeau, F. I. Woodward, I. C. Prentice, R. A. Betts, V. Brovkin, and P. M. Cox, et al. 2001. "Global Response of Terrestrial Ecosystem Structure and Function to CO_2 and Climate Change: Results from Six Dynamic Global Vegetation Models." *Global Change Biology* 7:357–73.

Cross, M. S., J. A. Hilty, G. M. Tabor, J. J. Lawler, L. J. Graumlich, and J. Berger. In press. "From Connect-the-dots to Dynamic Networks: Maintaining and Enhancing Connectivity as a Strategy to Address Climate Change Impacts on Wildlife." In *Wildlife Conservation in a Changing Climate,* J. Brodie, D. Doak, and E. Post, eds. Chicago: University of Chicago Press.

Cross, M. S. and C. Servheen. 2009. *Climate Change Impacts on Wolverines and Grizzly Bears in the Northern U.S. Rockies: Strategies for Conservation*. Final Workshop Summary Report

October 6–7, 2009. Wildlife Conservation Society and U.S.Fish and Wildlife Service. May 24, 2010.

Cross, M. S., E. S. Zavaleta, D. Bachelet, M. L. Brooks, C. A. F. Enquist, E. Fleishman, L. Graumlich, et al. In review. *The Adaptation for Conservation Targets (ACT) Framework: A Tool for Incorporating Climate Change into Natural Resource Management*.

Culver, M., C. Varas, P. M. Harveson, B. McKinney and L. A. Harveson. 2009. "Connecting Wildlife Habitats across the US–Mexico Border." In *Conservation of Shared Environments: Learning from the United States and Mexico*, L. López-Hoffman, E. McGovern, R. G. Varady, and K. W. Flessa, eds. Tucson: University of Arizona Press, 83–100.

Cumming, S. G., K. Lefevre, E. Bayne, T. Fontaine, F. K. A. Schmiegelow, and S. J. Song. 2010. "Toward Conservation of Canada's Boreal Forest Avifauna: Design and Application of Ecological Models at Continental Extents." *Avian Conservation and Ecology* 5:8.

DAAC (Distributed Active Archive Center). 2010. *MODIS Global Subsets: Data Subsetting and Visualization*. Distributed by Active Archive Center for Biogeochemical Dynamics. Oak Ridge, TN: Oak Ridge National Laboratory.

Daly, C., J. I. Smith, and R. McKane. 2007. "High-resolution Spatial Modeling of Daily Weather Elements for a Catchment in the Oregon Cascade Mountains, United States." *Journal of Applied Meteorology and Climatology* 46:1565–86.

DataBasin. 2011. *Boreal Region of Canada: Study Area of BEACONs Project*. app.databasin .org/app/pages/datasetPage.jsp?id=2d7bf937823f4d08acfef8741c33d594 (accessed January 9, 2012).

Daufresne, M. and P. Boët. 2007. "Climate Change Impacts on Structure and Diversity of Fish Communities in Rivers." *Global Change Biology* 13:2467–78.

Davis, Neil. 2001. *Permafrost: A Guide to Frozen Ground in Transition*. Fairbanks: University of Alaska.

Dawson, T. P., S. T. Jackson, J. I. House, I. C. Prentice, and G. M. Mace. 2011. "Beyond Predictions: Biodiversity Conservation in a Changing Climate." *Science* 332:53–58.

DeBano, L. F. and P. Ffolliott. 1995. *Biodiversity and Management of the Madrean Archipelago: The Sky Islands of Southwestern United States and Northwestern Mexico.* Fort Collins, CO: USDA Forest Service, Rocky Mountain Research Station, RM-GTR-264.

DeChaine, E. G. and A. P. Martin. 2004. "Historic Cycles of Fragmentation and Expansion in *Parnassius smintheus* (*Papilionidae*) Inferred Using Mitochondrial DNA." *Evolution* 58: 113–27.

Derksen, D. V., M. W. Weller, and W. D. Eldridge. 1979. "Distributional Ecology of Geese Molting near Teshekpuk Lake, National Petroleum Reserve—Alaska." In *Management and Biology of Pacific Flyway Geese*, R. L. Jarvis and J. C. Bartonek, eds. Corvallis, OR: Oregon State University Book Stores, 189–207.

Dixon, J., K. Hamilton, S. Pagiola, and L. Segnestam. 2001. *Tourism and the Environment in the Caribbean: An Economic Framework*. Washington, DC: World Bank. Environment Department Papers, 22740.

Dobrowski, S. Z. 2011. "A Climatic Basis for Microrefugia: The Influence of Terrain on Climate." *Global Change Biology* 17:1022–35.

Doherty, R. M., S. Sitch, B. Smith, S. L. Lewis, and P. K. Thornton. 2009. "Implications of Future Climate and Atmospheric CO_2 Content for Regional Biogeochemistry, Biogeography and Ecosystem Services across East Africa." *Global Change Biology* 16:617–40.

DOI (Department of the Interior). 2011. USDOI. *Climate Science Centers*. www.fws.gov/southeast/LCC/GulfPlains/pdf/GCPO%20Workshop_Jones%20&%20Shipp_Climate%20Science%20and%20LCCs_Mar10.pdf (accessed January 11, 2012).

Donato, D. C., J. B. Kauffman, D. Murdiyarso, S. Kurnianto, M. Stidham, and M. Kanninen. 2011. "Mangroves among the Most Carbon-rich Forests in the Tropics." *Nature Geoscience* 4:293–97.

Donner, S. D. 2009. "Coping with Commitment: Projected Thermal Stress on Coral Reefs under Different Future Scenarios." *PLoS ONE* 4:e5712.

Drielsma, M., G. Manion, and S. Ferrier. 2007. "The Spatial Links Tool: Automated Mapping of Habitat Linkages in Variegated Landscapes." *Ecological Modeling* 200:403–11.

Ducklow, H. W., K. Baker, D. G. Martinson, L. B. Quetin, R. M. Ross, R. C. Smith, S. E. Stammerjohn, M. Vernet, and W. R. Fraser. 2007. "Marine Pelagic Ecosystems: The West Antarctic Peninsula." *Philosophical Transactions of the Royal Society B* 362:67–94.

Dudley, N., S. Stolton, A. Belokurov, L. Krueger, N. Lopoukhine, K. MacKinnon, T. Sandwith, and N. Sekhran. 2010. *Natural Solutions: Protected Areas Helping People Cope with Climate Change*. Gland, Switzerland: IUCN-WCPA, TNC, UNDP, WCS, The World Bank, and WWF, 130 pp.

Dunlop, M. and P. R. Brown. 2008. *Implications of Climate Change for Australia's National Reserve System: A Preliminary Assessment*. Canberra, Australia: Report to the Department of Climate Change and the Department of the Environment, Water, Heritage and the Arts.

Durance, I. and S. J. Ormerod. 2007. "Climate Change Effects on Upland Stream Invertebrates over a 25 Year Period." *Global Change Biology* 13:942–57.

Duro, D., N. C. Coops, M. A. Wulder, and T. Han. 2007. "Development of a Large Area Biodiversity Monitoring System Driven by Remote Sensing." *Progress in Physical Geography* 31:235–61.

Edelman, M. 1995. "Rethinking the Hamburger Thesis: Deforestation and the Crisis of Central America's Beef Exports." In *The Social Causes of Environmental Destruction in Latin America*, M. Painter and W. H. Durham, eds. Ann Arbor: University of Michigan Press, 25–62.

Edwards, M. and A. J. Richardson. 2004. "Impact of Climate Change on Marine Pelagic Phenology and Trophic Mismatch." *Nature* 430(7002): 881–84.

Egli, D. P., T. Tui, S. D. Jupiter, and A. Caginitoba. 2010. *Perception Surveys of Coastal Resource Use and Changes Following Establishment of a Marine Protected Area Network in Kubulau, Fiji*. Suva, Fiji: Technical Report no. 07/10. Wildlife Conservation Society–Fiji.

El Raey, M., Kh. Dewidar, and M. El Hattab. 1999. "Adaptation to the Impacts of Sea Level Rise in Egypt." *Climate Research* 12:117–28.

Elsner, M. M., J. Littell, and L. W. Binder. 2009. *The Washington Climate Change Impacts Assessment*. Climate Impacts Group, Center for Science in the Earth System, Joint Institute for the Study of the Atmosphere and Oceans. Seattle: University of Washington.

Emers, M., J. C.. Jorgenson, and M. K. Raynolds. 1995. "Response of Arctic Tundra Plant Communities to Winter Vehicle Disturbance." *Canadian Journal of Botany* 73:905–17.

ENS (Environmental News Service). 2008. *Ancient Pronghorn Path Becomes First U.S. Wildlife Migration Corridor*. ENS. June 25. www.ens-newswire.com/ens/jun2008/2008-06-17-091.html (accessed January 11, 2012).

EPA (Environmental Protection Agency). 2009. *Synthesis of Adaptation Options for Coastal Areas*. Washington, DC: U.S. EPA, Climate Ready Estuaries Program. EPA 430-F-08-024, January 2009.

European Commission. 2007. *Combating Climate Change. The EU Leads the Way*. 24 pp. ec.europa.eu/publications (accessed January 8, 2012).

——. 2009. White paper. "*Adapting to Climate Change: Towards a European Framework for Action*." eur-lex.europa.eu/LexUriServ/LexUriServ.do?uri=COM:2009:0147:FIN:EN:PDF (accessed January 8, 2012).

——. 2010. *The 2010 Assessment of Implementing the EU Biodiversity Action Plan*. ec.europa.eu/environment/nature/biodiversity/comm2006/pdf/bap_2010/1_EN_ACT_part1_v2.pdf (accessed January 8, 2012).

European Topic Centre on Biological Diversity. 2008. *Habitats Directive Article 17 Report (2001–2006): Some Specific Analysis on Conservation Status*. Paris: European Topic Centre on Biological Diversity. biodiversity.eionet.europa.eu/article17 (accessed January 8, 2012).

Euskirchen, E. S., D. McGuire, F. S. Chapin III, S. Yi, and C. C. Thompson. 2009. "Changes in Vegetation in Northern Alaska under Scenarios of Climate Change, 2003–2100: Implications for Climate Feedbacks." *Ecological Applications* 19:1022–43.

Evenden, M. 2009. "Site C Forum: Considering the Prospect of Another Dam on the Peace River." *BC Studies* 161:93–114.

Falk, D. A., C. Miller, D. McKenzie, and A. E. Black. 2007. "Cross-scale Analysis of Fire Regimes." *Ecosystems* 10:809–26.

FAO (Food and Agricultural Organization). 2003. "Status and Trends in Mangrove Area Extent Worldwide." In *Forest Resources Assessment Working Paper No. 63*, M. L. Wilkie and S. Fortuna, eds. Rome: Forest Resources Division, UN FAO. www.fao.org/docrep/007/j1533e/j1533e00.htm (accessed January 11, 2012).

——. 2007. *Report of the Twenty-seventh Session of the Committee on Fisheries*. Rome: FAO Fisheries Report. No. 830. 74 pp.

Feely, R. A., C. L. Sabine, J. M. Hernandez-Ayon, D. Ianson, and B. Hales. 2008. "Evidence for Upwelling of Corrosive 'Acidified' Water onto the Continental Shelf." *Science* 320: 1490–92.

Felger, R. S. and M. F. Wilson. 1994. "Northern Sierra Madre Occidental and Its Apachian Outliers: A Neglected Center of Biodiversity." In *The Sky Islands of Southwestern United States and Northwestern Mexico*, L. F. Debano, G. J. Gottfried, R. H. Hamre, C. B. Edminster, P. F. Ffolliott, and A. Ortega-Rubio, eds. USDA Forest Service: General Technical Report RM-GTR-264, 6–18.

Fernandez-Gimenez, M. E. 2000. "The Role of Mongolian Nomadic Pastoralists' Ecological Knowledge in Rangeland Management." *Ecological Applications* 10:1318–26.

Field, J. C., R. C. Francis, and K. Aydin. 2006. "Top-down Modeling and Bottom-up Dynamics: Linking a Fisheries-based Ecosystem Model with Climate Hypotheses in the Northern California Current." *Progress in Oceanography* 68:238–70.

Fiji Government. 2005. *Climate Change the Fiji Islands Response: Fiji's First National Communication under the Framework Convention on Climate Change*. Prepared by the Pacific Island Climate Change Assistance Programme and Fiji Country Team. Suva, Fiji: Department of Environment.

Finlayson, C. M. and R. D'Cruz. 2005. "Millenium Ecosystem Assessment." Chap. 20. In *Inland Water Systems*. www.maweb.org/documents/document.289.aspx.pdf. (accessed January 8, 2012).

Finley-Brook, M. 2007. "Green Neoliberal Space: The Mesoamerican Biological Corridor." *Journal of Latin American Geography* 6:101–24.

Fish, M. R. and C. Drews. 2009. "Adaptation to Climate Change: Options for Marine Turtles." San José: WWF Central America. assets.panda.org/downloads/adaptation_to_climate_change_options_for_marine_turtles.pdf (accessed January 7, 2012).

Flannigan, M. D., K. A. Logan, B. D. Amiro, W. R. Skinner, and B. J. Stocks. 2005. "Future Area Burned in Canada." *Climatic Change* 72:1–16.

Flannigan, M. D., B. Stocks, M. Turetsky, and B. M. Wotton. 2009. "Impacts of Climate Change on Fire Activity and Fire Management in the Circumboreal Forest." *Global Change Biology* 15:549–60.

Flesch, A. D., C. W. Epps, J. W. Cain, M. Clark, P. R. Krausman, and J. R. Morgart. 2010. "Potential Effects of the United States–Mexico Border Fence on Wildlife." *Conservation Biology* 24:171–81.

FNSP (Far North Science Advisory Panel). 2010. "Science for a Changing Far North." Report submitted to the Ontario Ministry of Natural Resources; Queen's Printer for Ontario. www.web2.mnr.gov.on.ca/FarNorth/sciencereports/Far_North_Science_Panel_Report_Summary_June_2010.pdf (accessed January 12, 2012).

FOEN (Federal Office for the Environment). 2009a. "Integrated Water Management." http://www.bafu.admin.ch/wasser/01444 (accessed January 12, 2012).

———. 2009b. *Surface Waters*. http://www.bafu.admin.ch/gewaesserschutz/01267/index.html?lang=en (accessed January 6, 2012).

Ford, A. T., A. P. Clevenger, and K. Rettie. 2010. "The Banff Wildlife Crossings Project: An International Public-Private Partnership." In *Safe Passages: Highways, Wildlife, and Habitat Connectivity*, J. P. Beckmann, A. P. Clevenger, M. P. Huijser, and J. A. Hilty, eds. Washington, DC: Island Press, 157–72.

Ford, D. 2011. "Expanding South Nahanni National Park, Northwest Territories, Canada, to Include and Manage Some Remarkable Sub-arctic Karst Terrains." In *Karst Management*, Philip E. van Beynen, ed. New York: Springer, 415–37.

Foreman, D., B. Dugelby, J. Humphrey, B. Howard, and A. Holdsworth. 2000. "The Elements of a Wildlands Network Conservation Plan." *Wild Earth* 10:17–30.

Forman, R. T. T. 1995. *Land Mosaics: The Ecology of Landscapes and Regions*. New York: Cambridge University Press.

Forrest, S. C., H. Strand, W. H. Haskins, C. Freese, and E. Dinerstein. 2004. *Ocean of Grass: A Conservation Assessment for the Northern Great Plains*. Bozeman, MT: Northern Plains Conservation Network and Northern Great Plains Ecoregion, WWF-US.

Foster, D. and J. Aber. 2004. *Forests in Time: The Environmental Consequences of 1,000 Years of Change in New England*. New Haven, CT: Yale University Press.

Foster, D. R., B. M. Donahue, D. B. Kittredge, K. F. Lambert, M. L. Hunter, B. R. Hall, L. C. Irland, et al. 2010. *Wildlands and Woodlands: A Vision for the New England Landscape*. Petersham, MA: Harvard Forest. Cambridge, MA: Harvard University Press.

FPTGC (Federal, Provincial, and Territorial Governments of Canada). 2010. *Canadian Biodiversity: Ecosystem Status and Trends 2010*. Ottawa, Ontario: Canadian Councils of Resource Ministers, FPTGC.

France, Europe. 2010. *Open-Ended Intersessional Contact Group on "Non-Native Species" (NNS)—2009–2010 Report*. WP 9. 33rd ATCM. Punta del Este, Uruguay. May 3–14.

Francou, B., M. Vuille, P. Wagnon, J. Mendoza, and J. E. Sicart. 2003. "Tropical Climate Change Recorded by a Glacier of the Central Andes during the Last Decades of the 20th Century: Chacaltaya, Bolivia, 16°S." *Journal of Geographical Research* 108.

Franklin, J. F. and D. B. Lindenmayer. 2009. "Importance of Matrix Habitats in Maintaining Biological Diversity." *Proceedings of the National Academy of Sciences* 106:349–50.

Frazee, S. R., R. M. Cowling, R. L. Pressey, J. K. Turpie, and N. Lindenberg. 2003. "Estimating the Costs of Conserving a Biodiversity Hotspot: A Case-Study of the Cape Floristic Region, South Africa." *Biological Conservation* 112:275–290.

Frumhoff, P. C., J. J. McCarthy, J. M. Melillo, S. C. Moser, and D. J. Wuebbles. 2007. *Confronting Climate Change in the U.S. Northeast: Science, Impacts, and Solutions. Synthesis Report of the Northeast Climate Impacts Assessment*. Cambridge, MA: Union of Concerned Scientists.

FSC (Forest Steward Council). 2010. *Statistics on FSC Forest Management Standards*. www.fsccanada.org/fsccertifiedforests.htm#Standards (accessed January 12, 2012).

Fuentes, M., M. R. Fish, and J. A. Maynard. 2012. "Management Strategies to Mitigate the Impacts of Climate Change on Sea Turtle's Terrestrial Reproductive Phase." *Mitigation and Adaptation Strategies for Global Change*. 17:51–63.

Fundación PRISMA and Grupo CABAL. 2011. *Synthesis Report of Project: Designing a REDD Program that Benefits Forestry Communities in Mesoamerica*. San Salvador, El Salvador, 44 pp.

Galakhov, V. P. and R. M. Mukhammetov. 1999. *Glaciers of the Altai* [in Russian].

Galatowitsch, S., L. Frelich, and L. Phillips-Mao. 2009. "Regional Climate Change Adaptation Strategies for Biodiversity Conservation in a Midcontinental Region of North America." *Biological Conservation* 142:2012–22.

Game, E. T., C. Groves, M. Andersen, M. Cross, C. Enquist, Z. Ferdaña, E. Girvetz, et al. 2010. *Incorporating Climate Change Adaptation into Regional Conservation Assessments*. Arlington, VA: The Nature Conservancy. conserveonline.org/workspaces/climateadaptation/documents/incorporating-cc-adaptation-into-regional-0 (accessed January 12, 2012).

Gao, X. J. and F. Giorgi. 2008. "Increased Aridity in the Mediterranean Region under Greenhouse Gas Forcing Estimated from High Resolution Simulations with a Regional Climate Model." *Global and Planetary Change* 62:95–209.

Gardner, T. A., I. M. Côte, J. A. Gill, A. Grant, and A. R. Watkinson. 2003. "Long-term Region-wide Declines in Caribbean Corals." *Science* 301:958–60.

Gash, J. H. C. and C. A. Nobre. 1997. "Climatic Effects of Amazonian Deforestation: Some Results from ABRACOS." *Bulletin of the American Meteorological Society* 78:823–30.

Gelderblom, C. M., B. W. van Wilgen, J. L. Nel, T. Sandwith, M. Botha, and M. Hauck. 2003. "Turning Strategy into Action: Implementing a Conservation Action Plan in the Cape Floristic Region." *Biological Conservation* 112:291–97.

Gillett, N. P., A. J. Weaver, F. W. Zwiers, and M. D. Flannigan. 2004. "Detecting the Effect of Climate Change on Canadian Forest Fires." *Geophysical Research Letters* 31:L18211.

Girardin, M. P., A. A. Ali, C. Carcaillet, M. Mudelsee, I. Drobyshev, C. Hély, and Y. Bergeron. 2009. "Heterogeneous Response of Circumboreal Wildfire Risk to Climate Change since the Early 1900s." *Global Change Biology* 15:2751–69.

Girardin, M. P. and D. Sauchyn. 2008. "Three Centuries of Annual Area Burned Variability in Northwestern North America Inferred from Tree Rings." *Holocene* 18:205–14.

Girvetz, E. H., C. Zganjar, G. T. Raber, E. P. Maurer, P. Kareiva, and J. J. Lawler. 2009. "Applied Climate-change Analysis: The Climate Wizard Tool." *PLoS One* 4:12, e8320.

Glick, P., A. Staudt, and B. Stein. 2009. "A New Era for Conservation: Review of Climate Change Adaptation Literature." National Wildlife Federation report.

Glick, P., B. A. Stein, and N. A. Edelson. 2011. *Scanning the conservation horizon: A guide to climate change vulnerability assessment*. Washington: National Wildlife Federation. http://www.fs.fed.us/rm/pubs_other/rmrs_2011_glick_p001.pdf.

GLOCHAMORE. 2005. "Global Change and Mountain Regions, Research Strategy." A. B. Gorung, ed. Developed in the course of a Specific Support Action under the EU Framework Program 6 (Contract No. 506679). *Global Change and Mountain Regions: An Integrated Assessment of Causes and Consequences* (November 2003 – October 2005). www.epbrs.org/PDF/GLOCHAMORE_Research_Strategy_color.pdf (accessed January 10, 2012).

Glorioso, R. S. and L. A. G. Moss. 2007. "Amenity Migration to Mountain Regions: Current Knowledge and a Strategic Construct for Sustainable Management." *Social Change* 37:137–61.

Goldblatt, P. 1978. "An Analysis of the Flora of Southern Africa: Its Characteristics, Relationships, and Origins." *Annals of the Missouri Botanical Garden* 65(2): 369–436.

Goldblatt, P. and J. C. Manning. 2000. *Cape Plants: A Conspectus of the Cape Flora of Southern Africa*. National Botanical Institute, Cape Town and Missouri Botanical Garden, St. Louis.

——. 2002. "Plant Diversity of the Cape Region of Southern Africa." *Annals of the Missouri Botanical Garden* 89:281.

GoM (Government of Mexico). 2007. *National Strategy on Climate Change: Executive Summary*. Intersecretarial Commission on Climate Change.

Gonzalez, P., R. P. Neilson, J. M. Lenihan, and R. J. Drapek. 2010. "Global Patterns in the Vulnerability of Ecosystems to Vegetation Shifts Due to Climate Change." *Global Ecology and Biogeography* 19:755–68.

Gouvernement du Québec. 2009. *Plan Nord—For a Socially Responsible and Sustainable Form of Economic Development*. Working document, Québec: Ressources naturelles et Faune Québec. www.plannord.gouv.qc.ca/english/documents/plan-nord.pdf (accessed January 6, 2012).

Government of Ontario. 2010. *Bill 191—Far North Act*. Toronto, Ontario.

Graham, N. A. J., S. K. Wilson, S. Jennings, N. V. C. Polunin, J. P. Bijoux, and J. Robinson. 2006. "Dynamic Fragility of Oceanic Coral Reef Ecosystems." *Proceedings of the National Academy of Sciences* 103:8425–29.

Grandia, L. 2007. "Between Bolivar and Bureaucracy: The Mesoamerican Biological Corridor." *Conservation and Society* 5:478–503.

Grantham, H. S., E. McLeod, A. Brooks, S. D. Jupiter, J. Hardcastle, A. J. Richardson, E. S. Polozanska, et al. In press. "Ecosystem-based adaptation in marine ecosystems of tropical Oceania in response to climate change." *Pacific Conservation Biology*.

Graumlich, L. and W. L. Francis. 2010. *Moving toward Climate Change Adaptation: The Promise of the Yellowstone to Yukon Conservation Initiative for Addressing the Region's Vulnerabilities*. Canmore, Alberta: Yellowstone to Yukon Conservation Initiative. www.y2y.net/data/1/rec_docs/898_Y2Y_Climate_Adaptation_Report__FINAL_Web.pdf (accessed January 6, 2012)

Gravelle, G. and N. Mimura. 2008. "Vulnerability Assessment of Sea-level Rise in Viti Levu, Fiji Islands." *Sustainable Science* 3:171–80.

Gray, S. T. and G. J. McCabe 2010. "A Combined Water Balance and Tree Ring Approach to Understanding the Potential Hydrologic Effects of Climate Change in the Central Rocky Mountain Region." *Water Resources Research* 46:W05513.

Green, K. and C. M. Pickering. 2002. "A Scenario for Mammal and Bird Diversity in the Snowy Mountains in Relation to Climate Change." In *Mountain Biodiversity: A Global Assessment,* C. Koerner and E. M. Spehn, eds. London: Parthenon, 241–49.

Gripne, S. L. 2005. "Grassbanks: Bartering for Conservation." *Rangelands* 27:24–28.

Groves, C. R. 2003. *Drafting a Conservation Blueprint: A Practitioner's Guide to Planning for Biodiversity*. Washington, DC: Island Press.

Gunderson, L. H. 2000. "Resilience in Theory and Practice." *Annual Review of Ecology and Systematics* 31:425–39.

GUSAF (Giving USA Foundation). 2011. *Giving USA 2011: The Annual Report on Philanthropy for the Year 2010*. Chicago: Giving USA Foundation. www.givingusareports.org (accessed Febryary 8, 2012).

Hall, M. H. P. and D. B. Fagre. 2009. "Modeled Climate-induced Glacier Change in Glacier National Park, 1850– 2100." *BioScience* 53:131–40.

Halpin, P. N. 1997. "Global Climate Change and Natural-area Protection: Management Responses and Research Directions." *Ecological Applications* 7:828–43.

Hamann, A. and T. L. Wang. 2006. "Potential Effects of Climate Change on Tree Species and Ecosystem Distribution in British Columbia." *Ecology* 87:2773–86.

Hamilton, H., S. Caballero, A. G. Collins, and R. L. Brownell Jr. 2001. "Evolution of River Dolphins." *Proceedings of the Royal Society London* 268:549–56.

Hamilton, L. S. 2010. "Great Eastern Ranges (GER) Corridor (Australia)." In *Mountain Protected Areas Update, No 67, September 2010*. Gland, Switzerland: IUCN World Commission on Protected Areas, Mountains and Connectivity Conservation Biome.

Hamlet, A. F. and D. P. Lettenmaier. 1999. "Effects of Climate Change on Hydrology and Water Resources in the Columbia River Basin." *Journal of the American Water Resources Association* 35:1597–1623.

Hamlet, A. F., P. W. Mote, M. P. Clark, and D. P. Lettenmaier. 2005. "Effects of Temperature and Precipitation Variability on Snowpack Trends in the Western United States." *Journal of Climate* 18:4545–61.

——. 2007. "Twentieth-century Trends in Runoff, Evapotranspiration, and Soil Moisture in the Western United States." *Journal of Climate* 20:1468–81.

Hanekom, N., R. M. Randall, L. A. Russell, and B. Sachse. 1995. *The Agulhas Area. An Investigation of Its Potential for Proclamation as a National Park*. Internal report, SANParks, 15 pp.

Hannah, L. 2008. "Protected Areas and Climate Change." *Annals of the New York Academy of Sciences* 1134:201–12.

Hannah, L., ed. 2011. *Saving a Million Species: Extinction Risk from Climate Change*. Washington, DC: Island Press.

Hannah, L., G. Midgley, S. Andelman, M. Araujo, G. Hughes, E. Martinez-Meyer, R. Pearson, and P. Williams. 2007. "Protected Area Needs in a Changing Climate." *Frontiers in Ecology and the Environment* 5:131–38.

Hannah, L., G. Midgley, G. Hughes, and B. Bomhard. 2005. "The View from the Cape. Extinction Risk, Protected Areas, and Climate Change." *Bioscience* 55:231–42.

Hannah, L., G. F. Midgley, T. Lovejoy, W. J. Bond, M. Bush, J. C. Lovett, D. Scott, and F. I. Woodward. 2002. "Conservation of Biodiversity in a Changing Climate." *Conservation Biology* 16:264–68.

Hansen, A. J., R. Rasker, B. Maxwell, J. J. Rotella, J. D. Johnson, A. Wright Parmenter, U. Langner, W. B. Cohen, R. L. Lawrence, and M. P. V. Kraska. 2002. "Ecological Causes and Consequences of Demographic Change in the New West." *Bioscience* 52:151–62.

Hansen, L. J. and J. R. Hoffman. 2011. *Climate Savvy: Adapting Conservation and Resource Management to a Changing World*. Washington, DC: Island Press.

Hansen, L., J. Hoffman, C. Drew, and E. Mielbrecht. 2010. "Designing Climate-smart Conservation: Guidance and Case Studies." *Conservation Biology* 24:63–69.

Hansen, M., R. DeFries, J. R. Townshend, M. Carroll, C. Dimiceli, and R. Sohlberg. 2003. "Vegetation Continuous Fields MOD44B." *2001 Percent Tree Cover, Collection 3*. College Park: University of Maryland. glcf.umiacs.umd.edu/data/vcf/ (accessed January 10, 2012).

Harborne, A. R., P. J. Mumby, F. Micheli, C. T. Perry, C. P. Dahlgren, K. E. Holmes, and D. R. Brumbaugh. 2006. "The Functional Value of Caribbean Coral Reef, Seagrass and Mangrove Habitats to Ecosystem Processes." *Advances in Marine Biology* 50:57–189.

Hare, F. K. 1980. "The Planetary Environment: Fragile or Sturdy?" *Geographical Journal* 146:379–95.

Harley, C. D. G., A. R. Hughes, K. M. Hultgren, B. G. Miner, C. J. B. Sorte, C. S. Thornber, L. F. Rodriguez, L. Tomanek, and S. L. Williams. 2006. "The Impacts of Climate Change in Coastal Marine Systems." *Ecology Letters* 9:228–41.

Harmeling, S. 2009. *Global Climate Risk Index 2009: Weather-related Loss Events and Their Impacts on Countries in 2007 and in a Long-term Comparison*. Briefing paper. Bonn, Germany: Germanwatch E.V.

Harte, J. and J. Kitzes. 2011. "The Use and Misuse of Species Area Relationships in Predicting Climate Driven Extinction." In *Saving a Million Species: Extinction Risk from Climate Change*, L. Hannah, ed. Washington, DC: Island Press.

Hartman, G. and M. Farr. 2010. "Rosement Mine Benefits Are Small and Short-term, Negatives Are Lasting and Costly." *Arizona Daily Star*. March 9.

Harvey, A. 1998. *A Sense of Place: Issues, Attitudes and Resources in the Yellowstone to Yukon Ecoregion*. Canmore, AB, Canada: Yellowstone to Yukon Conservation Initiative.

Harvey, C. A., B. Dickson, and C. Kormos. 2010. "Opportunities for Achieving Biodiversity Conservation through REDD." *Conservation Letters* 3:53–61.

Hassan, R., R. Scholes, and N. Ash, eds. 2005. *Ecosystems and Human Well-being: Current State and Trends: Findings of the Condition and Trends Working Group of the Millennium Ecosystem Assessment.* Millennium Ecosystem Assessment Series. Vol. 1 of Ecosystems and Human Well-being, Millennium Ecosystem Assessment (Program). Washington, DC: Island Press.

Hauer, G., S. G. Cumming, F. K. A. Schmiegelow, A. Adamowicz, M. Weber, and R. Jagodzinski. 2010. "Tradeoffs between Forestry Resource and Conservation Values under Alternate Policy Regimes: A Spatial Analysis of the Western Canadian Boreal Plains." *Ecological Modelling* 221:2590–2603.

Hawkes, L. A., A. C. Broderick, M. H. Godfrey, and B. J. Godley. 2009. "Climate Change and Marine Turtles." *Endangered Species Research* 7:137–54.

Hayhoe, K., B. Jones, and J. Gross. 2011. "Peering into the Future: Climate and Ecological Models." In *Scanning the Conservation Horizon: A Guide to Climate Change Vulnerability Assessment,* P. Glick, B. A. Stein, and N. A. Edelson, eds. Washington, DC: National Wildlife Federation.

Heck, K. L., Jr, T. J. B. Carruthers, C. M. Duarte, A. R. Hughes, G. Kendrick, R. J. Orth, and S. W. Williams. 2008. "Trophic Transfers from Seagrass Meadows Subsidize Diverse Marine and Terrestrial Consumers." *Ecosystems* 11:1198–1210.

Heiner, M., G. Davaa, J. Kiesecker, B. McKenney, J. Evans, T. Enkhtsetseg, Z. Dash, et al. 2011. *Identifying Conservation Priorities in the Face of Future Development: Applying Development by Design in the Grasslands of Mongolia*. Arlington, VA: The Nature Conservancy.

Heino, J., R. Virkkala, and H. Toivonen. 2009. "Climate Change and Freshwater Biodiversity: Detected Patterns, Future Trends and Adaptations in Northern Regions." *Biological Reviews* 84:39–54.

Heinrichs, A. K., Y. Kohler, and A. Ulrichs. 2010. *Implementing a Pan-Alpine Ecological Network. A Compilation of Major Approaches, Tools and Activities.* Bonn, Germany: Bundesamt für Naturschutz (BfN).

Heinz Center. 2008. *Strategies for Managing the Effects of Climate Change on Wildlife and Ecosystems*. Washington, DC. 43 pp. www.heinzctr.org/Major_Reports_files/Strategies%20for%20Manageing%20the%20Effects%20of%20Climate%20Change%20on%20Wildlife%20and%20Ecosystems.pdf (accessed January 11, 2012).

Heller, N. E. and E. S. Zavaleta. 2009. "Biodiversity Management in the Face of Climate Change: A Review of 22 Years of Recommendations." *Biological Conservation* 142:14–32.

Hersteinsson, P. and D. W. MacDonald 1992. "Interspecific Competition and the Geographic Distribution of Red and Arctic Foxes *Vulpes vulpes* and *Alopex lagopus*." *Oikos* 54:505–15.

Hess, S., T. Boucher, A. Dabrovskyy, E. Ter Hoorn, and P. Van Beukering. 2010. *Evaluating the Effectiveness of Community-based Conservation in Mongolia's Gobi Desert*. Arlington, VA: The Nature Conservancy.

Heydenrych, B. J., R. Cowling, and A. Lombard. 1999. "Strategic Conservation Interventions in a Region of High Biodiversity and High Vulnerability: A Case Study from the Agulhas Plain at the Southern Tip of Africa." *Oryx* 33:256–69.

Higgins, K. F., D. E. Naugle, and K. J. Forman. 2002. "A Case Study of Changing Land Use Practices in the Northern Great Plains, USA: An Uncertain Future for Waterbird Conservation." *Waterbirds* 25, special publication no. 2:42–50.

Hilty, J. A., W. Z. Lidicker, and A. M. Merenlender. 2006. *Corridor Ecology: The Science and Practice of Linking Landscapes for Biodiversity Conservation*. Washington, DC: Island Press.

Hobbs, R. J., E. Higgs, and J. A. Harris. 2009. "Novel Ecosystems: Implications for Conservation and Restoration." *Trends in Ecology and Evolution* 24:599–605.

Hoegh-Guldberg, O. 1999. "Climate Change, Coral Bleaching and the Future of the World's Coral Reefs." *Marine and Freshwater Research* 50:839–66.

Hoegh-Guldberg, O., L. Hughes, S. McIntyre, D. B. Lindenmayer, C. Parmesan, H. P. Possingham, and C. D. Thomas. 2008. "Assisted Colonization and Rapid Climate Change." *Science* 321:345–46.

Hoegh-Guldberg, O., P. J. Mumby, A. J. Hooten, R. S. Steneck, P. Greenfield, E. Gomez, and C. D. Harvell, et al. 2007. "Coral Reefs under Rapid Climate Change and Ocean Acidification." *Science* 318:1737.

Hoekstra, J. M., T. M. Boucher, T. H. Ricketts, and C. Roberts. 2005. "Confronting a Biome Crisis: Global Disparities of Habitat Loss and Protection." *Ecology Letters* 8:23–29.

Hogg, E. H. T. and P. Y. Bernier. 2005. "Climate Change Impacts on Drought-prone Forests in Western Canada." *Forestry Chronicle* 81:675–82.

Holechek, J., R. D. Pieper, and C. H. Herbel. 1989. *Range Management: Principles and Practices*. Englewood Cliffs, NJ: Prentice Hall.

Holling, C. S. 1973. "Resilience and Stability of Ecological Systems." *Annual Review of Ecology and Systematics* 4:1–23.

Høye, T., E. Post, H. Meltofte1, N. M. Schmidt, and M. C. Forchhammer. 2007. "Rapid Advancement of Spring in the High Arctic." *Current Biology* 17:449–51.

Hudson, J. M. G. and G. H. R. Henry. 2009. "Increased Plant Biomass in a High Arctic Heath Community from 1981 to 2008." *Ecology* 90:2657–63.

Hughes, K. A. and P. Convey. 2010. "The Protection of Antarctic Terrestrial Ecosystems from Inter- and Intra-continental Transfer of Non-indigenous Species by Human Activities: A Review of Current Systems and Practices." *Global Environmental Change* 20:96–112.

Huntley, B. 2007. "Limitations on Adaptation: Evolutionary Response to Climatic Change?" *Heredity* 98:247–48.

IBRA (Interim Biogeographic Regionalisation of Australia). 2010. *Interim Biogeographic Regionalisation of Australia, Version 6.1*. Commonwealth of Australia. www.environment.gov.au/parks/nrs/science/pubs/regions.pdf (accessed January 6, 2012).

IISD (International Institute for Sustainable Development). 1990. *Wetland Areas Lost or Degraded since Settlement in the Prairie/Parkland Region (Potential Areas Available for Restoration, Creation, and Enhancement)*. IISD, citing Ducks Unlimited 1990 Continental Conservation Plan. www.iisd.org/wetlands/sci_abstrct1.htm (accessed January 12, 2012).

Immerzeel, W. W., L. P.H. van Beek, and M. F. P. Bierkens. 2010. "Climate Change Will Affect the Asian Water Towers." *Science* 328:1382–85.

Ims, R. A. and E. Fuglei. 2005. "Trophic Interaction Cycles in Tundra Ecosystems and the Impact of Climate Change." *BioScience* 55:311–22.

IPCC (Intergovernmental Panel on Climate Change). 2007. *Climate Change 2007: Synthesis Report*. Contribution of Working Groups I, II, and III to the fourth assessment report of the IPCC. Geneva.

IPCC-SRES (Intergovernmental Panel on Climate Change–Special Report on Emission Scenarios). 2000. *Special Report on Emissions Scenarios.* A Special Report of Working Group III of the IPCC. Cambridge, UK: Cambridge University Press.

IPCC-WGII (Intergovernmental Panel on Climate Change–Working Group II). 1990. *Climate Change. The IPCC Impacts Assessment*. Report prepared for IPCC by Working Group II.

IPCC-WG1 (Intergovernmental Panel on Climate Change-Working Group 1). 2001. "Climate Change 2001: The Scientific Basis." Cambridge, UK.

——. 2007. *Climate Change 2007: The Physical Science Basis*. Contribution of Working Group I to the fourth assessment report of the IPCC. New York: Cambridge University Press.

IPCC-WG2 (Intergovernmental Panel on Climate Change–Working Group 2). 2007. *Climate Change 2007: Impacts, Adaptation and Vulnerability*. Contribution of Working Group II to the fourth assessment report of the IPCC. New York: Cambridge University Press.

IUCN (International Union for Conservation of Nature). 2008. *Position Paper: Adaptation to Climate Change*. Gland, Switzerland: IUCN. Submitted to the fourteenth session of the Conference of the Parties to the UN Framework Convention on Climate Change (COP14), Poznan, Poland, December 1–12, 2008.

IUCN. 2008. *Position Paper: Adaptation to climate change*. Gland, Switzerland: International Union for the Conservation of Nature and Natural Resources. Submitted to the fourteenth session of the Conference of the Parties to the UN Framework Convention on Climate Change (COP14), Poznan, Poland, 1–12 December 2008. http://cmsdata.iucn.org/downloads/unfccc_cop14_adaptation_pp_19_nov_08_final_1.pdf.

Iverson, L. R. and A. M. Prasad. 1998. "Predicting Abundance of 80 Tree Species following Climate Change in the Eastern United States." *Ecological Monographs* 68:465–85.

Iverson, L., A. Prasad, and S. Matthews. 2008. "Modeling Potential Climate Change Impacts on the Trees of the Northeastern United States." *Mitigation and Adaptation Strategies for Global Change* 13:487–516.

Iverson, L.R., A. M. Prasad, S. N. Matthews, and M. Peters. 2008. "Estimating Potential Habitat for 134 Eastern U.S.Tree Species under Six Climate Scenarios." *Forest Ecology and Management* 254:390–406.

IWM (Institute of Water Modelling). 2003. *Sundarbans Biodiversity Conservation Project Surface Water Modelling*. TA No. 3158-Ban (Contract COCS/00-696). Final Report, Vol. 1.

Jacobson, G. L., I. J. Fernandez, P. A. Mayewski, and C. V. Schmitt, eds. 2009. *Maine's Climate Future: An Initial Assessment*. Orono: University of Maine.

Jäger, J. and R. Holderegger. 2005. "Thresholds of Landscape Fragmentation." *Gaia: Ecological Perspectives for Science and Society* 14:113–18.

Jallow, B. P., M. K. A. Barrow, and S. P. Leatherman. 1996. "Vulnerability of the Coastal Zone of The Gambia to Sea Level Rise and Development of Response Strategies and Adaptation Options." *Climate Research* 6:165–77.

Jenkins, A. P., S. D. Jupiter, I. Qauqau, and J. Atherton J. 2010. "The Importance of Ecosystem-based Management for Conserving Migratory Pathways on Tropical High Islands: A Case Study from Fiji." *Aquatic Conservation: Marine and Freshwater Ecosystems* 20:224–38.

Jenkins, J. 2010. *Climate Change in the Adirondacks: The Pathway to Sustainability.* Ithaca, NY: Cornell University Press.

Jenouvrier, S., H. Caswell, C. Barbraud, M. Holland, J. Stoeve, and H. Weimerskirch. 2009. "Demographic Models and IPCC Climate Projections Predict the Decline of an Emperor Penguin Population." *Proceedings of the National Academy of Sciences USA* 106:1844–47.

Johnson, J. A., R. B. Lanctot, B. A. Andres, J. R. Bart, S. C. Brown, S. J. Kendall, and D. C. Payer. 2007. "Distribution of Breeding Shorebirds on the Arctic Coastal Plain of Alaska." *Arctic* 60:277–93.

Johnson, W. C., B. V. Millet, T. Gilmanov, R. A. Voldseth, G. R. Guntenspergen, and D. E. Naugle. 2005. "Vulnerability of Northern Prairie Wetlands to Climate Change." *BioScience* 55:863–72.

Jones, B. M., K. M. Hinkel, C. D. Arp, and W. R. Eisner. 2008. "Modern Erosion Rates and Loss of Coastal Features and Sites, Beaufort Sea Coastline, Alaska." *Arctic* 61:361–72.

Jones, G. P., G. R. Almany, G. R. Russ, P. F. Sale, R. S. Steneck, M. J. H van Oppen, and B. L. Willis. 2009. "Larval Retention and Connectivity among Populations of Corals and Reef Fishes: History, Advances and Challenges." *Coral Reefs* 28:307–25.

Jones, P. F., M. Grue, and J. Landry-DeBoer. 2007. "Resource Selection by Pronghorn Antelope in the Grassland Natural Region of Alberta; Progress Report No. 3." Lethbridge, Alberta: Alberta Conservation Association. www.albertapronghorn.com/tasks/sites/default/assets/File/Antelope%202004-2005%20Summary_.pdf (accessed January 6, 2012).

Jørgensen, D. 2009. *Annual migrations of female prairie rattlesnakes*, Crotalus v. viridid, *in Alberta*. Master's thesis, University of Calgary. dspace.ucalgary.ca/bitstream/1880/47445/1/Jorgensen_2009.pdf.

Jørgensen, D., C. C. Gates, and D. P. Whiteside. 2008. "Movements, Migrations, and Mechanisms: A Review of Radiotelemetry Studies of Prairie (*Crotalus v. viridis*) and Western (*Crotalus oreganus*) Rattlesnakes." In *The Biology of Rattlesnakes*, W. K. Hayes, K. R. Beaman, M. D. Cardwell, and S. P. Bush, eds. Loma Linda, CA: Loma Linda University Press, 303–16.

Jørgensen, D. and J. Nicholson. 2007. *Reproductive Biology of Female Prairie Rattlesnakes (Crotalus viridis viridis) in Alberta*. Alberta Sustainable Resource Development, Fish and Wildlife Division, Alberta Species at Risk Report No. 103, Edmonton, AB.

Jupiter, S. D. and D. P. Egli. 2011. "Ecosystem-based Management in Fiji: Successes and Challenges after Five Years of Implementation." *Journal of Marine Biology* 2011: 14 pp.

Jupiter, S., G. Roff, G. Marion, M. Henderson, V. Schrameyer, M. McCulloch, and O. Hoegh-Guldberg. 2008. "Linkages between Coral Assemblages and Coral Proxies of Terrestrial Exposure along a Cross-shelf Gradient on the Southern Great Barrier Reef." *Coral Reefs* 27:887–903.

Kaimowitz, D. 1996. *Livestock and Deforestation: Central America in the 1980s and 1990s: A Policy Perspective.* Bogor, Indonesia: Center for International Forestry Research.

———. 2008. "The Prospects for Reduced Emissions from Deforestation and Degradation (REDD) in Mesoamerica." *International Forestry Review* 10:485–95.

Kaiser, J. 2001. "Bold Corridor Project Confronts Political Reality." *Science* 293:2196–99.

Kalin Arroyo, M. T. L. Cavieres, C. Marticorena, and M. Munoz-Schick. 1994. "Convergence in the Mediterranean Floras in Central Chile and California: Insights from Comparative Biology." In *Ecology and Biogeography of Mediterranean Ecosystems in Chile, California and Australia*, M. T. Kalin Arroyo, P. H. Zedler, and M. D. Fox, eds. Ecological Studies 108. New York: Springer-Verlag, 43–88.

Karl, J. W. and J. Hoth. 2005. *North American Central Grasslands Priority Conservation Areas: Technical Report and Documentation.* Montreal, Québec: Commission for Environmental Cooperation and The Nature Conservancy.

Karl, T. R., J. M. Melillo, and T. C. Peterson. 2009. *Global Climate Change Impacts in the United States.* U.S. Global Change Research Program: Cambridge University Press.

Kellert, S. R. 2005. *Building for Life: Designing and Understanding the Human-nature Connection.* Washington, DC: Island Press.

Kelly, B. P., A. Whiteley, and D. Tallmon. 2010. "The Arctic Melting Pot." *Nature* 468:891–91.

Kharlamova, N. 2010. "Trends of Climate Change in the Region." In *Climate Change and Connectivity Conservation in the Altai-Sayan Ecoregion.* Forthcoming [in Russian].

Kiesecker, J. M., H. Copeland, A. Pocewicz, and B. McKenney. 2010. "Development by Design: Blending Landscape-level Planning with the Mitigation Hierarchy." *Frontiers in Ecology and the Environment* 8:261–66.

Killeen, T., M. Douglas, T. Consiglio, P. M. Jorgensen, and J. Mejia. 2007. "Dry Spots and Wet Spots in the Andean Hotspot." *Journal of Biogeography* 34:1357–73.

Killeen, T. J. and L. A. Solorzano. 2008. "Conservation Strategies to Mitigate Impacts from Climate Change in Amazonia." *Philosophical Transactions of the Royal Society* B 363:1881–88.

Klyza, C. M. and S. C. Trombulak. 1999. *The Story of Vermont: A Natural and Cultural History.* Hanover, NH: University Press of New England.

Knight, R. 2009. "The Wisdom of the Sierra Madre: Aldo Leopold, the Apaches and the Land Ethic." In *Conservation of Shared Environments: Learning from the United States and Mexico*, L. López-Hoffman, E. McGovern, R. G. Varady, and K. W. Flessa, eds. Tucson: University of Arizona Press, 71–77.

Knutson, T. R., J. L. Mcbride, J. Chan, K. Emanuel, G. Holland, C. Landsea, I. Held, J. P. Kossin, A. K. Srivastava, and M. Sugi. 2010. "Tropical Cyclones and Climate Change." *Nature Geoscience* 3:157–63.

Kohler, Y. 2009. Ökologische Netzwerke in Zahlen. Szene Alpen, 90:24.

Konrad, S. K., S. N. Miller, W. K. Reeves, and N. S. Tietze. 2009. "Spatially Explicit West Nile Virus Risk Modeling in Santa Clara County, CA." *Vector-Borne and Zoonotic Diseases* 9:267–74.

Koshy, K. 2007. *Modeling Climage Change Impacts on Viti Levu (Fiji) and Aitutaki (Cook Islands).* A final report submitted to AIACC Project No. SIS09. International START Secretariat, Washington DC.

Kovacs, K. M., C. Lydersen, J. E. Overland, and S. E. Moore. 2011. "Impacts of Changing Sea-ice Conditions on Arctic Marine Mammals." *Marine Biodiversity* 41:181–194.

Krawchuk, M. A. and S. G. Cumming. 2011. "Effects of Biotic Feedback and Harvest Management on Boreal Forest Fire Activity under Climate Change." *Ecological Applications* 21: 122–36.

Krawchuk, M. A., M. A. Moritz, M.-A. Parisien, J. Van Dorn, and K. Hayhoe. 2009. "Global Pyrogeography: The Current and Future Distribution of Wildfire." *PLoS ONE* 4:e5102.

Krosby, M., J. Tewksbury, N. M. Haddad, and J. Hoekstra. 2010. "Ecological Connectivity for a Changing Climate." *Conservation Biology* 24:1686–89.

Kurz, W. A. and M. J. Apps. 1999. "A 70-year Retrospective Analysis of Carbon Fluxes in the Canadian Forest Sector." *Ecological Applications* 9:526–47.

LaGro, J. A., Jr. 2001. "Landscape Ecology." In *Encyclopedia of Life Sciences*. Hoboken, NJ: John Wiley.

Laird-Benner, W. and H. Ingram. 2011. "Sonoran Desert Network Weavers." *Environment Magazine* 53:7–16.

Lal, M. 2004. "Climate Change and Small Island Developing Countries of the South Pacific." *Fijian Studies* 2:15–31.

Lal, M., J. L. McGregor, and K. C. Nguyen. 2008. "Very High-resolution Climate Simulation over Fiji Using a Global Variable-resolution Model." *Climate Dynamics* 30:293–305.

Laliberte, A. S. and W. J. Ripple. 2004. "Range Contractions of North American Carnivores and Ungulates." *Bioscience* 54:123–38.

Lambert, J. D. 1997. *Preliminary Results of the Analysis of Biological Corridor Potential in Central America*. Case study online: www.conservationgis.org/test/stories/paseo (accessed March 31, 2011).

Lamers, M. and B. Amelung. 2007. "The Environmental Impacts of Tourism in Antarctica: A Global Perspective." In *Tourism and Climate Change Mitigation: Methods, Greenhouse Gas Reductions and Policies*, P. Peeters, ed. Proceedings of the E-Clat Climate Change and Tourism Conference, Westelbeers, June 11–14, 2006, Breda University of Applied Sciences, Breda, 51–62.

Larned, W., R. Stehn, and R. Platte. 2010. *Waterfowl Breeding Population Survey, Arctic Coastal Plain, Alaska, 2009*. Unpublished report. U.S. Fish and Wildlife Service.

Lassen and S. Savoia, eds. 2005. *Ecoregion Conservation Plan for the Alps*. Bellinzona, Switzerland: WWF European Alpine Program.

Lawler, J. J. 2009. "Climate Change Adaptation Strategies for Resource Management and Conservation Planning." *Annals of the New York Academy of Sciences* 1162:79–98.

Lawler, J. J. and J. Hepinstall-Cymerman. 2010. "Conservation Planning in a Changing Climate: Assessing the Impacts of Potential Range Shifts on a Reserve Network." In *Landscape-scale Conservation Planning*, S. C. Trombulak and R. F. Baldwin, eds. New York: Springer, 325–48.

Lawler, J .J., S. L. Shafer, D. White, P. Kareiva, E. P. Maurer, A. R. Blaustein, and P. J. Bartlein. 2009. "Projected Climate-induced Faunal Change in the Western Hemisphere." *Ecology* 90:588–97.

Lawler, J. J., T. H. Tear, C. Pyke, M. R. Shaw, P. Gonzalez, P. Kareiva, L. Hansen, et al. 2010. "Resource Management in a Changing and Uncertain Climate." *Frontiers in Ecology and the Environment* 8:35–43.

Lee, P. and R. Cheng. 2011. *Canada's Terrestrial Protected Areas Status Report 2010: Number, Area and Naturalness*. Edmonton, AB.

Lee, P. G., M. Hanneman, J. D. Gysbers, and R. Cheng. 2010. *Atlas of Canada's Intact Forest Landscapes*. Edmonton, Alberta: Global Forest Watch Canada 10th Anniversary Publication no.1.

Lee, T., M. Quinn, and D. Duke. 2010. "A Local Community Monitors Wildlife along a Major Transportation Corridor." In *Safe Passages: Highways, Wildlife, and Habitat Connectivity*, P. B. Jon, A. P. Clevenger, M. P. Huijser and J. A. Hilty, eds. Washington, DC: Island Press, 277–292.

Leemans, R. 1990. *Possible Changes in Natural Vegetation Patterns Due to a Global Warming*. Laxenburg, Austria: International Institute of Applied Systems Analysis. IIASA Working Paper WP90-08 and Publication Number 108 of the Biosphere Dynamics Project. www.ngdc.noaa.gov/ecosys/cdroms/ged_iia/datasets/a06/reprints/lh1.htm (accessed January 6, 2012).

Leisher, C., R. Brouwer, T. M. Boucher, R. Vogelij, W. R. Bainbridge, et al. 2011. "Striking a Balance: Socioeconomic Development and Conservation in Grassland through Community-Based Zoning." *PLoS ONE* 6:e28807.

Leisher, C., S. Hess, T. M. Boucher, P. van Beukering, and M. Sanjayan. 2012. "Measuring the Impacts of Community-based Grasslands Conservation in Mongolia's Gobi." *PLoS ONE* 7:e30991.

Lemieux, C. J., T. J. Beechey, D. J. Scott, and P.A. Gray. 2010. *Protected Areas and Climate Change in Canada: Challenges and Opportunities for Adaptation*. Canadian Council on Ecological Areas (CCEA) Occasional Paper No. 19. Ottawa, Ontario, Canada: CCEA Secretariat.

Lemieux, C. J. and D. J. Scott. 2005. "Climate Change, Biodiversity Conservation and Protected Area Planning in Canada." *Canadian Geographer* 49:384–99.

Lemos, M. C. and B. J. Morehouse. 2005. "The Co-production of Science and Policy in Integrated Climate Assessments." *Global Environmental Change* 15:57–68.

Lenton, T. M., H. Held, E. Kriegler, J. Hall, W. Lucht, S. Rahmstorf, and H. Schellnhuber. 2008. "Tipping Elements in the Earth's Climate System." *PNAS* 105:1786–93.

Leroux, S. J., F. K. A. Schmiegelow, R. B. Lessard, and S. G. Cumming. 2007. "Minimum Dynamic Reserves: A Conceptual Framework for Reserve Size." *Biological Conservation* 138:464–73.

Liebezeit, J. R., S. J. Kendall, S. Brown, C. B. Johnson, P. Martin, T. L. McDonald, C. L. Payer, et. al. 2009. "Influence of Human Development and Predators on Nest Survival of Tundra Birds, Arctic Coastal Plain, Alaska." *Ecological Applications* 19:628–44.

Liebezeit, J. and S. Zack 2008. "Point Counts Underestimate the Importance of Arctic Foxes as Avian Nest Predators: Evidence from Remote Video Cameras in Arctic Alaskan Oil Fields." *Arctic* 61:153–61.

Lin, J. L., E. Mapes, K. M. Weickmann, G. N. Kiladis, S. D. Schubert, M. J. Suarez, J. T. Bacmeister, and M. Lee. 2007. "North American Monsoon and Convectively Coupled Equatorial Waves Simulated by IPCC AR4 Coupled GCMs." *Journal of Climate* 21: 2919–37.

Linder, H. P., M. E. Meadows, and R. M. Cowling. 1992. "History of the Cape Flora." In *The Ecology of Fynbos: Nutrients, Fire and Diversity*, R. M. Cowling, ed. Cape Town: Oxford University Press, 113–34.

Loarie, S. R., P. B. Duffy, H. Hamilton, G. P. Asner, C. B. Field, and D. D. Ackerly. 2009. "The Velocity of Climate Change." *Nature* 462:1052–55.

Locatelli B., V. Evans, A. Wardell, A. Andrade, and R. Vignola. 2011. "Forests and Climate Change in Latin America: Linking Adaptation and Mitigation." *Forests* 2:431–50.

Lochner, P., A. Weaver, C. Gelderblom, R. Peart, T. Sandwith, and S. Fowkes. 2003. "Aligning the Diverse: The Development of a Biodiversity Conservation Strategy for the Cape Floristic Region." *Biological Conservation* 112:29–43.

Logan, J. A., W. W. Macfarlane, and L. Willcox. 2010. "Whitebark Pine Vulnerability to Climate-driven Mountain Pine Beetle Disturbance in the Greater Yellowstone Ecosystem." *Ecological Applications* 20:895–902.

Logan, J. A. and J. A. Powell 2001. "Ghost Forests, Global Warming, and the Mountain Pine Beetle (Coleoptera: Scolytidae)." *American Entomologist* 43:160–72.

Lombard, A. T., R. M. Cowling, R. L. Pressey, and P. J. Mustart. 1997. "Reserve Selection in a Species-rich and Fragmented Landscape on the Agulhas Plain, South Africa." *Conservation Biology* 11:1101–16.

Lombard, A. T. , A. G. Rebelo, R. L. Pressey, and R. M. Cowling. 1999. "Reserve Selection in the Succulent Karoo, South Africa: Coping with High Compositional Turnover." *Plant Ecology* 142:35–55.

Lomolino, M. V., B. R. Riddle, R. J. Whittaker, and J. H. Brown. 2010. *Biogeography*. 4th edition. Sunderland, MA: Sinauer Associates.

López-Hoffman, L., E. D. McGovern, R. G. Varady, and K. W. Flessa. 2009. *Conservation of Shared Environments: Learning from the United States and Mexico*. First volume in a new series, *Edge: Environmental Science, Law and Policy*, M. Miller, B. Morehouse and J. Overpeck, eds. Tucson: University of Arizona Press.

Loss, S. R., L. A. Terwilliger, and A. C. Peterson. 2011. "Assisted Colonization: Integrating Conservation Strategies in the Face of Climate Change." *Biological Conservation* 144:92–100.

Loucks, C., S. Barber-Meyer, Md. A. A. Hossain, A. Barlow, and R. M. Chowdhury. 2010. "Sea-level Rise and Tigers: Predicted Impacts to Bangladesh's Mangroves." *Climate Change* 98:291–98.

Lovejoy, T. E. and L. J. Hannah, eds. 2005. *Climate Change and Biodiversity*. New Haven, CT: Yale University Press.

Lovell, E. and H. Sykes. 2008. "Rapid Recovery from Bleaching Events—Fiji Global Coral Reef Monitoring Network Assessment of Hard Coral Cover from 1999–2007." *Proceedings of the 11th International Coral Reef Symposium*, Ft. Lauderdale, FL, 836–40.

Luckman, B. H. 1990. "Mountain Areas and Global Change: A View from the Canadian Rockies." *Mountain Research and Development* 10:183–95.

Mackey, B. and S. Hugh. 2010. *Spatial Analysis of Conservation Priorities in the Great Eastern Ranges: Project 3. Productivity Analysis and Drought Refuge*. Final Report to the NSW Department of Environment, Climate Change and Water.

Mackey, B., J. Watson, and G. L. Worboys. 2010. *Connectivity Conservation and the Great Eastern Ranges Corridor*. An Independent Report to the Interstate Agency Working Group, Alps to Atherton Connectivity Conservation Working Group convened under the Environment Protection and Heritage Council/Natural Resources Management Ministerial Council. Sydney: NSW Department of Environment Climate Change and Water.

Mahmoud, M., Y. Liu, H. Hartmann, S. Stewart, T. Wagener, D. Semmens, R. Stewart, et al. 2009. "A Formal Framework for Scenario Development in Support of Environmental Decision-making." *Environmental Modelling and Software* 24:798–808.

Maina, J., T. R. McClanahan, V. Venus, M. Ateweberhan, and J. Madin. 2011. "Global Gradients of Coral Exposure to Environmental Stresses and Implications for Local Management." *PLoS ONE* 6:e23064.

Malhi, Y., J. T. Roberts, R. A. Betts, T. J. Killeen, W. Li, and C. A. Nobre. 2008. "Climate Change, Deforestation, and the Fate of the Amazon." *Science* 319:169–72.

Marengo J., C. A. Nobre, R. A. Betts, P. M. Cox, G. Sampaio, and L. Salazar. 2009. "Global Warming and Climate Change in Amazonia: Climate-vegetation Feedback and Impacts on Water Resources." In *Amazonia and Global Change,* M. Keller, M. Bustamante, J. Gash, and P. S. Dias, eds. Washington, DC American Geophysical Union, 273–92.

Mars, J. C. and D. W. Houseknecht. 2007. "Quantitative Remote Sensing Study Indicates Doubling of Coastal Erosion Rate in Past 50 Years along a Segment of the Arctic Coast of Alaska." *Geology* 35:583–86.

Marshall, R. M., D. Turner, A. Gondor, D. Gori, C. Enquist, G. Luna, and R. Paredes Aguilar. 2004. *An Ecoregional Analysis of Conservation Priorities in the Apache Highlands Ecoregion*. Prepared by The Nature Conservancy of Arizona, Instituto del Medio Ambiente y el Desarrollo Sustentable del Estado de Sonora.

Martin, P. J., Jenkins, F. J. Adams, M. T. Jorgenson, A. C. Matz, D. C. Payer, P. E. Reynolds, A. C. Tidwell, and J. R. Zelenak. 2009. *Wildlife Response to Environmental Arctic Change: Predicting Future Habitats of Arctic Alaska*. Report of the Wildlife Response to Environmental Arctic Change (WildREACH), Predicting Future Habitats of Arctic Alaska Workshop, November 17–18, 2008. Fairbanks, AK: USFWS.

Maurer, E. P., L. Brekke, T. Pruitt, and P. B. Duffy. 2007. "Fine-resolution Climate Projections Enhance Regional Climate Change Impact Studies." *Eos, Transactions, American Geophysical Union* 88:504.

Mawdsley, J. R., R. O'Malley, and D. S. Ojima. 2009. "A Review of Climate-change Adaptation Strategies for Wildlife Management and Biodiversity Conservation." *Conservation Biology* 23:1080–89.

McAlpine, C. A., J. Syktus, R. C. Deo, P. J. Lawrence, H. A. McGowan, I. G. Watterson, and S. R. Phinn. 2007. "Modeling the Impact of Historical Land Cover Changes on Australia's Regional Climate." *Geophysical Research Letters* 34:L22711.1–L22711.6.

McCarty, J. 2001. "Ecological Consequences of Recent Climate Change." *Conservation Biology* 15:320–31.

McClanahan, T. R., J. E. Cinner, N. A. J. Graham, T. M. Daw, J. Maina, S. M. Stead, A. Wamukota, K. Brown, V. Venus, and N. V. C. Polunin. 2009. "Identifying Reefs of Hope and Hopeful Actions: Contextualizing Environmental, Ecological, and Social Parameters to Respond Effectively to Climate Change." *Conservation Biology* 23:662–71.

McClanahan, T. R., J. E. Cinner, J. Maina, N. A. J. Graham, T. M. Daw, S. M. Stead, A. Wamukota, K. Brown, M. Atweberhan, V. Venus, and N. V. C. Polunin. 2008. "Conservation Action in a Changing Climate." *Conservation Letters* 1:53–59.

McCullough, D. A. 1999. *A Review and Synthesis of Effects of Alterations to the Water Temperature Regime on Freshwater Life Stages of Salmonids, with Special Reference to Chinook Salmon*. EPA 910-R-99-010. Prepared for the U.S. Environmental Protection Agency (EPA), Region 10. Seattle, WA.

McDonald, R. I., J. Fargione, J. Kiesecker, W. M. Miller, and J. Powell. 2009. "Energy Sprawl or Energy Efficiency: Climate Policy Impacts on Natural Habitat for the United States of America." *PLoS One* 4:e6802.

McGranahan, D. A. and C. A. Beale. 2002. "Understanding Rural Population Loss." *Rural America* 17:2–11.

McKelvey, K., J. Copeland, M. Schwartz, J. Littell, K. Aubry, J. Squires, S. Parks, M. Elsner, and G. Mauger. 2011. "Climate Change Predicted to Shift Wolverine Distributions, Connectivity, and Dispersal Corridors." *Ecological Applications* 21:2882–97.

McKenzie, D., Z. Gedalof, D. L. Peterson, and P. Mote. 2004. "Climatic Change, Wildfire, and Conservation." *Conservation Biology* 18:890–902.

Meehl, G. A., C. Covey, T. Delworth, M. Latif, B. McAvaney, J. F. B. Mitchell, R. J. Stouffer, and K. E. Taylor. 2007. "The WCRP CMIP3 Multi-model Dataset: A New Era in Climate Change Research." *Bulletin of the American Meteorological Society* 88:1383–94.

Memmott, J., P. G. Craze, N. M. Waser, and M. V. Price. 2007. "Global Warming and the Disruption of Plant-pollinator Interactions." *Ecology Letters* 10:710–17.

MFWP (Montana Fish Wildlife and Parks). 2010. *Crucial Areas Planning System*. Helena, MT: MFWP. fwp.mt.gov/wildthings/conservationInAction/crucialAreas.html (accessed January 6, 2012).

Midgley, G. F., L. Hannah, D. Millar, M. C. Rutherford, and L. W. Powrie. 2002. "Assessing the Vulnerability of Species Richness to Anthropogenic Climate Change in a Biodiversity Hotspot." *Global Ecology and Biogeography* 11:445–51.

Miles, E. L., A. K. Snover, A. F. Hamlet, B. M. Callahan, and D. L. Fluharty. 2000. "Pacific Northwest Regional Assessment: The Impacts of Climate Variability and Climate Change on the Water Resources of the Columbia River Basin." *Journal of the American Water Resources Association* 36:399–420.

Millar, C. I., N. L. Stephenson, and S. L. Stephens. 2007. "Climate Change and Forests of the Future: Managing in the Face of Uncertainty." *Ecological Applications* 17:2145–51.

Miller, K., E. Chang, and N. Johnson. 2001. *Defining Common Ground for the Mesoamerican Biological Corridor.* Washington, DC: World Resources Institute.

Miller, S., M. Cross, and A. Schrag. 2008. *Anticipating Climate Change in Montana: A Report on the Workshop with Montana Fish, Wildlife and Parks Focused on the Sagebrush-steppe and Yellowstone River Systems*. NWF, WCS, WWF, and MTWFP. www.worldwildlife.org/what/wherewework/ngp/publications.html (accessed January 6, 2012).

Milly, P. C. D., K. A. Dunne, A. V. Vecchia. 2005. "Global Pattern of Trends in Streamflow and Water Availability in a Changing Climate." *Nature* 438:347–50.

Mimura, N. 1999. "Vulnerability of Island Countries in the South Pacific to Sea Level Rise and Climate Change." *Climate Research* 12:137–43.

Mirza, M. and M. Qadar. 1998. "Diversion of the Ganges Water at Farakka and Its Effects on Salinity in Bangladesh." *Environmental Management* 22:711–22.

Mitchell, T. D. and P. D. Jones. 2005. "An Improved Method of Constructing a Database of Monthly Climate Observations and Associated High-resolution Grids." *International Journal of Climatology* 25:693–712.

Mittermeier, R. A., P. R. Gil, M. Hoffman, J. Pilgrim, T. Brooks, C. G. Mittermeier, J. Lamoreux, and G. A. B. da Fonseca. 2004. *Hotspots Revisited*. Mexico City: CEMEX.

Mittermeier, R. A. and C. G. Mittermeier. 1997. *Megadiversity: Earth's Biologically Wealthiest Nations*. Mexico City: CEMEX.

MNET (Ministry of Nature, Environment, and Tourism). 2009. "Mongolia Assessment Report on Climate Change 2009." Ulaanbaatar, Mongolia: MNET.

MoEF (Ministry of Environment and Forests). 2009a. *Bangladesh Climate Change Strategy and Action Plan 2009*. Dhaka, Bangladesh: MoEF. www.moef.gov.bd/climate_change_strategy 2009.pdf (accessed January 12, 2012).

——. 2009b. *National Adaptation Programme of Action*. Dhaka: Ministry of Environment and Forests, Government of the People's Republic of Bangladesh. www.undp.org.bd/info/pub /NAPA_English.pdf (accessed January 12, 2012).

Moore, J., S. Antenen, G. Davaa, C Ferree, D. Batbold, M. DePhilip, C. Pague, Y. Onon, D. Sanjmyatav, and R. McCready. 2010. *Biodiversity Gap Analysis of the Grasslands and Forest Steppe of Central and Eastern Mongolia: Setting the Stage for Establishment of an Ecologically Representative Protected Areas Network for Mongolia*. Ulaanbaatar, Mongolia: The Nature Conservancy.

Morgan, G., H. Dowlatabadi, M. Henrion, D. Keith, R. Lempert, S. McBride, M. Small, and T. Wilbanks. 2009. *Best Practice Approaches for Characterizing, Communicating, and Incorporating Scientific Uncertainty in Decisionmaking*. A Report by the U.S. Climate Change Science Program and the Subcommittee on Global Change Research. Synthesis and Assessment Product 5.2. Washington, DC: NOAA.

Moss, R. H., J. A. Edmonds, K. A. Hibbard, M. R. Manning, S. K. Rose, D. P. van Vuuren, T. R. Carter, S. Emori, M. Kainuma, T. Krums, et al. 2010. "The Next Generation of Scenarios for Climate Change Research and Assessment." *Nature* 463:747–56.

Mote, P. W. 2003. "Trends in Temperature and Precipitation in the Pacific Northwest during the Twentieth Century." *Northwest Science* 77:271–82.

——. 2006. "Climate-driven Variability and Trends in Mountain Snowpack in Western North America." *Journal of Climate* 19:6209–20.

Mote, P. W., A. F. Hamlet, M. P. Clark, and D. P. Lettenmaier. 2005. "Declining Mountain Snowpack in Western North America." *Bulletin of the American Meteorological Society* 86:39–49.

Mote, P. W., E. A. Parson, A. F. Hamlet, W. S. Keeton, D. Lettenmaier, N. Mantua, and E. L. Miles. 2003. "Preparing for Climatic Change: The Water, Salmon, and Forests of the Pacific Northwest." *Climatic Change* 61:45–88.

Mueller, T., K. A. Olson, T. K. Fuller, G. B. Schaller, M. G. Murray, and P. Leimgruber. 2008. "In Search of Forage: Predicting Dynamic Habitats of Mongolian Gazelles Using Satellite-based Estimates of Vegetation Productivity." *Journal of Applied Ecology* 45:649–58.

Muhar, S., M. Kainz, M. Kaufmann, M. Schwarz, and M. Jungwirth. 1998. *Ausweisung flußtypspezifisch erhaltener Fließgewässerabschnitte in Österreich. Fließgewässer mit einem Einzugsgebiet > 500 km² ohne Bundesflüsse*. Wasserwirtschaftskataster, Bundesministerium für Land- und Forstwirtschaft. Wien.

Mumby, P. J., A. J. Edwards, J. E. Arias-Gonzalez, K. C. Lindeman, P. G. Blackwell, A. Gall, M. I. Gorczynska, et al. 2004. "Mangroves Enhance the Biomass of Coral Reef Fish Communities in the Caribbean." *Nature* 427:533–36.

Myagmarsuren, D. 2008. *Mongolia's Protected Areas*. Ulaanbaatar, Mongolia: Mongolian Environmental Protection Agency.

Myers, N. 1981. "The Hamburger Connection." *Ambio* 10:3–8.

Myers, N. 1990. "The biodiversity challenge: Expanded hot-spots analysis." *The Environmentalist* 10, 243–256.

Myers, N., R. A. Mittermeier, C. G. Mittermeier, G. A. da Fonseca, and J. Kent. 2000. "Biodiversity Hotspots for Conservation Priorities." *Nature* 403:853–58.

Nabhan, G. P., L. López-Hoffman, C. K. Presnall, R. Knight, J. Goldstein, H. Gosnell, L. Gwen, D. Thilmany, and S. Charnley. Forthcoming. "Payments for Ecosystem Services: Keeping Working Lands in Working Hands." In *Saving the Wide-Open Spaces*, T. Sheridan, S. Charnley, and G. P. Nabhan, eds. Tucson: University of Arizona Press.

Narozhny, Y. K., S. A. Nikitin, and P. S. Borodavko. 2006. "Glaciers of Belukha Massif (Altai): Mass Balance, Dynamics and Distribution of Ice." *Proceedings of Glaciological Studies* 101:117–27 [in Russian].

NAS (National Academy of Sciences). 2003. "Cumulative Environmental Effects of Oil and Gas Activities on Alaska's North Slope." National Research Council of the NAS. Washington, DC: National Academies Press.

NCCSC (North Central Climate Science Center). 2010. *North Central Climate Science Center*. U. S. Geological Survey. www.doi.gov/csc/northcentral/about.cfm (accessed January 10, 2012).

Neilson, R. P., L. F. Pitelka, A. M. Solomon, R. Nathan, G. F. Midgley, J. M. V. Fragoso, H. Lischke, and K. Thompson. 2005. "Forecasting Regional to Global Plant Migration in Response to Climate Change." *Bioscience* 55:749–59.

New Zealand. 2005. *A Review of the Antarctic Protected Areas System*. IP 29. 28 ATCM. June 6–17, 2005, Stockholm, Sweden.

———. 2009a. *Updated Analysis of Representation of Annex 5 Categories and Environmental Domains in the System of Antarctic Specially Protected and Managed Areas*. WP 31. 32nd ATCM. April 6–17, 2009, Baltimore, USA.

———. 2009b. *A Framework for Analyzing and Managing Non-native Species Risks in Antarctica*. IP 36. 32nd ATCM. April 6–17, 2009, Baltimore, USA.

NFLC (Northern Forest Lands Council). 1994. "Finding Common Ground: Conserving the Northern Forest." Concord, NH: NFLC. 178 pp.

Nicholls, N. 2004. "The Changing Nature of Australian Droughts." *Climate Change* 63:323–36.

Nicholls, R. J., V. Burkett, J. Codignotto, J. Hay, R. McLean, S. Ragoonaden, and C. D. Woodroffe. 2007. "Coastal Systems and Low-lying Areas." In *Climate Change 2007: Impacts, Adaptation and Vulnerability.* Contribution of Working Group II to the fourth assessment report of the Intergovernmental Panel on Climate Change, M. L. Parry, O. F. Canziani, J. P. Palutikof, P. J. van der Linden and C. E. Hanson, eds. Cambridge and New York: Cambridge University Press, 315–356.

Nicol, S., A. Bowie, S. Jarman, D. Lannuzel, K. M. Meiners, and P. van der Merwe. 2010. "Southern Ocean Iron Fertilization by Baleen Whales and Antarctic Krill." *Fish and Fisheries* 11:203–9.

Nilsson, C., C. A. Reidy, M. Dynesius, and C. Revenga. 2005. "Fragmentation and Flow Regulation of the World's Large River Systems." *Science* 308:405–8.

Nkem, J., H. Santoso, M. Idinoba, J. Saborio, and H. Herawati. 2009. *Tropical Forests and Climate Change Adaptation: Criteria and Indicators for Adaptive Management for Reduced Vulnerability and Long-term Sustainability.* Final Technical Report for Center for International Forestry Research and Centro Agronómico Tropical de Investigación y Enseñanza. www.cifor.org/trofcca/_ref/home/index.htm (accessed January 10, 2012).

Nobre, C. A. and L. S. Borma. 2009. "Tipping Points for the Amazon Forest." *Current Opinion in Environmental Sustainability* 1:28–36.

Nobre, C. A., P. J. Sellers, and J. Shulka. 1991. "Amazonian Deforestation and Regional Climate Change." *Journal of Climatology* 4:957–88.

Norse, E. A. and R. E. McManus. 1980. "Ecology and Living Resources: Biological Diversity." In *Environmental Quality 1980: The Eleventh Annual Report of the Council on Environmental Quality*. Council on Environmental Quality: 31–80. Washington, DC: U.S. Government Printing Office.

Norway and United Kingdom. 2010. *Report from Antarctic Treaty Meeting of Experts on Implications of Climate Change for Antarctic Management and Governance. Co-chairs' Executive Summary with Advice for Actions.* Working Paper 63. 33rd Antarctic Treaty Consultative Meeting, May 3–14, 2010. Punta del Este, Uruguay.

Noss, R. F. 1983. "A Regional Landscape Approach to Maintain Diversity." *BioScience* 33:700–706.

———. 2001. "Beyond Kyoto: Forest Management in a Time of Rapid Climate Change." *Conservation Biology* 15:578–90.

Noss, R. F. and L. D. Harris. 1986. "Nodes, Networks, and MUMs: Preserving Diversity at All Scales." *Environmental Management* 10:299–309.

NRC (Natural Resources Canada). 2009. *The Atlas of Canada: Wetlands.* Natural Resources Canada. atlas.nrcan.gc.ca/site/english/maps/freshwater/distribution/wetlands/1 (accessed January 10, 2012).

NRDC (Natural Resources Defense Council). 2010. "Climate Change, Water and Risk: Current Water Demands Are Not Sustainable." Washington, DC: NRDC. www.nrdc.org/globalwarming/watersustainability/files/WaterRisk.pdf (accessed January 10, 2012).

NRPPC (Natural Resource Policies and Program Committee). 2009. *Australia's Strategy for the National Reserve System 2009–2030*. Department of the Environment, Water, Heritage and the Arts, Australian Government, Canberra.

NRTEE (National Round Table on the Environment and Economy). 2005. *Boreal Futures: Governance, Conservation and Development in Canada's Boreal: State of the Debate Report*. Ottawa, ON: NRTEE.

NSDE (Novia Scotia Department of the Environment). 2009. *Toward a Greener Future: Nova Scotia's Climate Change Action Plan*.

Nuñez, T. 2011. *Connectivity Clanning to Facilitate Species Movements in Response to Climate Change*. MS thesis. University of Washington, College of the Environment. Seattle, Washington. 46 pp.

Nunn, P. D. and W. R. Peltier. 2001. "Far-field Test of the ICE-4G Model of Global Isostatic Response to Deglaciation Using Empirical and Theoretical Holocene Sea-level Reconstructions for the Fiji Islands, Southwestern Pacific." *Quaternary Research* 55:203–14.

Obura, D. O. and G. Grimsditch. 2009. *Resilience Assessment of Coral Reefs: Assessment Protocol for Coral Reefs, Focusing on Coral Bleaching and Thermal Stress*. IUCN Working Group on Climate Change and Coral Reefs, Gland, Switzerland.

Ohlson, D. W., G. A. McKinnon, and K. G. Hirsch. 2005. "A Structured Decision-making Approach to Climate Change Adaptation in the Forest Sector." *Forestry Chronicle* 81:97–103.

Olson, D., L. Farley, A. Patrick, D. Watling, M. Tuiwawa, V. Masibalavu, L. Lenoa, et al. 2009. "Priority Forests for Conservation in Fiji: Landscapes, Hotspots and Ecological Processes." *Oryx* 44:57–70.

Olson, D. M., E. Dinerstein, E. D. Wikramanayake, N. D. Burgess, G. V. N. Powell, E. C. Underwood, J. A. D'Amico, et al. 2001. "Terrestrial Ecoregions of the Worlds: A New Map of Life on Earth." *Bioscience* 51:933–38.

Olson, K. A. 2008. "Distribution and Ecology of Mongolian Gazelles (*Procapra gutturosa Pallas* 1777) in Mongolia's Eastern Steppe." Amherst: University of Massachusetts.

Orlove, Ben. 2009. "Glacier Retreat: Reviewing the Limits of Human Adaptation to Climate Change." *Environment* 51:22–34.

Ormerod, S. J. 2009. "Climate Change, River Conservation and the Adaptation Challenge." *Aquatic Conservation: Marine and Freshwater Ecosystems* 19:609–13.

Orr, J. C., V. J. Fabry, O. Aumont, L. Bopp, S. C. Doney, R. A. Feely, A. Gnanadesikan, et al. 2005. "Anthropogenic Ocean Acidification over the Twenty-first Century and Its Impact on Calcifying Organisms." *Nature* 437:681–86.

Österblom, H., S. Hansson, U. Larsson, O. Hjerne, F. Wulff, R. Elmgren, and C. Folk. 2007. "Human-induced Trophic Cascades and Ecological Regime Shifts in the Baltic Sea." *Ecosystems* 10:877–89.

Overpeck, J. and B. Udall. 2010. "Dry Times Ahead." *Science* 328:1642–43.

Oyama, M. D. and C. A. Nobre. 2003. "A New Climate-vegetation Equilibrium State for Tropical South America" *Geophysical Research Letters* 30:2199.

Ozgul, A., D. Z. Childs, M. K. Oli, K. B. Armitage, D. T. Blumstein, L. E. Olson, S. Tuljapurkar, and T. Coulson. 2010. "Coupled Dynamics of Body Mass and Population Growth in Response to Environmental Change." *Nature* 466:482–85.

Palmer, M. A., D. P. Lettenmaier, N. LeRoy Poff, S. L. Postel, B. Richter, and R. Warner. 2009. "Climate Change and River Ecosystems: Protection and Adaptation Options." *Environmental Management* 44:1053–68.

Pamperin, N. J., E. H. Follmann, and B. Petersen. 2006. "Interspecific Killing of an Arctic Fox by a Red Fox at Prudhoe Bay, Alaska." *Arctic* 59:361–64.

Papworth, S. K., J. Rist, L. Coad, and E. J. Milner-Gulland. 2009. "Evidence for Shifting Baseline Syndrome in Conservation." *Conservation Letters* 2:93–100.

Park, R., M. Trehan, P. Mausel, and R. Howe. 1989. "The Effects of Sea Level Rise on US Coastal Wetlands." In *The Potential Effects of Global Climate Change on the United States. Appendix B–Sea Level Rise,* J. Smith, and D. Tirpak, eds, Washington, DC: U.S. Environmental Protection Agency.

Parkinson, C. L. 2002. "Trends in the length of the Southern Ocean sea-ice season, 1979–99." *Annals of Glaciology* 34, 435–440.

Parmesan, C. 2006. "Ecological and Evolutionary Responses to Recent Climate Change." *Annual Review of Ecology, Evolution, and Systematics* 37:637–69.

Parmesan, C., C. Duarte, E. Poloczanska, A. J. Richardson, and M. C. Singer. 2011. "Overstretching Attribution." *Nature Climate Change* 1:2–4.

Parmesan, C., N. Ryrholm, C. Stefanescu, J. K. Hill, C. D. Thomas, H. Descimon, and B. Huntley, et al. 1999. "Poleward Shift of Butterfly Species' Ranges Associated with Regional Warming." *Nature* 399:579–83.

Patkar, M. 2004. *River Linking: A Millennium Folly*. Mumbai, India: National Alliance of People's Movements and Initiatives.

Patrusheva, T. 2010. "Dendrological Indication of Climate Change in the High-altitudinal Zone of the Altai." In *Climate Change and Connectivity Conservation in the Altai-Sayan Ecoregion*. Forthcoming [in Russian].

Peacock, S. 2011. "Projected 21st Century Climate Change for Wolverine Habitats within the Contiguous United States." *Environmental Research Letters* 6:014007.

Pearson, R. G. and T. P. Dawson. 2003. "Predicting the Impacts of Climate Change on the Distribution of Species: Are Bioclimate Envelope Models Useful?" *Global Ecology and Biogeography* 12:361–71.

Permanent Secretariat of the Alpine Convention. 2009. *Water and Water Management Issues*. Bolzano, Italy: Permanent Secretariat of the Alpine Convention.

Peters, R. L. 1988. *Executive Overview of World Wildlife's Conference on Consequences of the Greenhouse Effect for Biological Diversity*. Washington, DC: World Wildlife Fund–US and National Technical Information Service, U.S. Department of Commerce.

Peters, R. L. and J. D. S. Darling. 1985. "The Greenhouse Effect and Nature Reserves." *BioScience* 35:707–17.

Peterson, A. T. 2003. "Projected Climate Change Effects on Rocky Mountain and Great Plains Birds: Generalities of Biodiversity Consequences." *Global Change Biology* 9:647–55.

Peterson, G., G. A. De Leo, J. J. Hellmann, M. A. Janssen, A. Kinzig, J. R. Malcolm, K. L. O'Brien, et al. 1997. "Uncertainty, Climate Change, and Adaptive Management." *Conservation Ecology* 1:4.

Peterson, G. D., G. S. Cumming, and S. R. Carpenter. 2003. "Scenario Planning: A Tool for Conservation in an Uncertain World." *Conservation Biology* 17:358–66.

Phillips, S. J., P. Williams, G. Midgley, and A. Archer. 2008. "Optimizing Dispersal Corridors for the Cape Proteaceae Using Network Flow." *Ecological Applications* 18:1200–1211.

Picton Phillipps, G. and A. Seimon. 2009. *Potential Climate Change Impacts in Conservation Landscapes of the Albertine Rift*. Wildlife Conservation Society, Whitepaper.

Pierce, S., R. Cowling, A. Knight, A. Lombard, M. Rouget, and T. Wolf. 2005. "Systematic Conservation Planning Products for Land-use Planning: Interpretation for Implementation." *Biological Conservation* 125:441–58.

Plumptre, A. J. 2012. *The Ecological Impact of Long-term Changes in Africa's Rift Valley*. New York: Nova Science Publishers.

Plumptre, A. J., M. Behangana, E. Ndomba, T. Davenport, C. Kahindo, R. Kityo, P. Ssegawa, G. Eilu, D. Nkuutu, and I. Owiunji. 2003. *The Biodiversity of the Albertine Rift*. Albertine Rift Technical Reports no. 3. Wildlife Conservation Society.

Plumptre, A. J., T. R. B. Davenport, M. Behangana, R. Kityo, G. Eilu, P. Ssegawa, C. Ewango, et al. 2007. "The Biodiversity of the Albertine Rift." *Biological Conservation* 134:178–94.

Plumptre, A. J., A. Kayitare, H. Rainer, M. Gray, I. Munanura, N. Barakabuye, S. Asuma, M. Sivha, and A. Namara. 2004. *The Socio-economic Status of People Living near Protected Areas in the Central Albertine Rift*. Albertine Rift Technical Reports, no. 4. www.albertine rift.org (accessed January 10, 2012).

Plumptre, A. J., D. Kujirakwinja, and S. Nampindo. 2009. "Conservation of Landscapes in the Albertine Rift." In *Protected Areas, Governance and Scale*, K. H. Redford and C. Grippo, eds. WCS Working Paper no. 36, 28–34.

Plumptre, A. J., D. Kujirakwinja, A. Treves, I. Owiunji, and H. Rainer. 2007. "Transboundary Conservation in the Greater Virunga Landscape: Its Importance for Landscape Species." *Biological Conservation* 134:279–87.

Poff, N. L., M. M. Brinson, and J. W. Day Jr. 2002. *Aquatic Ecosystems and Global Climate Change: Potential Impacts on Inland Freshwater and Coastal Wetland Ecosystems in the United States*. Washington, DC: Pew Center on Global Climate Change.

Poiani, K. A., R. L. Goldman, J. Hobson, J. M. Hoekstra, and K. S. Nelson. 2011. "Redesigning Biodiversity Conservation Projects for Climate Change: Examples from the Field." *Biological Conservation* 20:185–201.

Post, E., M. C. Forchhammer, M. S. Bret-Harte, T. V. Callaghan, T. R. Christensen, B. Elberling, A. Fox, et al. 2009. "Ecological Dynamics across the Arctic Associated with Recent Climate Change." *Science* 325:1355–58.

Postel, S. L. and B. H. Thompson. 2005. "Watershed Protection: Capturing the Benefits of Nature's Water Supply Services." *Natural Resources Forum* 29:98–108.

Pouyaud, B., M. Zapata, J. Yerren, J. Gómez, G. Rosas, W. Suarez, and P. Ribstein. 2005. "Avenir des ressources en eau glaciaire de la Cordillére Blanche." *Hydrological Sciences Journal* 50:999–1022.

Povilitis, T. 1994. "A Nature Reserve System for the Gila River–Sky Island Region of Arizona and New Mexico; Some Preliminary Suggestions." In The *Sky Islands of Southwestern United States and Northwestern Mexico*, L. F. Debano, G. J. Gottfried, R. H. Hamre, C. B. Edminster, P. F. Ffolliott, and A. Ortega-Rubio, eds. USDA Forest Service: General Technical Report RM-GTR-264, 6–18.

Prasad, P. V., S. R. Pisipati, Z. Ristic, U. Bukovnik, and A. K. Fritz. 2008. "Impact of Nighttime Temperature on Physiology and Growth of Spring Wheat." *Crop Physiology and Metabolism* 48:2372–80.

Pressey, R. L., R. M. Cowling, and M. Rouget. 2003. "Formulating Conservation Targets for Biodiversity Pattern and Process in the Cape Floristic Region, South Africa." *Biological Conservation* 112:99–127.

Pressey, R. L., S. Ferrier, T. C. Hager, C. A. Woods, S. L. Tully, and K. M. Weinman. 1996. "How Well Protected Are the Forests of North-eastern New South Wales?" Analyses of Forest Environments in Relation to Formal Protection Measures, Land Tenure, and Vulnerability to Clearing." *Forest Ecology and Management* 85:311–33.

Pressey, R. L., I. R. Johnson, and P. D. Wilson. 1994. "Shades of Irreplaceability: Towards a Measure of the Contribution of Sites to a Reservation Goal." *Biodiversity and Conservation* 3:242–62.

Programa Estado de la Nación. 2008. *Informe Estado de la Región en Desarrollo Humano Sostenible*. San José, Costa Rica: Programa Estado de la Nación.

Pulsford, I., G. L. Worboys, K. Evans, and G. Howling. 2010. "Great Eastern Ranges Corridor: A Continental Scale Vision to Protect Our Richest Biodiversity." *Journal of the National Parks Association* (NSW), Oct.–Dec.

Pulsford, I., G. L. Worboys, J. Gough, and T. Shepherd. 2004. "The Australian Alps and the Great Escarpment of Eastern Australia Conservation Corridors." In *Managing Mountain Protected Areas: Challenges and Responses for the 21st Century*, D. D. Harmon and G. L. Worboys, eds. Collendara: Andromeda Editrice, 106–14.

Pulsford, I., G. L. Worboys, and G. Howling. 2010. "Australian Alps to Atherton Connectivity Conservation Corridor." In *Connectivity Conservation Management: A Global Guide,* G. L. Worboys, W. L. Francis, and M. Lockwood, eds. London: Earthscan, 96–105.

Quijada-Mascareñas, A., D. Falk, J. Weiss, M. McClaran, J. Koprowski, M. Culver, S. Drake, S. Marsh, W. van Leeuwen, and M. Skroch. In review. "Potential Multi-scale Ecological Effects of Climate Change in Sky Islands." *Climate Research*.

Quijada-Mascareñas, J. A., C. Van Riper, D. James, L. López-Hoffman, C. Sharp, R. Gimblett, M. Scott, L. Norman, et al. 2011. "An Ecosystem Approach for Monitoring the Border Wall." In *The Border Wall between Mexico and the United States. Venues, Mechanisms and Stakeholders for a Constructive Dialogue*, C. de la Parra, E. Peters, and A. Cordova y Vazquez, eds. Instituto Nacional de Ecología, Mexico: SEMARNAT, Mexico DF.

Racine, C. and R. Jandt. 2008. "The 2007 'Anaktuvuk River' Tundra Fire on the Arctic Slope of Alaska: A New Phenomenon?" In *Ninth International Conference on Permafrost, Extended Abstracts*, D. L. Kane and K. M. Hinkel, eds. University of Alaska– Fairbanks, Institute of Northern Engineering, 247–48.

Rahmstorf, S. 2007. "A Semi-empirical Approach to Projecting Future Sea Level Rise." *Science* 315:368–70.

Ramírez, D., J. L. Ordaz, and J. Mora. 2009. *Istmo Centroamericano: Efectos del Cambio Climático sobre la Agricultura*. Technical document from project "La economía del cambio climático en Centroamérica." Comisión Económica para America Latina y el Caribe (CEPAL). Mexico: Subregional Program Office.

Ramirez-Llodra, E., P. A. Tyler, M. C. Baker, O. A. Bergstad, M. R. Clark, E. Escobar, and L. A. Levin. 2011. "Man and the Last Great Wilderness: Human Impact on the Deep Sea." *PLoS* 6:1–25.

Rasker, R. 1994. "A New Look at Old Vistas: The Economic Role of Environmental Quality of Western Public Lands." *University of Colorado Law Review* 65:369–99.

Rebelo, A. G. 2001. *Proteas: A Field Guide to the Proteas of Southern Africa.* Cape Town: Fernwood Press.

Rebelo, A. G. and W. R. Siegfried. 1992. "Where Should Nature Reserves Be Located in the Cape Floristic Region, South Africa? Models for the Spatial Configuration of a Reserve Network Aimed at Maximizing the Protection of Floral Diversity." *Conservation Biology* 6:243–52.

Reeves, Brian O. K. 1998. "Sacred Geography: First Nations of the Yellowstone to Yukon." In *A Sense of Place: Issues, Attitudes and Resources in the Yellowstone to Yukon Ecoregion*, Ann Harvey, ed. Canmore, Alberta: Yellowstone to Yukon Conservation Initiative, 31–50.

Reiners, W. A., J. S. Baron, D. M. Debinski, S. A. Elias, D. B. Fagre, J. S. Findley, and L. O. Mearns. 2003. "Natural Ecosystems 1: The Rocky Mountains." In *Rocky Mountain/Great Basin Regional Climate-change Assessment*, F. H. Wagner, ed. Logan: Utah State University report for the U. S. Global Change Research Program, 145–84.

Reuters. 2011. *Correction-Update-Grupo Mexico: Cananea at Capacity in March.* 8 February. www.reuters.com/article/2011/02/08/grupomexico-idUSN0829272520110208 (accessed January 10, 2012).

Reyers, B., M. Rouget, Z. Jonas, R. M. Cowling, A. Driver, K. Maze, and P. Desmet. 2007. "Developing Products for Conservation Decision-making: Lessons from a Spatial Biodiversity Assessment for South Africa." *Diversity and Distributions* 13:608–19.

Reyers, B., D. J. Roux, R. M. Cowling, A. E. Ginsburg, J. L. Nel, and P. O. Farrell. 2010. "Conservation Planning as a Transdisciplinary Process." *Conservation Biology* 24:957–65.

Ricciardi, A. and D. Simberloff. 2009. "Assisted Colonization Is Not a Viable Conservation Strategy." *Trends in Ecology and Evolution* 24:248–53.

Richardson, J. E., F. M. Weitz, M. F. Fay, Q. C. Cronk, H. P. Linder, G. Reeves, and M. W. Chase. 2001. "Rapid and Recent Origin of Species Richness in the Cape Flora of South Africa." *Nature* 412:181–83.

Richardson, M. 2004. *Making Sense of Rural Poverty in Central America: Lessons from the Rural Development Institute*. RUTA, San José, Costa Rica. unpan1.un.org/intradoc-cgi/idc_cgi_isapi.dll?IdcService=DOC_INFO&dID=38565 (accessed January 10, 2012).

Ricketts, T. H., E. Dinerstein, D. M. Olson, and C. J. Loucks. 1999. *Terrestrial Ecoregions of North America: A Conservation Assessment.* Washington, DC: Island Press.

Riemann, H. and E. Ezcurra. 2007. "Endemic Regions of the Vascular Flora of the Peninsula of Baja California, Mexico." *Journal of Vegetation Science* 18:327–36.

RMRS (Rocky Mountain Research Station). 2009. *2009 Climate Change Research Strategy.* U.S. Forest Service, RMRS. www.fs.fed.us/rmrs/docs/climate-change/climate-change-research-strategy.pdf (accessed January 10, 2012).

Robertson, C., T. A. Nelson, D. E. Jelinski, M. A. Wulder, and B. Boots. 2009. "Spatial–temporal Analysis of Species Range Expansion: The Case of the Mountain Pine Beetle, *Dendroctonus ponderosae*." *Journal of Biogeography* 36:1446–58.

Rodenhouse, N., S. Matthews, K. McFarland, J. Lambert, L. Iverson, A. Prasad, T. Sillett, and R. Holmes. 2008. "Potential Effects of Climate Change on Birds of the Northeast." *Mitigation and Adaptation Strategies for Global Change* 13:517–40.

Roots, E. F. 1989. "Climate Change: High Latitude Regions." *Climatic Change* 15:223–53.

Rose, N. A. and P. J. Burton. 2009. "Using Bioclimatic Envelopes to Identify Temporal Corridors in Support of Conservation Planning in a Changing Climate." *Forest Ecology and Management* 258:S64–S74.

Rosero-Bixby, L., and A. Palloni. 1998. "Population and Deforestation in Costa Rica." *Population and Environment* 20:149–85.

Rouget, M., D. M. Richardson, R. M. Cowling, J. W. Lloyd, and A. T. Lombard. 2003. "Current Patterns of Habitat Transformation and Future Threats to Biodiversity in Terrestrial Ecosystems of the Cape Floristic Region, South Africa." *Biological Conservation* 112:63–85.

Roura, R. and A. D. Hemmings. Forthcoming. "Realising Strategic Environmental Assessment in Antarctica." *Journal of Environmental Assessment Policy and Management.*

Ruckstuhl K. E., E. A. Johnson, and K. Miyanishi. 2008. "Introduction. The Boreal Forest and Global Change." *Philosophical Transactions of the Royal Society B* 363:2245–49.

Rudel, T. K., O. T. Coomes, E. Moran, F. Achard, A. Angelsen, J. Xiu, and E. Lambin. 2005. "Forest Transitions: Towards a Global Understanding of Land Use Change." *Global Environmental Change* 15:23–31.

Running, S. W. and L. S. Mills. 2009. *Terrestrial Ecosystem Adaptation*. Washington, DC: Climate Policy Program, Resources for the Future. www.cas.umt.edu/facultydatabase/FILES_Faculty/1135/RFF-Rpt-Adaptation-RunningMills.pdf (accessed January 10, 2012).

Rutherford, M. C., L. W. Powrie, and R. E. Schulze. 1999. "Climate Change in Conservation Areas of South Africa and Its Potential Impact on Floristic Areas of South Africa." *Diversity* 5:253–62.

Sader, S. A., D. J. Hayes, J. A. Hepinstall, and C. Soza. 2001. "Forest Change Monitoring of a Remote Biosphere Reserve." *International Journal of Remote Sensing* 22:1937–50.

Sader, S. A. and A. T. Joyce. 1988. "Deforestation Rates and Trends in Costa Rica, 1940 to 1983." *Biotropica* 20:11–19.

Salati, E., A. Dall'Olio, E. Matsui, and J. R. Gat. 1979. "Recycling of Water in the Amazon Basin: An Isotopic Study." *Water Resources Research* 15:1250–58.

Salinger, M. J., R. Basher, B. Fitzharris, J. Hay, P. Jones, J. McVeigh, and I. Schmidely-Leleu. 1995. "Climate Trends in the South-West Pacific." *International Journal of Climatology* 15:285–302.

Sampaio, G., C. A. Nobre, M. H. Costa, P. Satyamurty, B. S. Soares-Filho, and M. Cardoso. 2007. "Regional Climate Change over Eastern Amazonia Caused by Pasture and Soybean Cropland Expansion." *Geophysical Research Letters* 34:L17709.

Sanderson, E. W., M. Jaiteh, M. A. Levy, K. H. Redford, A. V. Wannebo, and G. Woolmer. 2002. "The Human Footprint and the Last of the Wild." *BioScience* 52:891–904.

SCBD (Secretariat of the Convention on Biological Diversity). 2009. *Connecting Biodiversity and Climate Change Mitigation and Adaptation: Report of the Second Ad Hoc Technical Expert Group on Biodiversity and Climate Change*. Montreal: Secretariat of the Convention on Biological Diversity, Technical Series No. 41.

Schauffler, M. and G. L. Jacobson Jr. 2002. "Persistence of Coastal Spruce Refugia during the Holocene in Northern New England, USA, Detected by Stand-scale Pollen Stratigraphies." *Journal of Ecology* 90:235–50.

Schindler, D. W. and P. G. Lee. 2010. "Comprehensive Conservation Planning to Protect Biodiversity and Ecosystem Services in Canadian Boreal Regions under a Warming Climate and Increasing Exploitation." *Biological Conservation* 143:1571–86.

Schlacher, T. A., D. S. Schoeman, J. Dugan, M. Lastra, A. Jones, F. Scapini, and A. McLachlan. 2008. "Sandy Beach Ecosystems: Key Features, Sampling Issues, Management Challenges and Climate Change Impacts." *Marine Ecology* 29:70–90.

Schmiegelow, F. K. A., S. J. Leroux, S. G. Cumming, L G. Anderson, M. A. Krawchuk, K. Lisgo, R. F. Noss, F. Saucier, and P. Vernier. In prep. "Conservation beyond Crisis Management: A New Model for Comprehensive Planning for Sustainability."

Schneider, R. and S. Dyer. 2006. *Death by a thousand cuts: The impacts of in situ oil sands development on Alberta's boreal forest*. Edmonton: Canadian Parks and Wilderness Society and the Pembina Institute for Appropriate Development. 27 February. http://pubs.pembina.org/reports/1000-cuts.pdf.

Schneider, R. R., A. Hamann, D. Farr, X. Wang, and S. Boutin. 2009. "Potential effects of climate change on ecosystem distribution in Alberta." *Canadian Journal of Forest Research* 39:1001–1010.

Schrag, A. M. 2011. "Addendum: Climate Change Impacts and Adaptation Strategies." In *Ocean of Grass: A Conservation Assessment for the Northern Great Plains*, S. C. Forrest, H. Strand, W.H. Haskins, C. Freese, J. Proctor, and E. Dinerstein, eds. Bozeman, MT: Northern Plains Conservation Network and Northern Great Plains Ecoregion, World Wildlife Fund–US.

Schrag, A., S. Konrad, S. Miller, B. Walker, and S. Forrest. 2011. "Climate-change Impacts on Sagebrush Habitat and West Nile Virus Transmission Risk and Conservation Implications for Greater Sage-grouse." *GeoJournal* 76:561–75.

Schuur, E. A. G., J. G. Vogel, K. G. Crummer, H. Lee, J. O. Sickman, and T. E. Osterkamp. 2009. "The Effect of Permafrost Thaw on Old Carbon Release and Net Carbon Exchange from Tundra." *Nature* 459:556–59.

Schwartz, C. C., P. H. Gude, L. Landenburger, M. A. Haroldson, and S. Podruzny. Forthcoming. "Impact of Rural Residential Development on Grizzly Bear Habitat in the Greater Yellowstone Ecosystems." *Journal of Wildlife Management.*

Seimon, A. and G. Picton Phillipps. 2011. "Regional Climatology of the Albertine Rift." In *Long-term Changes in Africa's Rift Valley*, A. J. Plumptre, ed. New York: Nova Science Publishers, 9–30.

Seimon, A., J. Watson, R. Dave, E. Gray, and J. Oglethorpe. 2011. *A Review of Climate Change Adaptation Initiatives within the Africa Biodiversity Collaborative Group Members*. New York: Wildlife Conservation Society, and Washington, DC: Africa Biodiversity Collaborative Group. frameweb.org/adl/en-US/8202/file/1090/abcg-climatechangeadaptation.pdf (accessed January 9, 2012).

SEMARNAT-CONANP. 2010. Estrategia de cambio climático para áreas protegidas (ECCAP); SEMARNAT (Secretaria de Medio Ambiente y Recursos Naturales), CONANP (Comisión Nacional de Areas Naturales Protegidas): Mexico D.F., Mexico, 41.

Servheen, C. and M. S. Cross. 2010. *Climate Change Impacts on Grizzly Bears and Wolverines in Northern US and Transboundary Rockies: Strategies for Conservation*. Report on a workshop held September 13–15, 2010, in Fernie, British Columbia.

SFMN (Sustainable Forest Management Network). 2011. *About Us*. Alberta: Alberta School of Forest Science and Management, University of Alberta. www.ales.ualberta.ca/forestry/Sustainable_Forest_Management.aspx (accessed June 21, 2011).

Shaw, M. R., E. S. Zavaleta, N. R. Chiariello, E. E. Cleland, H. A. Mooney, and C. B. Field. 2002. "Grassland Responses to Global Environmental Changes Suppressed by Elevated CO_2." *Science* 298:1987–90.

Sheppard, P. R., A. C. Comrie, G. D. Packin, K. Angersbach, and M. K. Hughes. 2002. "The Climate of the Southwest." *Climate Research* 21:219–38.

Shirsat, S. V. and H. F. Graf. 2009. "An Emission Inventory of Sulfur from Anthropogenic Sources in Antarctica." *Atmospheric Chemistry and Physics* 9:3397–3408.

Sills, E., G. Hartshorn, P. Ferraro, and B. Spergel. 2005. *Evaluation of the World Bank–GEF Ecomarkets Project in Costa Rica*. North Carolina State University, Blue Ribbon Panel. 47 pp.

Simmons, M. T. and R. M. Cowling. 1996. "Why Is the Cape Peninsula So Rich in Plant Species? An Analysis of the Independent Diversity Components." *Biodiversity and Conservation* 5:551–73.

SINAC (Sistema National de Areas de Conservacion)). 2007. Gruas II. *Análisis de vacíos de conservación en Costa Rica* vol. 1: "Gap Analysis in the Representation and Integrity of Terrestrial Biodiversity." San José, Costa Rica.

Siniff, D. B., R. A. Garrott, J. J. Rotella, W. R. Fraser, and D. G. Ainley. 2008. "Projecting the Effects of Environmental Change on Antarctic Seals." *Antarctic Science* 20:425–35.

Sivakumar, M. V. K. 2007. "Interactions between Climate and Desertification." *Agricultural and Forest Meteorology* 142:143–55.

Skagen, S. K. 2009. *Avian Conservation in the Prairie Pothole Region, Northern Great Plains: Understanding the Links between Climate, Ecosystem Processes, Wetland Management and Bird Communities*. Project summary, USGS National Climate Change and Wildlife Science Center.

Sklar, F. H. and J. A. Browder. 1998. "Coastal Environmental Impacts Brought about by Alterations to Freshwater Flow in the Gulf of Mexico." *Environmental Management* 22:547–62.

Smit, B. and J. Wandel. 2006. "Adaptation, Adaptive Capacity and Vulnerability." *Global Environmental Change* 16:282–92.

Smith, A. C. M. 2010. *Birds in the Great Eastern Ranges: Movement and Connectivity*. An analysis by Birds Australia for the NSW Department of Environment, Climate Change and Water. Sydney.

Smith, B. D. 1993. "1990 Status and Conservation of the Ganges River Dolphin (*Platanista gangetica*) in the Karnali River, Nepal." *Biological Conservation* 66:159–70.

Smith, B. D., B. Ahmed, R. M. Mowgli, and S. Strindberg. 2008. "Species Occurrence and Distributional Ecology of Nearshore Cetaceans in the Bay of Bengal, Bangladesh, with Abundance Estimates for Irrawaddy Dolphins, *Orcaella brevirostris*, and Finless Porpoises, *Neophocaena phocaenoides*." *Journal of Cetacean Research and Management* 10:45–58.

Smith, B. D., G. Braulik, S. Strindberg, B. Ahmed, and R. Mansur. 2006. "Abundance of Irrawaddy Dolphins (*Orcaella brevirostris*) and Ganges River Dolphins (*Platanista gangetica gangetica*) Estimated Using Concurrent Counts from Independent Teams in Waterways of the Sundarbans Mangrove Forest in Bangladesh." *Marine Mammal Science* 22:527–47.

Smith, B. D., G. Braulik, S. Strindberg, R. Mansur, M. A. A. Diyan, and B. Ahmed. 2009. "Habitat Selection of Freshwater Cetaceans and the Potential Effects of Declining Freshwater Flows and Sea-level Rise in Waterways of the Sundarbans Mangrove Forest, Bangladesh." *Aquatic Conservation: Marine and Freshwater Ecosystems* 19:209–25.

Smith, B. D., M. A. A. Diyan, R. M. Mansur, E. F. Mansur, and B. Ahmed. 2010. "Identification and Channel Characteristics of Cetacean 'Hotspots' in Waterways of the Eastern Sundarbans Mangrove Forest, Bangladesh." *Oryx* 44:241–47.

Smith, B. D. and R. R. Reeves. 2000. "Report of the Workshop on the Effects of Water Development on River Cetaceans in Asia, Rajendrapur, Bangladesh, 26–28 February 1997." In *Biology and Conservation of Freshwater Cetaceans in Asia,* IUCN Occasional Papers Series no. 23, R. R. Reeves, B. D. Smith, and T. Kasuya, eds. Gland, Switzerland: IUCN, 15–21.

——. 2010. "Freshwater-dependent Cetaceans: Integrating Climate Change-related Impacts from Mountain to Sea." *Whalewatcher* 39:25–29.

——. Forthcoming. "River Cetaceans and Habitat Change: Generalist Resilience or Specialist Vulnerability?" *Journal of Marine Biology.*

Smith, B. D., R. K. Sinha, Z. Kaiya, A. A. Chaudhry, L. Renjun, W. Ding, B. Ahmed, A. A. K. M. Haque, K. Sapkota, and R. S. L. Mohan. 2000. "Register of Water Development Projects Affecting Asian River Cetaceans." In *Biology and Conservation of Freshwater Cetaceans in Asia,* IUCN Occasional Papers Series no. 23, R. R. Reeves, B. D. Smith, and T. Kasuya, eds. Gland, Switzerland: IUCN, 22–39.

Smith, L. C., Y. Sheng, G. M. MacDonald, and L. D. Hinzman. 2005. "Disappearing Arctic Lakes." *Science* 308:1429.

Smith, W. O. Jr., D. G. Ainley, R. Cattaneo-Vietti, and E. E. Hofmann. 2010. "The Ross Sea Continental Shelf: Regional Biogeochemical Cycles, Trophic Interactions, and Potential Future Changes." *In Antarctica: An Extreme Environment in a Changing World.* London: John Wiley.

Smol, J. P. and M. S. V. Douglas. 2007. "Crossing the Final Ecological Threshold in High Arctic Ponds." *Proceedings of the National Academy of Sciences* 104:12395–97.

Snäll, T., R. B. O'Hara, C. Ray, and S. K. Collinge. 2008. "Climate-driven Spatial Dynamics of Plague among Prairie Dog Colonies." *American Naturalist* 171:238–48.

SNAP (Scenario Network for Alaska Planning). 2011. www.snap.uaf.edu (accessed January 10, 2012).

Solomon, S., G. K. Plattner, R. Knutti, and P. Friedlingstein. 2009. "Irreversible Climate Change Due to Carbon Dioxide Emissions." *Proceedings of the National Academy of Sciences* 106:1704–9.

Spoerl, M. and J. C. Ravesloot. 1994. "From Casas Grandes to Casa Grande: Prehistoric Human Impacts in the Sky Islands of Southern Arizona and Northwestern Mexico." In *The Sky Islands of Southwestern United States and Northwestern Mexico*, L. F. Debano, G. J. Gottfried, R. H. Hamre, C. B. Edminster, P. F. Ffolliott, and A. Ortega-Rubio, eds. USDA Forest Service: General Technical Report RM-GTR-264, 6–18.

Stager, J. C. and M. Thill. 2010. *Climate Change in the Champlain Basin*. Prepared for the Nature Conservancy. www.nature.org/ourinitiatives/regions/northamerica/unitedstates

/vermont/howwework/champlain_climate_report_5_2010.pdf (accessed on January 10, 2012).

Steffen, W., A. A. Burbidge, L. Hughes, R. Kitching, D. Lindenmayer, W. Musgrave, M. Stafford Smith, and P. A. Werner. 2009. *Australia's Biodiversity and Climate Change*. Canberra: CSIRO Publishing and the Australian Government.

Strassburg, B. B. N., A. Kelly, A. Balmford, R. G. Davies, H. K. Gibbs, A. Lovett, and L. Miles. 2010. "Global Congruence of Carbon Storage and Biodiversity in Terrestrial Ecosystems." *Conservation Letters* 3:98–105.

Sud, Y. C., W. Chao, and G. Walker. 1993. "Dependence of Rainfall on Vegetation: Theoretical Considerations, Simulation Experiments, Observations, and Inferences from Simulated Atmospheric Soundings." *Journal of Arid Environments* 25:5–18.

Suding, K. N. and D. Goldberg. 2001. "Do Disturbances Alter Competitive Hierarchies? Mechanisms of Change Following Gap Creation." *Ecology* 82:2133–49.

Sweeney, B. W., J. K. Jackson, J. D. Newbold, and D. H. Funk. 1992. "Climate Change and the Life Histories and Biogeography of Aquatic Insects in Eastern North America." In *Global Climate Change and Freshwater Ecosystems*, P. Firth and S. G. Fisher, eds. New York: Springer, 143–76.

Swiderska, K. 2002. *Mainstreaming Biodiversity in Development Policy and Planning: A Review of Country Experience*. Biodiversity and Livelihoods Group, International Institute for Environment and Development. www.iied.org/pubs/pdfs/G01228.pdf (accessed January 11, 2012).

Tack, J. D. 2006. *Sage Grouse and the Human Footprint: Implications for Conservation of Small and Declining Populations*. MS thesis, University of Montana, Missoula.

Taggart, M. 2009. *The Feasibility of a Yukon Carbon Offset Fund*. Marsh Lake, Yukon: Research Northwest for the Yukon Conservation Society. March 27. www.yukonconservation.org /energydocuments/The_Feasibility_of_a_Yukon_Carbon_Fund_FINAL_REPORT_2009 _03_27-1%5B1%5D.pdf (accessed January 11, 2012).

Tape, K., M. Sturm, and C. Racine. 2006. "The Evidence for Shrub Expansion in Northern Alaska and the Pan-Arctic." *Global Change Biology* 12:686–702.

Tarnocai, C., J. G. Canadell, E. A. G. Schuur, P. Kuhry, G. Mazhitova, and S. Zimov. 2009. "Soil Organic Carbon Pools in the Northern Circumpolar Permafrost Region." *Global Biogeochemical Cycles* 23:GB2023.

Taylor, A. R., R. B. Lanctot, A. N. Powell, F. Huettmann, D. A. Nigro, and S. J. Kendall. 2010. "Distribution and Community Characteristics of Staging Shorebirds on the Northern Coast of Alaska." *Arctic* 63:451–67.

Taylor, M. A., J. Centella, J. Charlery, I. Borrajero, J. Campbell, R. Rivero, T. S. Stephenson, E. Whyte, and R. Watson. 2007. *Glimpses of the Future: A Briefing from the PRECIS Caribbean Climate Change Project*. Belmopan, Belize: Caribbean Community Climate Change Centre, 24 pp.

Taylor, S. W. and A. L. Carroll. 2004. "Disturbance, Forest Age, and Mountain Pine Beetle Outbreak Dynamics in BC: A Historical Perspective." In *Mountain Pine Beetle Symposium: Challenges and Solutions, October 30–31, 2003, Kelowna, British Columbia, Canada*, T. L. Shore, J. E. Brooks, and J. E. Stone, eds. Victoria, BC: Pacific Forestry Centre, 41–51.

Teh, L. C. L., L. S. L. Teh, B. Starkhouse, and U. R. Sumaila 2009. "An Overview of Socio-economic and Ecological Perspectives of Fiji's Inshore Reef Fisheries." *Marine Policy* 33:807–17.

Thackway, R. and I. T. Cresswell. 1995. *An Interim Biogeographic Regionalisation for Australia: A Framework for Establishing the National System of Reserves, Version 4*. Canberra: Australian Nature Conservation Agency.

Thomas, C. D., A. Cameron, R. E. Green, M. Bakkenes, L. J. Beaumont, Y. C. Collingham, and B. F. N. Erasmus, et al. 2004. "Extinction Risk from Climate Change." *Nature* 427:145–48.

Thornton, P. K., P. G. Jones, G. Alagarswamy, and J. Andresen. 2009. "Spatial Variation of Crop Yield Response to Climate Change in East Africa." *Global Environmental Change* 19:54–65.

Thorpe, J. 2010. *Climate Change and the South of the Divide Project*. Presentation at Climate Adaptation Workshop, Cypress Hills, Saskatchewan, June 23–24.

Thuiller, W., S. Lavorel, G. Midgley, S. Lavergne, and T. Rebelo. 2004. "Relating Plant Traits and Species Distributions along Bioclimatic Gradients for 88 Leucadendron Taxa." *Ecology* 85:688–99.

Tin, T., B. K. Sovacool, D. Blake, P. Magill, S. El Naggar, S. Lidstrom, K. Ishizawa, and J. Berte. 2010. "Energy Efficiency and Renewable Energy under Extreme Conditions: Case Studies from Antarctica." *Renewable Energy* 35:1715–23.

TNC (The Nature Conservancy). 2000. *Ecoregional Planning in the Northern Great Plains Steppe*. TNC Northern Great Plains Ecoregional Planning Team.

Tonn, W. M. 1990. "Climate Change and Fish Communities—A Conceptual-Framework." *Transactions of the American Fisheries Society* 119:337–52.

Toropova, C., I. Meliane, D. Laffoley, E. Matthews, and M. Spalding, eds. 2010. *Global Ocean Protection: Present Status and Future Possibilities*. Brest, France: Agence des aires marines protégées, Gland, Switzerland, Washington, DC, and New York, USA; IUCN WCPA, Cambridge, UK; UNEP-WCMC, Arlington, USA; TNC, Tokyo, Japan; UNU, New York, USA; WCS.

Tortell, P. 2010. "C.A.P.E. Biodiversity Conservation and Sustainable Development Project (BCSD) Terminal Evaluation Report." Washington, DC: World Bank.

Trathan, P. N. and D. Agnew. 2010. "Climate Change and the Antarctic Marine Ecosystem: An Essay on Management Implications." *Antarctic Science* 22:387–98.

Treves, A., A. J. Plumptre, L. Hunter, and J. Ziwa. 2009. "Identifying a Potential Lion, *Panthera leo*, Stronghold in Queen Elizabeth National Park, Uganda and Parc National des Virunga, Democratic Republic of Congo." *Oryx 43:1–8.*

Trombulak, S. C. 1994. "The Northern Forest: Conservation Biology, Public Policy, and the Failure of Democracy." *Endangered Species UPDATE* 11:7–16.

Trombulak, S. C., M. G. Anderson, R. F. Baldwin, K. Beazley, J. C. Ray, C. Reining, G. Woolmer, C. Bettigole, G. Forbes, and L. Gratton. 2008. "The Northern Appalachian/Acadian Ecoregion: Priority Locations for Conservation Action." *Two Countries, One Forest/Deux Pays, Une Forêt,* Special Report 1.

Trombulak, S. C. and R. F. Baldwin. 2010. "Introduction: Creating a Context for Landscape-scale Conservation Planning." In *Landscape-scale Conservation Planning*, S. C. Trombulak and R. F. Baldwin, eds. New York: Springer, 1–16.

Troy, D. M. 2000. "Shorebirds." In *The Natural History of an Arctic Oil Field—Development and Biota*, J. C. Truett and S. R. Johnson, eds. San Diego: Academic Press, chap. 14.

Tulp, I. and H. Schekkerman. 2008. "Has Prey Availability for Arctic Birds Advanced with Climate Change? Hindcasting the Abundance of Tundra Arthropods Using Weather and Seasonal Variation." *Arctic* 61:48–60.

Turner, J., R. Bindschadler, P. Convey, G. di Prissco, E. Fahrbach, J. Gutt, D. Hodgson, P. Mayewski, and C. Summerhayes, eds. 2009. *Antarctic Climate Change and the Environment.* Scientific Committee on Antarctic Research: Cambridge, UK. www.scar.org/publications/occasionals/acce.html (accessed January 10, 2012).

Turner, M. G. 2010. "Disturbance and Landscape Dynamics in a Changing World." *Ecology* 91:2833–49.

Tynan, C. and J. Russell. 2008. *Assessing the Impacts of Future 2°C Global Warming on Southern Ocean Cetaceans*. International Whaling Commission, Information Paper SC/60/E3.

U.S. Census Bureau. 2010. State and County QuickFacts: North Slope Borough, Alaska. quickfacts.census.gov/qfd/states/02/02185.html (accessed January 11, 2012).

UNEP (United Nations Environment Program). 1992. *World Atlas of Desertification*. London: UNEP.

UNESCO (United Nations Educational, Scientific, and Cultural Organization). 2011. *The Sundarbans*. http://whc.unesco.org/en/list/798 (accessed February 8, 2012).

UNFCCC (United Nations Framework Convention on Climate Change). No date. "NAPAs Received by the Secretariat." unfccc.int/cooperation_support/least_developed_countries_portal/submitted_napas/items/4585.php (accessed January 11, 2012).

United Kingdom. 2010a. *Draft Procedures for Vehicle Cleaning to Prevent Transfer of Non-native Species into and around Antarctica*. Working Paper 8. 33rd Antarctic Treaty Consultative Meeting, May 3–14. Punta del Este, Uruguay.

——. 2010b. *Guidance for Visitors and Environmental Managers Following the Discovery of a Suspected Non-native Species in the Terrestrial and Freshwater Antarctic Environment*. Working Paper 15. 33rd Antarctic Treaty Consultative Meeting, May 3–14. Punta del Este, Uruguay.

——. 2010c. *The Implications of Climate Change for the Antarctic Protected Areas System*. Working Paper 16. 33rd Antarctic Treaty Consultative Meeting, May 3–14. Punta del Este, Uruguay.

UNPD (United Nations Population Division). 2009. *World Urbanization Prospects and Revision of World Population Prospects*. UNDP, Dept. of Economic and Social Affairs. http://esa.un.org/unpd/wup/index.htm (accessed March 3, 2011).

USAID (United States Aid for International Development). 2007. *Adapting to Climate Variability and Change: A Guidance Manual for Development Planning*. Washington, DC: USAID.

USDOI (United States Department of Interior). 2010. *Climate Change*. U.S. Department of the Interior. www.doi.gov/whatwedo/climate/index.cfm (accessed January 11, 2012).

USFWS (United States Fish and Wildlife Service). 2010. *Plains and Prairie Potholes Landscape Conservation Cooperative Map*. www.fws.gov/midwest/climate/LCC/PPP/areamap.htm (accessed January 11, 2012).

——. 2011. *Landscape Conservation Cooperatives: Shared Science for a Sustainable Future*. www.fws.gov/science/shc/pdf/LCC_Fact_Sheet.pdf (accessed January 11, 2012).

USGS (United States Geological Survey). 2011. *Protected Areas Database of the United States (PAD-US)*. USGS Gap Analysis Program. gapanalysis.usgs.gov/data/padus-data (accessed January 11, 2012).

Valkenburg, P. 1998. "Herd Size, Distribution, Harvest, Management Issues, and Research Priorities Relevant to Caribou Herds in Alaska." *Rangifer* 10:125–29.

Van Rijn, L. C. 1993. *Principles of Sediment Transport in Rivers, Estuaries and Coastal Seas*. Amsterdam: Aqua Publications.

Varady, R. and E. V. Ward. 2009. "Transboundary Conservation in Context: What Drives Environmental Change?" In *Conservation of Shared Environments: Learning from the United States and Mexico*, L. López-Hoffman, E. McGovern, R. G. Varady, and K. W. Flessa, eds. Tucson: University of Arizona Press, 9–23.

Veit, H. 2002. *Die Alpen—Geoökologie und Landschaftsentwicklung*. UTB Ulmer Verlag, Stuttgart.

Vergara, W. 2009. *Assessing the Potential Consequences of Climate Destabilization in Latin America*. Latin America and the Caribbean Region, Sustainable Development Working Paper no. 32. Washington, DC: World Bank.

Vergara, W. and S. M. Scholz, eds. 2010. *Assessment of the Risk of Amazonian Dieback*. Washington, DC: World Bank.

Vic DSE (Victoria Department of Sustainability and Environment). 2009. *Securing Our Natural Future, A White Paper for Land and Biodiversity at a Time of Climate Change*. Melbourne: Victoria Department of Sustainability and Environment.

Vignola, R., B. Locatelli, C. Martinez, and P. Imbach. 2009. "Ecosystem-based Adaptation to Climate Change: What Role for Policy-makers, Society and Scientists?" *Mitigation and Adaptation Strategies for Global Change* 14:691–96.

Von Hase, A., M. Rouget, and R. M. Cowling. 2010. "Evaluating Private Land Conservation in the Cape Lowlands, South Africa." *Conservation Biology* 24:1182–89.

Vörösmarty, C. K., P. Green, J. Salisbury, and R. B. Lammers. 2000. "Global Water Resources Vulnerability from Climate and Population Growth." *Science* 289:284–88.

Vos, C. C., P. Berry, P. Opdam, H. Baveco, B. Nijhof, J. O'Hanley, C. Bell, and H. Kuipers. 2008. "Adapting Landscapes to Climate Change: Examples of Climate-proof Ecosystem Networks and Priority Adaptation Ones." *Journal of Applied Ecology* 45:1722–31.

Walters, C. J. 1986. *Adaptive Management of Renewable Resources*. New York: Macmillan Press.

Walther, G., E. Post, P. Convey, A. Menzel, C. Parmesan, T. J. C. Beebee, J. Fromentin, O. Hoegh-Guldberg, and F. Bairlein. 2002. "Ecological Responses to Recent Climate Change." *Nature* 416:389–95.

Walther, G. R. 2010. "Community and Ecosystem Responses to Recent Climate Change." *Philosophical Transactions of the Royal Society B: Biological Sciences* 365:2019–24.

Warshall, P. 1994. "The Madrean Sky Island Archipelago: A Planetary Overview." In *The Sky Islands of Southwestern United States and Northwestern Mexico*, L. F. Debano, G. J. Gottfried, R. H. Hamre, C. B. Edminster, P. F. Ffolliott, and A. Ortega-Rubio, eds. USDA Forest Service: General Technical Report RM-GTR-264, 6–18.

Watermeyer, K. E., L. J. Shannon, and C. L. Griffiths. 2008. "Changes in the Trophic Structure of the Southern Benguela before and after the Onset of Industrial Fishing." *African Journal of Marine Science* 30:351–82.

WCAT (Washington Climate Advisory Team). 2008. "Chapter 3: The Climate Change Challenge." In *Leading the Way on Climate Change: The Challenge of Our Time*. Washington State Departments of Ecology and Community, Trade and Economic Development. Ecology Publication #08-01-008. www.ecy.wa.gov/climatechange/interimreport.htm (accessed January 10, 2012).

WCS (Wildlife Conservation Society). 2009. *Ecosystem-based Management Plan: Kubulau District, Vanua Levu, Fiji*. Suva, Fiji: WCS.

———. 2010. *New Conservation Priorities in a Changing Arctic Alaska*. A workshop summary by the Wildlife Conservation Society. http://www.wcsnorthamerica.org/WildPlaces/ArcticAlaska/ClimateChangeImpactsinArcticAlaska/tabid/3648/Default.aspx (accessed January 6, 2012).

———. 2011. *Ocean*. Wildlife Conservation Society. www.wcs.org/saving-wild-places/ocean.aspx (accessed January 11, 2012).

WCS Canada (Wildlife Conservation Society Canada). 2009. *On the Golden Pond*. Pamphlet. www.wcscanada.org/LinkClick.aspx?fileticket=VUpxOxePLdo%3dandtabid=3071 (accessed January 11, 2012).

Weaver, J. 2011. "Conservation Value of Roadless Areas for Vulnerable Fish and Wildlife Species in the Crown of the Continent Ecosystem, Montana." Working Paper, no. 40. April.

Webb, A. P. and P. S. Kench. 2010. "The Dynamic Response of Reef Islands to Sea-level Rise: Evidence from Mult-decadal Analysis of Island Change in the Central Pacific." *Global and Planetary Change* 72:234–46.

Weber, E. U. and P. C. Stern. 2011. "Public Understanding of Climate Change in the United States." *American Psychologist* 66:315–28.

Weber, M. G. and M. D. Flannigan. 1997. "Canadian Boreal Forest Ecosystems Structure and Function in a Changing Climate: Impact on Fire Regimes." *Environmental Reviews* 5:145–66.

Weissmann, H. Z., C. Künitzer, and A. Bertiller. 2009. *Strukturen der Fliessgewässer in der Schweiz—Zustand von Sohle, Ufer und Umland (Ökomorphologie); Ergebnisse der ükomorphologischen Kartierung*. Bern, Switzerland: Bundesamt für Umwelt (BAFU).

Wenger, E. 1999. *Communities of Practice: Learning, Meaning and Identity*. Cambridge, UK: Cambridge University Press.

West, J. M., S. H. Julius, P. Kareiva, C. Enquist, J. J. Lawler, B. Petersen, A. E. Johnson, and M. R. Shaw. 2009. "U.S. Natural Resources and Climate Change: Concepts and Approaches for Management Adaptation." *Environmental Management* 44:1001–21.

Westerling, A. L., H. G. Hidalgo, D. R. Cayan, and T. W. Swetnam. 2006. "Warming and Earlier Spring Increase Western U.S. Forest Wildfire Activity." *Science* 313:940–43.

WGA (Western Governors' Association). 2008. *Western Wildlife Habitat Council Established*. Jackson, Wyoming: WGA. June 29. www.westgov.org/wga/publicat/wildlife08.pdf (accessed January 11, 2012).

———. 2010. *Climate Adaptation Priorities for the Western States: Scoping Report*. Denver, CO: Western Governors' Assocation.

WHCWG (Wildife Habitat Connectivity Working Group). 2010. "Washington Connected Landscapes Project: Statewide Analysis." Washington Departments of Fish and Wildlife, and Transportation, Olympia. www.waconnected.org (accessed January 11, 2012).

——. 2011. *Washington Connected Landscapes Project: Climate-gradient Corridors Report*. Washington Departments of Fish and Wildlife, and Transportation, Olympia. www.waconnected.org (accessed January 10, 2012).

Wiens, J. A. and D. Bachelet. 2010. "Matching the Multiple Scales of Conservation with the Multiple Scales of Climate Change." *Conservation Biology* 24:51–62.

Wilder, M., C. A. Scott, M. Pineda Pablos, R. G. Varady, G. M. Garfin, and J. McEvoy. 2010. "Adapting across Boundaries: Climate Change, Social Learning, and Resilience in the US–Mexico Border Region." *Annals of the Association of American Geographers* 100:1–11.

Williams, J. E., A. L. Haak, H. M. Neville, and W. T. Colyer. 2009. "Potential Consequences of Climate Change to Persistence of Cutthroat Trout Populations." *North American Journal of Fisheries Management* 29:533–48.

Williams, J. W., S. T. Jackson, and J. E. Kutzbacht. 2007. "Projected Distributions of Novel and Disappearing Climates by 2100 AD." *Proceedings of the National Academy of Sciences* 104:5738–42.

Williams, P., L. Hannah, S. Andelman, G. Midgley, M. Araújo, G. Hughes, L. Manne, E. Martinez-Meyer, and R. Pearson. 2005. "Planning for Climate Change: Identifying Minimum-dispersal Corridors for the Cape Proteaceae." *Conservation Biology* 19: 1063–74.

Williamson, T. B., S. J. Colombo, P. N. Duinker, P. A. Gray, R. J. Hennessey, D. Houle, M. H. Johnston, A. E. Ogden, and D. L. Spittlehouse. 2009. *Climate Change and Canada's Forests: From Impacts to Adaptation*. Edmonton, Alberta: Sustainable Forest Management Network and Natural Resources Canada.

Willis, K. J. and S. A. Bhagwat. 2009. "Biodiversity and Climate Change." *Science* 326:5954, 806–7.

Willows, R. I. and R. K. Connell. 2003. "Climate Adaptation: Risk, Uncertainty and Decision-making." UKCIP Technical Report. Oxford: United Kingdom Climate Impacts Program.

Wilson, K. A., J. Carwardine, and H. P. Possingham. 2009. "Setting Conservation Priorities." *Annals of the New York Academy of Sciences: The Year in Ecology and Conservation Biology* 1162:237–64.

Winters, A. 2001. "Three Tools for Rangeland Management." In *Rangeland Management Pilot Project, Land Use and Range Management Workshop*. Sergelen Soum, Mongolia.

Wood, A.W., L. R. Leung, V. Sridhar, and D. P. Lettenmaier. 2004. "Hydrologic Implications of Dynamical and Statistical Approaches to Downscaling Climate Model Outputs." *Climatic Change* 62:189–216.

Woodburne, M. O. 2010. "The Great American Biotic Interchange: Dispersals, Tectonics, Climate, Sea Level and Holding Pens." *Journal of Mammalian Evolution* 17:245–64.

Woodford, J. 2007. "2800km Coastal Route for Wildlife to Escape Warming." *Sydney Morning Herald*, February 24. www.smh.com.au/news/environment/route-for-wildlife-to-escape-warming/2007/02/23/1171734021093.html (accessed January 11, 2012).

Woolmer, G., S. C. Trombulak, J. C. Ray, P. J. Doran, M. G. Anderson, R. F. Baldwin, A. Morgan, and E. W. Sanderson. 2008. "Rescaling the Human Footprint: A Tool for Conservation Planning at an Ecoregional Scale." *Landscape and Urban Planning* 87:42–53.

Worboys, G. L. 1996. *Conservation Corridors and the NSW Section of the Great Escarpment of Eastern Australia*. Paper presented to the IUCN World Conservation Congress, October 13–23, Montreal, Canada.

———. 2005. "The South East Forest National Park of NSW." In *Protected Area Management, Principles and Practice*, 2nd ed., G. L. Worboys, M. Lockwood, and T. De Lacey, eds. Melbourne: Oxford University Press, 25.

———. 2010. *The Altai-Sayan Mega Connectivity Conservation Corridor, An Adaptation Response to Climate Change in the Heart of Asia*. An IUCN WCPA unpublished mission report of a workshop conducted in July 2010 at Ust Koksa, Altai Republic, Russia. Gland: International Union for the Conservation of Nature.

Worboys, G. L., R. B. Good, and A. Spate. 2010. *Caring for Our Australian Alps Catchments: A Climate Change Action Strategy for the Australian Alps to Conserve the Natural Condition of the Catchments and to Help Minimise Threats to High Quality Water Yields*. Canberra: Australian Alps Liaison Committee, Department of Climate Change.

Worboys, G. L., W. Francis, and M. Lockwood. 2010. *Connectivity Conservation Management: A Global Guide*. London: Earthscan.

World Bank. 2006. *Mongolia. A Review of Environmental and Social Impacts in the Mining Sector.* Discussion Paper.

———. 2010. *Convenient Solutions to an Inconvenient Truth: Ecosystem Based Approaches to Climate Change*. Washington, DC.

World Bank, Development Research Group. 2011. *PovCalNet Calculation of Poverty Headcount Ratio at $1.25 a Day PPP*.

WWF (World Wildlife Fund). 2001. *Ecoregional Climate Change and Biodiversity Decline: Altai-Sayan Ecoregion*. Moscow: WWF for Nature–Russia. Issue 1 of the Climate Passport series. www.wwf.ru/resources/publ/book/eng/19 (accessed January 11, 2012).

———. 2004. *Setting Priorities for Marine Conservation in the Fiji Islands Marine Ecoregion*. Suva, Fiji: WWF South Pacific Programme.

———. 2008. *Are We on Track? Marine Protected Areas for Antarctica and the Southern Ocean*. WWF Australia.

———. 2009. *The Golden Mountains at the Centre of Eurasia: Ensuring Long-term Conservation of the Altai-Sayan Ecoregion*. Krasnoyarsk, Russia: Litera-print Printing House for WWF for Nature.

———. 2009. *Adaptation to Climate Change Toolkit: Coasts*. wwf.panda.org/what_we_do/endangered_species/marine_turtles/lac_marine_turtle_programme/projects/climate_turtles/act_toolkit.

Xenopoulos, M., D. M. Lodge, J. Alcamo, M. Märker, K. Schulze, and D. P. Van Vuuren. 2005. "Scenarios of Freshwater Fish Extinctions from Climate Change and Water Withdrawal." *Global Change Biology* 11:1557–64.

Xu, J. R., E. Grumbine, A. Shrestha, M. Eriksson, X. Yang, Y. Wang, and A. Wilkes. 2009. "The Melting Himalayas: Cascading Effects of Climate Change on Water, Biodiversity, and Livelihoods." *Conservation Biology* 23:520–30.

Y2Y (Yellowstone to Yukon). 2011. *Cabinet-Purcells Mountain Corridor: Collaborative Projects*. Yellowstone to Yukon Conservation Initiative. www.y2y.net/Default.aspx?cid=4-67-532andpre=view (accessed January 11, 2012).

Younge, A. and S. Fowkes. 2003. "The Cape Action Plan for the Environment: Overview of an Ecoregional Planning Process." *Biological Conservation* 112:15–28.

Yu, F., K. P. Price, J. Ellis, J. J. Feddema, and P. Shi. 2004. "Interannual Variations of the Grassland Boundaries Bordering the Eastern Edges of the Gobi Desert in Central Asia." *International Journal of Remote Sensing* 25:327–46.

Zhang, G., Y. Kang, G. Han, and K. Sakurai. 2011. "Effect of Climate Change over the Past Half Century on the Distribution, Extent and NPP of Ecosystems of Inner Mongolia." *Global Change Biology* 17:377–89.

Zhang, X. Y., S. L. Gong, T. L. Zhao, R. Arimoto, Y. Q. Wang, and Z. J. Zhou. 2003. "Sources of Asian Dust and Role of Climate Change Versus Desertification in Asian Dust Emission." *Geophysical Research Letters* 30:8-1-8-4.

Zisenis, M. and P. Kristensen. 2010. *10 Messages for 2010—Freshwater Ecosystems.* Copenhagen, Denmark: European Environment Agency.

CONTRIBUTORS

David Ainley conducted his PhD research on Adélie penguins at Cape Crozier, Ross Island. He has since made approximately thirty trips to the Antarctic. Currently, he investigates penguin demography as well as effects of cetacean foraging on penguin prey availability. He founded the PRBO Conservation Science marine research program on the Farallon Islands in the California Current, and has been a main player in an attempt to have the Ross Sea designated a Marine Protected Area. He has written four books, twelve monographs, and approximately 200 papers about the ecology of marine top predators, including seabirds, mammals, and sharks.

Mark G. Anderson is director of conservation science of The Nature Conservancy's Eastern US Region. He provides ecological analysis and develops landscape-scale assessment tools for conservation efforts across eight ecoregions. He has worked as an ecologist for over twenty years and is coauthor of the National Vegetation Classification as well as numerous journal articles on biodiversity conservation. He holds a PhD in ecology from University of New Hampshire, where his research focused on the viability and spatial assessment of ecological communities in the Northern Appalachians.

Yuri Badenkov is a mountain geographer with more than fifty years' field experience working in the Far East, Central Asia, Causasus, and Altai-Sayan mountain regions. Since 1981, he has been with the Institute of Geography, Russian Academy of Science, as head of the Mountain Geosystems Laboratory, and deputy director and leader of the United Nations Educational, Scientific and Cultural Organization–Man and the Biosphere Programme (UNESCO–MAB-6) Mountain Group. He was involved in writing the Mountain Chapter in Agenda 21 (1992), and in 2004–6 he was a member of UNESCO–MAB International Advisory Committee for Biosphere Reserves. He has published more than 300 papers and books.

Robert F. Baldwin is a landscape ecologist/conservation biologist whose career has primarily been focused on the Appalachian ecoregions. He began his career at the National Zoological Park and received degrees from Colby College in Maine, George Mason University, and University of Maine. He worked closely with Two Countries, One Forest to develop and integrate threat analyses. He is currently on the faculty at

Clemson University in the Upper Piedmont near the Blue Ridge escarpment. His current research includes landscape-scale conservation planning and habitat connectivity in wetland/aquatic landscapes and influences of land use, forest management, and climate on forest-dwelling vertebrates.

THOMAS BARRETT is a landscape ecologist and conservation-planning specialist. For the past sixteen years Tom has worked for both the Australian government and nongovernmental organizations undertaking systematic conservation-planning projects as well as developing and applying geographic information system-based decision support tools. He works in a research and development group within the New South Wales (NSW) Office of Environment and Heritage and is an adjunct research fellow in the University of New England School of Environmental and Rural Science. He is currently modeling fauna habitat values for investment in natural resource management projects for the NSW Catchment Management Authorities.

CHARLES C. CHESTER teaches global environmental politics at Brandeis University and the Fletcher School at Tufts University, where he is an adjunct assistant professor of international environmental policy. He is the author of *Conservation across Borders: Biodiversity in an Interdependent World* (Island Press 2006), which focuses on case studies of transborder conservation in North America. Chester has consulted for the Union of Concerned Scientists, the Henry P. Kendall Foundation, and other environmental organizations. He is currently cochair of the board of the Yellowstone to Yukon Conservation Initiative and has served on the boards of Bat Conservation International and Root Capital.

MOLLY S. CROSS is the climate change adaptation coordinator for the North America Program of the Wildlife Conservation Society. Her work brings together experts in climate change, conservation planning, and natural resource management to translate broad-brush climate change-adaptation strategies into on-the-ground conservation actions. She has worked with numerous partners to lead scenario-based climate change-planning efforts involving diverse stakeholders at more than eight landscapes across North America, focused on a range of targets from individual species to more complex ecosystems. Previously, her research focused on ecosystem responses to climate warming and plant diversity loss in the Colorado Rocky Mountains.

STEVE CUMMING is Canada Research Chair in Boreal Ecosystems Modelling at Université Laval, Québec. His research focuses on quantitative analysis and spatial simulation of ecological processes in boreal forests, including natural disturbance regimes, stand dynamics, and the distribution and abundances of boreal fauna. His current efforts are directed at modelling methodologies to support spatial simulation at national extents, in order to evaluate trade-offs between forest management and other economic activities on the one hand, and conservation objectives on the other, while taking into account uncertainty associated with future climatic and disturbance regimes.

GALBADRAKH DAVAA is director of conservation for The Nature Conservancy's Mongolia Program Office. He has worked with the Colorado chapter of The Nature Conservancy, UNDP/GEF Mongolia Biodiversity Project, UNDP/GEF Eastern Steppe Biodiversity Project, and the German Technical Cooperation Agency's Nature Conservation and Buffer Zone Development projects. He is a founding board member of the Mongolian Society for Environmental Education, and holds a master's of environmental management degree from Yale University and a master's of environmental sciences and policy degree from Central European University.

MARC DOUROJEANNI is the former dean of the faculty of forestry and emeritus professor of the Universidad Nacional Agraria of La Molina (Peru), vice president of the Universidad San Martin de Porres, and professor at the High Military Studies Center of Peru. In the 1970s, he served as deputy minister for Forestry of Peru and as vice president of the World Conservation Union (IUCN) and deputy chairman of the World Commission on Protected Areas (WCPA). He also was a senior advisor at the World Bank and first chief of the environment division of the Inter-American Development Bank. He has authored fourteen books.

MATTHEW DURNIN is the Asia-Pacific Region conservation science director for The Nature Conservancy. Since 1994, he has been living in, and conducting research on wildlife in, China. Prior to joining The Nature Conservancy, he was a MacArthur Foundation postdoctoral fellow at the California Academy of Sciences and lead mammalogist on a project cataloging the biodiversity of the Gaoligongshan area in western Yunnan province. He completed his PhD in wildlife ecology at the University of California, Berkeley, in 2005.

CAROLYN A. F. ENQUIST focuses on the conservation and management implications of climate change. She recently launched the Southwest Climate Change Initiative, a regional collaboration focused on adaptation planning, and has contributed to national reports including the *Preliminary Review of Adaptation Options for Climate Sensitive Ecosystems* led by the US Climate Change Science Program, and *Scanning the Conservation Horizon: A Guide to Climate Change Vulnerability Assessment* led by the National Wildlife Foundation. Carolyn currently coordinates the USA–National Phenology Network's science activities on behalf of The Wildlife Society and is a member of several teams contributing to the National Climate Assessment.

MARIANNE FISH is the marine and coastal adaptation leader for World Wildlife Fund–Latin America and Caribbean. She has worked for WWF since 2008, providing technical support to adaptation projects throughout the region and coordinating the Adaptation to Climate Change for Marine Turtles (ACT) initiative. She has ten years of experience in climate adaptation, including climate impacts modeling, vulnerability assessment, development of climate adaptation strategies, training, and outreach. She has a BS in biology from the University of Leeds, UK; a MS (applied ecology and conservation); and a PhD (coastal ecology, management, and climate change) from the University of East Anglia, UK.

STEVE FORREST is a consulting wildlife biologist with a BS in forestry from Oregon State University, an MS in environmental studies from the Yale University School of Forestry, and a JD from the University of Washington School of Law. Steve is an authority on the biology and conservation of the endangered black-footed ferret, serves as member of the IUCN Bison Specialist Group, and conducts annual censuses of seabird population trends in the Antarctic Peninsula linked to climate change. Steve formerly served as manager of restoration science for the World Wildlife Fund's Northern Great Plains Program.

WENDY L. FRANCIS began her career in conservation twenty-five years ago as a volunteer with the Canadian Parks and Wilderness Society (CPAWS) in Calgary, eventually becoming its founding conservation director. In 1999, Wendy launched her own consulting practice where she served foundations, government agencies, and nonprofits, providing research and analysis, organizational leadership, and strategic-planning assistance. In 2005, Wendy became director of conservation and science for Ontario Nature. In 2007, Wendy returned to the mountains and joined the staff of the Yellowstone to Yukon Conservation Initiative, where she now serves as its program director.

LEOPOLD FÜREDER is a professor at the Institute of Ecology, University of Innsbruck, Austria, and head of the research group on "River Ecology and Invertebrate Biology." He is involved in basic and applied international research projects, with a central focus on ecohydrology, climate change, biodiversity, food webs, environmental change, and aquatic conservation. His study areas are primarily in the Alps, in the High Arctics (Svalbard), and in the Tropics (Sri Lanka, Costa Rica).

EVAN H. GIRVETZ is a senior scientist with The Nature Conservancy's Global Climate Change Program. He provides expert support on climate change impacts and ecosystem-based adaptation to Conservancy programs and project teams globally. Girvetz has extensive experience in environmental decision support, conservation planning, and climate change-impact assessment. He coleads the development of the Climate Wizard, an online climate change mapping and analyses tool for practitioners. Girvetz received his PhD in ecology from the University of California, Davis, and holds an affiliate assistant professor position at the University of Washington, School of Environment Forest Sciences.

LEE HANNAH is senior researcher in climate change biology at Conservation International, where he leads efforts to develop conservation responses to climate change. Hannah coedited the award-winning book *Climate Change and Biodiversity* and authored the first undergraduate textbook on the biological impacts of climate change, *Climate Change Biology*. Hannah has led research collaborations to improve conservation-planning tools for climate change adaptation, including the development of conservation-planning software and very high resolution climatologies suitable for species distribution modeling. His broader research interests include the role of climate change in conservation planning and methods of corridor design.

MICHAEL HEINER is a conservation scientist with The Nature Conservancy, based in Fort Collins, Colorado. His work focuses on the use of geographic information systems and conservation planning to advance conservation strategies in the western United States and Mongolia.

JEFFREY HEPINSTALL-CYMERMAN is an associate professor in the Warnell School of Forestry and Natural Resources at the University of Georgia. He is a landscape ecologist and teaches applied GIS, GPS, and remote sensing. His current research includes, among other interests, exploring the implications of land use and climate change on birds across the urban-rural gradient in Georgia and North Carolina. He is principal investigator on a NASA Global Climate Change Education grant to train undergraduates in designing and conducting ecological field research to explore species-climate interactions and how species may be responding to a changing climate.

JODI A. HILTY has served as the director of the North America Program for the Wildlife Conservation Society since October 2007 and is based in Bozeman, Montana. As director, Dr. Hilty provides leadership on scientific applications to natural resource management and conservation. This includes leading efforts across seven landscapes and addressing four major conservation challenges including natural resource extraction, livelihoods, connectivity, and climate change. Trained as a conservation biologist at the University of California, Berkeley, her passion is focused on finding creative science-based solutions to resolve critical conflicts between humans and the natural world.

Jennifer R. Hoffman is a cofounder and directing scientist at EcoAdapt, a nonprofit focused on adapting natural resource management to climate change. In addition to her peer-reviewed scientific papers, Jennie is author or coauthor of several books and reports, most recently *Climate Savvy: Adapting Conservation and Resource Management to a Changing World*, published by Island Press in 2011, and *Scanning the Conservation Horizon: A Guide to Climate Change Vulnerability Assessment*, published by the National Wildlife Federation. When environmental problems seem daunting, she calls on her undergraduate degree in geology for a long-term perspective that keeps her chipper.

Margaret Buck Holland is an assistant professor with the Department of Geography and Environmental Systems at the University of Maryland, Baltimore County (UMBC). Her research focuses on human dimensions of global environmental change. She uses geospatial technologies to model the complex relationships between social conditions, ecosystem services, and environmental management. After receiving her PhD from the University of Wisconsin–Madison, Maggie did a postdoctoral fellowship supported jointly by the Land Tenure Center at UW–Madison and Conservation International. Her current research focuses on the linkages between land tenure, community welfare, governance, and land-use change in Ecuador and Central America.

Gary Howling is a senior policy officer with the New South Wales Government, Australia. Since early 2008, Gary has been principal conservation analyst with the Great Eastern Ranges Conservation Initiative. In this role, he has provided government agencies, nongovernment organizations, and private conservation partners with specialist scientific, technical, and conservation assessment advice to guide the development of a wide-ranging portfolio of projects. Gary has a wide background in regional and landscape-scale conservation planning, biodiversity and native vegetation conservation policy, community engagement, and conservation science brokering.

Stacy Jupiter has been working with the Wildlife Conservation Society since 2008, first as an associate conservation scientist and currently as the Fiji Country Program Director. Her research focuses on understanding human impacts to marine and freshwater ecosystems in order to design best approaches for management. With the WCS Fiji team, she is helping communities use ecosystem-based approaches to adapt to climate change. In addition, she is working at the national scale in Fiji to aid the government in expanding protected-area coverage and developing integrated coastal management plans to preserve ecosystem services, livelihoods, and human health.

Joseph Kiesecker is a lead scientist for The Nature Conservancy. He has published over 100 articles, on topics ranging from climate change to the effectiveness of conservation strategies. He holds a PhD in zoology from Oregon State University and has held faculty appointments at Yale University, Penn State University, and the University of Wyoming. He pioneered the Conservancy's Development by Design strategy to improve impact mitigation through the incorporation of predictive modeling to provide solutions that benefit both conservation and development, and he is currently testing this process through a series of infrastructure (i.e., energy development and mining) pilot projects.

MEG KRAWCHUK is an assistant professor in the Department of Geography at Simon Fraser University in British Columbia, Canada. Meg leads the Landscape and Conservation Science Research Group, with research focused on understanding the drivers and outcomes of ecological disturbances and their interactions, vulnerability of ecosystems to climate and land cover change, and conservation challenges in terrestrial ecosystems. Much of Meg's research integrates a theme of understanding spatial patterns of fire, or pyrogeography, including work in Canada's boreal, the People's Republic of China, California, and using a global perspective.

MEADE KROSBY is a postdoctoral research associate at the University of Washington and chair of the Climate Change Subgroup of the Washington Wildlife Habitat Connectivity Working Group. She works with scientists, land managers, and policy makers to develop rigorous methods for integrating climate change and landscape connectivity into large-scale conservation planning efforts in the Pacific Northwest, United States.

JOSHUA J. LAWLER is an associate professor in the School of Environmental and Forest Sciences and the College of the Environment at the University of Washington. His research interests generally lie in the fields of landscape ecology and conservation biology. He is most interested in how anthropogenic factors affect species distributions, population dynamics, and community composition at regional and continental scales. His current research involves investigating the effects of climate change on species distributions and populations, exploring the influence of landscape pattern on animal populations and communities, and addressing the issue of climate change for conservation planning and natural resource management.

SHAWN LEROUX is a Natural Science and Engineering Research Council of Canada postdoctoral fellow working at the Canadian facility for Ecoinformatics Research at the University of Ottawa and in collaboration with the Boreal Ecosystems Analysis for Conservation Networks project. He is an ecosystem ecologist who uses mathematical and field-based techniques to understand the effects of spatial flows of energy, materials, and organisms in natural (e.g., aquatic-terrestrial) and human-modified (e.g., protected areas surrounding landscapes) ecosystems.

JOE LIEBEZEIT has worked as an associate conservation biologist for the Wildlife Conservation Society Arctic Program since 2001. He develops and implements collaborative research projects investigating how energy development and climate change are impacting wildlife on the Arctic Coastal Plain of Alaska. Joe works closely with diverse stakeholders, including governmental agencies, other NGOs, and private industry in order to achieve project objectives and conservation goals. Joe is a recent Alcoa Practitioner Fellow and an advisor to the US shorebird conservation plan council. He received his BA from the University of New Hampshire and his MS from Humboldt State University.

KIM LISGO is an ecologist on the BEACONs Project at the University of Alberta, working on large-scale, conservation-planning methods and tools for intact landscapes in the boreal region of Canada. Prior to this, Kim studied and coordinated research on the effects of industrial development on the ecology of boreal wildlife, including birds, weasels, and caribou.

LAURA LÓPEZ-HOFFMAN earned a BS from Princeton University and a PhD from Stanford University. She is an assistant research professor of environmental policy at the Udall Center for Studies in Public Policy and an assistant professor of natural resource studies at the School of Natural Resources and Environment at the University of Arizona. Her research fields include the nature-human dimensions of conservation biology, conservation policy, and climate change impacts on ecosystems. Her current research focuses are the ecology and policy of transboundary systems under global change and the opportunities and challenges of using the concept of ecosystem services to frame natural resource governance.

THOMAS LOVEJOY became the first recipient of the newly created Heinz Center Biodiversity Chair in August 2008. Previously he served as president of the Heinz Center as the World Bank's chief biodiversity advisor and lead specialist for environment for Latin America and as the Caribbean and senior advisor to the president of the United Nations Foundation. He has held positions with the Smithsonian Institution, the US Department of the Interior, and the World Wildlife Fund–US. He conceived the idea for the Minimum Critical Size of Ecosystems project and originated the concept of debt-for-nature swaps.

ELISABETH FAHRNI MANSUR currently serves as director of training and education for the Wildlife Conservation Society's Bangladesh Cetacean Diversity Project (BCDP). She has previously worked as a nature guide, wildlife photographer, and CEO of a nature tourism company. Elisabeth has written and illustrated several educational publications on cetaceans and coauthored a comprehensive field guide and a book of photographs on the Sundarbans mangrove forest. Elisabeth's work with the BCDP consists of convening interactive conservation workshops; coordinating training and internship programs; maintaining a conservation network of local community members, institutions and researchers; and building constituencies in support of cetacean conservation in Bangladesh.

Elizabeth Matthews is assistant director of the Wildlife Conservation Society's Marine Program and supports marine conservation programs in Indonesia, Papua New Guinea, Fiji, Madagascar, and Kenya. Her extensive experience fostering marine and coastal conservation in the Pacific Islands includes leading the research and marine conservation programs at the Palau Conservation Society and conducting research projects with the Palau Division of Marine Resources. With a Fulbright fellowship at the University of the South Pacific and a PhD from the University of Rhode Island, her particular interests are in ensuring a sustainable and equitable balance between the needs of local people and biodiversity conservation.

Caleb McClennen serves as marine conservation director for the Wildlife Conservation Society, which works around the world to improve fisheries management, establish effective marine reserves, and mitigate the impact of industry to conserve some of the world's most important marine biodiversity. Caleb has served as an environmental advisor to the Republic of the Marshall Islands, a GIS analyst, and a blue water oceanographer. Caleb holds an undergraduate degree from Middlebury College in environmental studies and geography and a master's degree and doctorate in international environmental policy and development economics from the Fletcher School of Law and Diplomacy at Tufts University.

Robert McDonald is senior scientist for sustainable land use at The Nature Conservancy. McDonald works on issues related to energy, agriculture, and ecosystem services. He has recently led a working group with the National Center for Ecological Analysis and Synthesis into how global urban growth and climate change will affect urban water availability and air quality. He also researches the effect of US energy policy on natural habitat and water use. Prior to joining the Conservancy, he was a Smith Conservation Biology Fellow at Harvard University, studying the impact global urban growth will have on biodiversity and conservation.

Brad H. McRae is a landscape ecologist with The Nature Conservancy. He has authored papers on landscape and conservation genetics, population biology, climate change, and habitat connectivity conservation. He has also developed GIS tools for conservation planning, including Circuitscape and Linkage Mapper software.

Guy Midgley has worked for the South African National Biodiversity Institute since 1983, starting as a dryland ecologist, shifting to a climate change focus in the late 1980s, and currently leading the Climate Change and BioAdaptation Program. Through international collaborative work on global change research issues, he has coauthored more than 100 publications, including the popular book *A Climate for Life*, published in 2008. He was a colead author for the Intergovernmental Panel on Climate Change's (IPCC) Third, Fourth, and Fifth (in progress) Assessment Reports, and cochaired the Ad Hoc Technical Expert Group on climate change and biodiversity for the Convention on Biological Diversity.

Agenor Mundim earned a degree in electrical engineering from the Federal University of Minas Gerais, Brazil. Presently, he is an energy project coordinator at the Brazilian Foundation for Sustainable Development, where he conducts studies on renewable energy and energy efficiency and other studies related to climate change. Formerly, he was a head of the High Voltage and Power Department of the Brazilian Electrical Research Center where he conducted several studies on electrical equipment.

Enkhtuya Oidov has been the Mongolia Country program director for The Nature Conservancy since 2008. Previously she was Mongolia's secretary general for the National Council for the Millennium Challenge Account, where she negotiated a $285 million grant from the United States. From 1996 to 2000, she was a member of the Mongolian National Parliament. In the early 1990s she founded the first ever nongovernmental organization in Mongolia for pioneering advocacy of the emerging civil society. She earned a BA in economics from College of Economy, Plauen, Germany, an MA from American University, and was a Hubert H. Humphrey Fellow and a Reagan-Fascell Fellow.

Chris Pague is a senior conservation scientist with The Nature Conservancy Colorado program. With more than thirty-five years of on-the-ground experience, Pague leads the science team for the Conservancy in Colorado, providing science leadership and support for conservation efforts, and assisting in the development and actions of the Center for Conservation Science and Strategy. He has traveled to more than fifteen countries to provide assistance in conservation planning. His current international work focuses on the eastern steppe of Mongolia and the Patagonian grasslands of Argentina.

Dave Panitz is a researcher and manager for conservation planning and monitoring projects with special focus on climate change impacts and adaptation. He has worked as manager of metrics for ecosystem services for the Tropical Ecology Assessment and

Monitoring Network at Conservation International, and as a climate change researcher for the University of California, Santa Barbara (USCB), and the Universidad Nacional de Costa Rica. Panitz has a MESM in environmental science and management from the Bren School at UCSB, and a BA in science, technology, and society from Stanford University.

ANDREW PLUMPTRE has been working in Africa for the past twenty-three years, focusing on conservation in Africa's Albertine Rift region. Prior experience includes studying large herbivores in the Virunga Volcanoes, establishing the Budongo Forest Conservation Project in western Uganda, and working as assistant director for Africa Programs of the Wildlife Conservation Society. In 2000 he established WCS's Albertine Rift Program, which focuses on regional efforts to build the capacity of protected areas authorities, to undertake research, to develop strategic plans for conservation of these sites, to establish national monitoring programs, to develop species action plans, and to support transboundary collaboration.

IAN PULSFORD is a specialist in protected areas and linking landscapes, with over thirty years of experience in conservation policy and practice with the New South Wales government including selection, design, and management of protected areas. His discussion paper to the New South Wales government resulted in the establishment of the Great Eastern Ranges corridor, Australia's first continental-scale conservation corridor, and he was the founding manager from 2007 to 2010. He is a member of the International Union for the Conservation of Nature World Commission on Protected Areas and has served on various government committees, including an independent expert panel advising the Australian government on its proposed National Wildlife Corridors Plan.

ADRIAN QUIJADA-MASCAREÑAS is Assistant Researcher at the Institute of Environment and Adjunct Professor in the School of Natural Resources at the University of Arizona. He is also a member of Mexico's National System of Investigators. His present research program is mostly located in the US Southwest–Northern Mexico region, focusing on ecology and conservation genetics. He also is focused in biological consequences of climate change in the border region and transborder adaptation tools for conservationists, decision makers and stakeholders.

ENEAS SALATI is an agronomist engineer from Luiz de Queiroz College of Agriculture, São Paulo University, with a PhD in agronomy. He has published more than 120 scientific works in international journals on themes such as Amazon hydrology and ecology and global climate change, and he has been director of the National Amazon Research Institute, the São Carlos Institute of Physics and Chemistry, and the Center of Nuclear Energy in Agriculture. He was the chief researcher responsible for the description of the water cycle in the Brazilian Amazon, and was awarded the Medal of Brazilian Scientific Merit.

GILVAN SAMPAIO is with the Group on Biosphere-Atmosphere Interactions, Division of Natural Systems, Earth Systems Science Center, National Institute of Space Research. His focal research areas are in the geosciences with an emphasis on biosphere-atmosphere interactions, climatic modeling, climatic forecasting, and climatic phenomena.

THOMAS SCHEURER earned his PhD in geography, geology, and anthropology at the University of Berne. He was a scientific collaborator in the Swiss United Nations Educational, Scientific and Cultural Organization (UNESCO) Man and Biosphere Project, "Socio-economic development and ecological capacity in mountain regions." Since 1986 he has served several mandates of the Swiss Academies of Sciences: scientific coordinator of Swiss National Park research, collaborator of the Swiss Commission for Integrated Environmental Monitoring, managing director of the Swiss Interacademic Commission for Alpine Studies (ICAS) and of the International Scientific Committee for Alpine Research (ISCAR). Since 2007, he has held several lectureships at Swiss universities.

FIONA SCHMIEGELOW is a professor at the University of Alberta, and director of the Northern Environmental and Conservation Sciences Program. Her base is the Yukon Territory, where for over twenty years she has studied boreal systems with a focus on wildlife responses to industrial development, and associated large-scale conservation planning. In northern regions of boreal Canada, the effects of climate change are already conspicuous, accelerating efforts to understand and address underlying factors. Increasingly, Fiona's interests lie at the interface of science and policy, and she is heavily involved in several initiatives that strive to bridge that divide.

ANNE M. SCHRAG is the landscape ecologist for World Wildlife Fund's Northern Great Plains Program, based in Bozeman, Montana. In this capacity, Anne directs the program's efforts related to climate adaptation, as well as assisting with large-landscape conservation planning. Anne received a BS in ecology and evolutionary biology and a BA in Spanish literature from the University of Kansas; she also holds an MS in environmental sciences from Montana State University. Anne's thesis research focused on the treeline forests of Yellowstone and Grand Teton National Parks, but she has since returned to working on her native prairie ecosystem.

ANTON SEIMON is an applied climate scientist at the Wildlife Conservation Society where he leads the Albertine Rift Climate Assessment, a multiyear program on climate change adaptation funded by the MacArthur Foundation. He also contributes to developing new climate change initiatives within WCS's Global Conservation Program, and in 2011 led a survey of member organization programs on climate change adaptation for the Africa Biodiversity Collaborative Group. His research portfolio spans a broad range of topics, including ecological and species response to climate change in tropical mountains, tornadoes and other meteorological hazards, and historical climate reconstruction developed from low-latitude ice cores.

BRIAN D. SMITH is director of the Wildlife Conservation Society's Asian Freshwater and Coastal Cetacean Program. His work focuses on working with local partners to apply conservation science to, and establish effective protection for, threatened cetaceans. Brian has nearly twenty years of experience conducting cetacean research and conservation in Asian countries, including Bangladesh, India, Indonesia, Myanmar, Nepal, Thailand, Vietnam, and the Philippines, resulting in more than thirty peer-reviewed publications. He serves as the Asia coordinator of the IUCN Species Survival Commission Cetacean Specialist Group, and he is a member of the Conservation Committee of the Society for Marine Mammalogy.

TINA TIN conducted her PhD research on the thickness of Antarctic sea ice based out of the University of Alaska, Fairbanks. Since then she has worked with environmental nonprofits internationally, including the World Wildlife Fund and the Antarctic and Southern Ocean Coalition, with a focus on promoting climate change science and policy in Europe and North America and on the protection of the Antarctic wilderness. She has been participating in the annual Antarctic Treaty Consultative Meetings as a member of the ASOC delegation since 2006.

STEPHEN C. TROMBULAK holds the professorship of environmental and biosphere studies at Middlebury College in Vermont, where he has been on the faculty since 1985. His teaching and research interests are in the fields of natural history and conservation biology, particularly in the Northern Appalachian Mountains. He is the author or editor of several articles and books, including *The Story of Vermont: A Natural and Cultural History* and, most recently, *Landscape-scale Conservation Planning*. He is a founding member of the board of the Natural History Network and is the editor of the *Journal of Natural History Education and Experience*.

AURELIA ULLRICH-SCHNEIDER is a landscape ecologist educated at the Westfälische Wilhelms-Universität in Münster, Germany. Since 2002 she has been working as a project coordinator for the International Commission for the Protection of the Alps (CIPRA International) in Schaan, Liechtenstein. Her main field of work concentrates on CIPRA's activities toward the establishment of an Alpine-wide ecological network. In addition, she is in charge of the sustainable construction topic and is collaborating on climate change projects. She has many years of experience in knowledge management, information transfer, public relations, and transnational networking.

Pierre Vernier is a spatial analyst with the BEACONs Project at the University of Alberta. He has been involved in conservation and forest management research in boreal and coastal ecosystems since the early 1990s. He has extensive experience in habitat modeling, biodiversity monitoring, and landscape planning using spatial statistical analysis and tools.

Graeme Worboys is vice chair of mountains and connectivity conservation for International Union for the Conservation of Nature (IUCN) World Commission on Protected Areas. He is a protected area-management specialist with thirty-eight years of practitioner and policy experience, a lead author of IUCN's 2010 book *Connectivity Conservation, a Global Guide,* coeditor of IUCN's 2006 book *Managing Protected Areas, a Global Guide,* a University of Tasmania guest lecturer, a board member of the National Heritage-listed Tidbinbilla Nature Reserve, and managing director of Jagumba Consulting Propriety Limited. He works nationally and internationally and has recently completed World Heritage assignments in South Africa, China, and Italy.

Tatyana Yashina has a professional background in physical geography and landscape ecology. For the last ten years, she has been deputy director at the Katunskiy Biosphere Reserve, Altai Republic of the Russian Federation; she is responsible for coordination of the research and monitoring projects and international cooperation. In 2008 she received the United Nations Educational, Scientific and Cultural Organization–Man and the Biosphere Programme (UNESCO–MAB) Young Scientists Award. Since 2010, Tatjana has been engaged with the UNDP-GEF-ICI Project "Biodiversity Conservation in the Russian Portion of the Altai-Sayan Ecoregion" as coordinator of its climate component. Her professional interests include global change research in the Altai-Sayan ecoregion, applied landscape planning, and protected area management.

Steve Zack is a coordinator of bird conservation for the Wildlife Conservation Society, joining WCS in 1997, and has been in charge of studies of wildlife and conservation in Arctic Alaska since 2001. He earned his BS from Oregon State University and his PhD from the University of New Mexico. He was on the biology faculty at Yale University prior to joining WCS. He has also done extensive studies of birds in Kenya, Venezuela, Madagascar, and in the western United States. He lives in Portland, Oregon, and migrates with the birds to his various projects.

INDEX

Page numbers followed by "b", "f" and "t" indicate boxes, figures and tables.

Island Press | Board of Directors